UNDERSTANDING
BRAIN AGING
AND DEMENTIA

UNDERSTANDING BRAIN AGING AND DEMENTIA

A LIFE COURSE APPROACH

LAWRENCE J. WHALLEY

COLUMBIA UNIVERSITY PRESS *NEW YORK*

Columbia University Press
Publishers Since 1893
New York   Chichester, West Sussex
Copyright © 2015 Columbia University Press
All rights reserved

Library of Congress Cataloging-in-Publication Data
Whalley, Lawrence J., author.
Understanding brain aging and dementia : a life course approach /
Lawrence Whalley.
p. ; cm.
Includes bibliographical references and index.
ISBN 978-0-231-16382-8 (cloth : alk. paper) — ISBN 978-0-231-16383-5
(pbk. : alk. paper) — ISBN 978-0-231-53637-0 (ebook)
I. Title.
[DNLM: 1. Aging. 2. Brain—growth & development.
3. Dementia—prevention & control. WM 220]
RC521
616.8′3—dc23
2014036252

Columbia University Press books are printed on permanent
and durable acid-free paper.
This book is printed on paper with recycled content.
Printed in the United States of America

c 10 9 8 7 6 5 4 3 2 1
p 10 9 8 7 6 5 4 3 2

Cover image: © Science Photo Library/Corbis
Cover design: Adam Bohannon
Book design: Lisa Hamm

References to Internet Web sites (URLs) were accurate at the time of writing.
Neither the author nor Columbia University Press is responsible for URLs
that may have expired or changed since the manuscript was prepared.

TO

HELEN

*A test of a people is how it behaves towards the old. It is easy
to love children. Even tyrants and dictators make a point of being
fond of children. But the affection and care for the old,
the incurable, the helpless are the true gold mines of a culture.*

ABRAHAM J. HESCHEL, POLISH AMERICAN RABBI (1907–1972)

Youth is a gift of nature, but age is a work of art.

STANISLAW JERZY LEC, POLISH POET (1906–1966)

CONTENTS

List of Tables xvii
List of Illustrations xix
Prologue xxix
Acknowledgments xxxiii

1. INTRODUCTION
1

2. THE LIFE COURSE APPROACH
18

3. THE WELL-CONNECTED BRAIN
45

4. EVOLUTION, AGING, AND DEMENTIA
75

5. THE AGING BRAIN
107

6. THE BIOLOGY OF THE DEMENTIAS
134

7. THE DISCONNECTED MIND
163

8. EMOTIONAL AGING
207

9. DEMENTIA SYNDROMES
233

10. DEMENTIA RISK REDUCTION, 1: CONCEPTS, RESERVE,
AND EARLY LIFE OPPORTUNITIES
274

11. DEMENTIA RISK REDUCTION, 2: MIDLIFE OPPORTUNITIES
TO DELAY DEMENTIA ONSET
304

12. DEMENTIA RISK REDUCTION, 3:
MULTIDOMAIN APPROACHES
331

Epilogue 352
Notes 357
Index 389

CONTENTS

List of Tables xvii
List of Illustrations xix
Prologue xxix
Acknowledgments xxxiii

1. INTRODUCTION 1

I. Life Histories and the Life Course Approach 1
II. Greater Numbers of the Elderly 2
III. The Role of Epigenetics in Brain Aging and Dementia 3
IV. Social Class and the Life Course 4
V. The Study of Individual Differences 5
VI. Resilience and Social Capital 5
VII. Developmental Origins of Adult Disease 7
VIII. Limitations of Association Studies and the Perils of Reductionism 8
IX. Systems Biology, Reductionism, and the Value of Visualization 9
X. The Biology of the Aging Mind and Brain 13
XI. Social Cognition, Aging, and Dementia 16

2. THE LIFE COURSE APPROACH 18

I. Influential Ideas 18
II. Developmental Stages 24
III. Critical Periods 27
IV. Programmed Development and Aging 31

V. Three Lives 36

VI. The Developmental Origins of Aging and Adult Disease 38

VII. Bringing It All Together 41

3. THE WELL-CONNECTED BRAIN 45

I. A Brief Introduction to the Nervous System 45

II. Functional Organization 53

III. Diffusion Tensor Imaging 61

IV. Brain Structures and Functions Relevant to Aging 62

V. Blueprints for a Well-Connected Brain 65

VI. The Brain and Language 70

VII. Brain Maps 72

VIII. Bringing It All Together 73

4. EVOLUTION, AGING, AND DEMENTIA 75

I. Introduction 75

II. Is AD Specific to Humans? 80

III. Human Evolution 83

IV. Are the Most Evolved Neural Networks More Vulnerable
to AD? 86

V. Is Reactivation of Developmental Programs in the Course of AD
Incompatible with the Mature Brain? 91

VI. Brain Aging and the Evolution of Neurological Disease 103

VII. Bringing It All Together 105

5. THE AGING BRAIN 107

I. Introduction 107

II. The Frontal Lobes and Aging 120

III. Brain Blood Vessels 121

IV. Genes, Diet, and Behavior 126

V. Stress Responses and the Aging Brain 128

VI. Bringing It All Together 132

6. THE BIOLOGY OF THE DEMENTIAS 134

I. Historical Trends 134
II. The Modern Era: Focus on β-Amyloid 141
III. Molecular Biology of Tau 147
IV. Vascular Cognitive Impairment and Dementia 149
V. The Modern Era: Focus on the Biology of Aging 151
VI. Breakdown of Neural Networks in Dementia 158
VII. Bringing It All Together 161

7. THE DISCONNECTED MIND 163

I. Introduction 163
II. Cognitive Aging 164
III. Sensory Systems 171
IV. What Influences Rates of Cognitive Aging? 175
V. Attention 176
VI. Information Processing 181
VII. General Mental Ability 181
VIII. Memory 195
IX. Vulnerable People 199
X. Personality and Cognitive Aging 205
XI. Bringing It All Together 205

8. EMOTIONAL AGING 207

I. Emotional Life 208
II. Older Adults as Emotional Experts 211
III. Emotional States 213
IV. The Aging Brain and Emotional Aging 214
V. The Anatomy of Emotion 215
VI. Emotional Intelligence 216
VII. Older Adults as Actors 219
VIII. Life Narratives, Self-concept, and Possible
Selves in Late Life 222

IX. Social Support, Social Cohesion, and Social Pain 226

X. How Much Time Is Left? 228

XI. Bringing It All Together 231

9. DEMENTIA SYNDROMES 233

I. Introduction 233

II. Alzheimer's Disease 244

III. Brain Imaging in Early Dementia 255

IV. Frontotemporal Dementias 263

V. Parkinson's Disease with Dementia 265

VI. Dementias Associated with Brain Blood Vessel Disease 270

VII. Bringing It All Together 272

10. DEMENTIA RISK REDUCTION, 1: CONCEPTS, RESERVE, AND EARLY LIFE OPPORTUNITIES 274

I. The Dementia Epidemic 274

II. Cognitive Reserve 279

III. Biological Plausibility of Strategies to Reduce the Risk of Dementia 283

IV. Early Life Opportunities to Prevent Dementia 287

V. Bringing It All Together 301

11. DEMENTIA RISK REDUCTION, 2: MIDLIFE OPPORTUNITIES TO DELAY DEMENTIA ONSET 304

I. The Neurovascular Hypothesis 306

II. The Inflammatory and Metabolic Hypotheses 315

III. The Amyloid and Related Therapeutic Approaches Hypothesis 319

IV. Stress, Depression, and the Role of Growth Factors 321

V. The Brain Activity Hypothesis 325

VI. Bringing It All Together 327

12. DEMENTIA RISK REDUCTION, 3: MULTIDOMAIN APPROACHES 331

I. Background 331
II. Nutrients and Dementia 337
III. The Multidomain Hypothesis 345
IV. Bringing It All Together 351

Epilogue 352
Notes 357
Index 389

LIST OF TABLES

TABLE 6.1 Acetylcholinesterase Inhibitor Drugs for Alzheimer's Disease   139

TABLE 6.2 Genetic Mutations and Variants Linked to Alzheimer's Disease   143

TABLE 7.1 Piaget's Four Stages of Cognitive Development During Childhood   191

TABLE 9.1 The Major Subdivisions of Dementia with Representative Examples   243

TABLE 9.2 Recent Advances in the Early Diagnosis of Dementia   259

TABLE 9.3 Percentage of 426 Individuals (median age 81 years at death) with Different Levels of Severity of Brain Pathology Grouped by Presence of Dementia Before Death   261

TABLE 10.1 A Two-Decade Comparison of Prevalence of Dementia in Individuals Sixty-Five Years of Age and Older from Three Geographic Areas of England   292

TABLE 11.1 Drugs in Development for Treatment and Possible Prevention   320

TABLE 12.1 Overview of Health Care, Environmental, Genetic, and Social Factors Shown in Observational Studies and Randomized Controlled Trials to Influence Dementia Risk in Populations Mostly of European Ancestry   334

LIST OF ILLUSTRATIONS

FIGURE 2.1 Proposed pathways from socioeconomic status (SES) in childhood that can affect linguistic development and language skills and from SES through greater exposure to stress and changes in socioemotional processing and cognitive control. Intervening brain structures are identified on these pathways and explained more fully later in the book. 20

FIGURE 2.2 Growth in body weight over a woman's life course in relation to changes in reproductive status. 23

FIGURE 2.3 How grandmother's nutritional status and family size may affect the growth and health of her daughter and later the birth weight of her grandchild. 23

FIGURE 2.4 This figure illustrates the links between what happened during growth in the womb with birth weight. A possible link between birth weight and the regulation of blood pressure in adulthood is shown as a direct line between birth weight and blood pressure. 24

FIGURE 2.5 This illustration shows in diagrammatic form multiple successive critical periods in the acquisition of sensory functions, the use of language, thinking strategies, and the development of higher mental abilities. These are just four systems that could be represented in this way and are overlapping systems without abrupt onsets and offsets. 28

FIGURE 2.6 Cortisol is a stress hormone released from the adrenal cortex. Its release is stimulated by adrenocorticotrophic hormone (ACTH) released from the anterior part of the pituitary gland lying on the base of the brain. Above the pituitary, the hypothalamus integrates many stress responses, including the release of corticotrophin releasing factor (CRF), which stimulates ACTH release. The hypothalamus senses increased cortisol release, and this "feedback" inhibits further release of CRF. 35

FIGURE 3.1 This figure shows aspects of the external and internal structure of the brain. A diagram in the lower right panel shows how the outer surface of the brain is made up of a deeply folded layer of brain tissue (the cortex) with prominent gyri separated by clefts (sulci). The upper right panel shows an MRI scan of the head with structures that lie along the midline (through the tip of the nose backward between the eyes) and points to two structures: the frontal lobes and the cerebellum. The left-hand panels show how the MRI scan can slice through the head through planes chosen to reveal specific brain structures. The upper left panel shows the brain along a slice through the top of the skull and both ears to reveal fluid-filled spaces in the middle of the brain. The lower left panel slices through a horizontal plane at the level of the eyeballs to expose relationships between the cortex and deeper brain structures.   46

FIGURE 3.2 This figure shows a diagram of the mature central nervous system. The neural tube which formed from the neural groove in the embryo has become the tube-like structure of the spinal cord. The highest (anterior) part of this tube has developed into a recognizable brain now seen as a greatly enlarged swelling encased in the skull.   47

FIGURE 3.3 Examples of different types of neuron.   48

FIGURE 3.4 Schematic diagram of the components of the neuron. Incoming electrical information arrives through dendrites to pass through synapses as chemical messages. The neuron responds to the sum total of incoming information and generates an outgoing electrical message through its axon.   49

FIGURE 3.5 The cortex includes six layers each containing neurons that are typical of each layer. The relative numbers of cell types vary between cortical regions.   50

FIGURE 3.6 Schematic diagram of the divisions of the cortex into lobes with areas linked to specific brain functions. The brain is made up of two symmetrical halves, and the diagram shows the outer surface of the left half. The brain faces to the left where the anterior parts are the frontal lobes. The temporal lobes lie under the part of the skull between the eyes and ears (the temples). At the posterior part of the brain lies the occipital lobes, and the parietal lobes occupy the space between the frontal and occipital lobes.   55

FIGURE 4.1 A branch from the "tree of life" that shows evolutionary relationships between modern humans and many living primates.   82

FIGURE 4.2 This figure shows the relationships between modern humans and their most closely related primates in terms of time, as we shared a common ancestor (millions of years ago); percent of genes that differ between humans and primate species; and brain volume (uncorrected for body size).   85

FIGURE 4.3 The cognitive development of children and young adults closely follows the pattern of progressive myelination of cortical axons. Regions myelinated first include the spinal cord and brainstem. Myelination continues toward the frontal cortex, with sensory pathways myelinating before motor pathways, and downstream projection pathways myelinating before cortical association pathways.

The prefrontal cortex myelinates last. Although initiated prenatally in humans, most tracts and regions are myelinated completely during the first year of life. In humans, some regions do not complete myelination until the second and third decades of life.   90

FIGURE 4.4 Embryos of selected vertebrate species. The striking similarities between species are most marked during early development. As development progresses, species begin to show major differences.   91

FIGURE 4.5 Life course pathways control improved cellular maintenance, which leads to slowed aging (e.g., slowed normal cognitive ageing) and protection against diseases of aging (e.g., neurodegenerative diseases of ageing, such as Alzheimer's and Parkinson's disease, and cancer). Aging can evolve via selection to reduce investment in energetically costly somatic maintenance processes and instead to increase early fitness traits, such as growth and reproduction.   93

FIGURE 4.6 During differentiation, cells pass through stages of development. These are shown as a landscape falling toward the front of this schematic. At first, each cell has great potential to differentiate into one of any different type of cell (i.e., it is totipotent). Next, its eventual form is limited, as it becomes pluripotent. As it develops further, choices are further restricted so it can become multipotent, and then finally it can only be one type of cell that is unipotent. This is the "epigenetic landscape" model of cell differentiation. The figure shows a graded landscape down which the cell passes but cannot return "uphill" to a former state without "reprogramming."   99

FIGURE 5.1 The principal pathways toward aging. This is a simplified schematic and omits many diverse possible contributions from the environment (e.g., diet) and interactions between factors that modify rates of aging.   109

FIGURE 5.2 The appearances of two brains. Figure 5.2a is that of a young adult in which the sulci are tightly folded between gyri. Figure 5.2b is the brain of an adult who is about eighty years old with no history of dementia or brain disease. The gyri appear narrower and the sulci wider than those of a young adult.   112

FIGURE 5.3 These show MRI findings from examinations of two individuals.

(A) combines two images from one individual to indicate how the hippocampus and some adjacent tissue has shrunk from the age of sixty-eight to seventy-three.

(B) combines images from a second individual to show where white matter hyperintensities (WMH) developed between ages of sixty-eight and seventy-three. The white areas show where WMH were detected at age sixty-eight.   114

FIGURE 5.4 A photomicrograph of thin slices of brain tissue stained to reveal presence of amyloid (lower panel) or neurofibrillary tangles (upper panel). The images are not to the same scale: the solid bar represents 100 microns on either image.   115

FIGURE 5.5 The problem of counting brain cells in the aging brain. A brain specimen is made up of a mixture of brain cells each of different sizes. The cells form part of structure with boundaries that can be measured. In this example, the density of the structure differs from surrounding tissue, so that as density shifts from outside to inside the structure, the density changes abruptly. This is the boundary. When the structure is set in a hardening compound (e.g., wax) very thin slices can be made across multiple planes (a)–(d).

This figure shows how brain cells of different sizes can be distributed across planes (a)–(d) in the center of the specimen. The (a)–(d) pass through single brain cells. Some adjacent sections misidentify a single large cell as two separate cells in the same structure. If a structure contains more large cells than small cells, cell counting may be biased toward overestimation of cell number in the whole structure. Conversely, as brain cells shrink with age, adjacent sections may miss small cells completely. 119

FIGURE 5.6 The schematic shows a loop in a single brain blood vessel. When blood pressure is continuously raised, this exerts mechanical pressure on the surrounding brain tissue. The brain area contained within the loop is compressed simultaneously from two sides. Eventually, these brain cells die, leaving a space—a lacunar infarct—within. 124

FIGURE 5.7 White matter hyperintensities detected using MRI in the brain of an older adult without dementia. 125

FIGURE 5.8 An area of lacunar infarction detected at MRI examination. 126

FIGURE 5.9 The effects of aging on the brain. On the right are aging-driven processes that contribute to increased damage to cell membranes, DNA, and other large biogregulatory molecules, including advanced glycation end products (AGEs). Not shown on this diagram is the stimulation of inflammatory responses in microglia through actions of AGEs. On the left hand side of the diagram are the powerful antiaging effects of intrinsic repair and anti-inflammatory systems. Brain responses include compensatory remodeling of neural networks. This schematic diagram summarizes the balance between harmful age-related damage to brain function and the helpful effects of systems that can counter these effects. These systems include pathways that help remodel neural networks damaged by aging or dementia. 133

FIGURE 6.1 A cholinergic synapse. Acetylcholine (Ach) is synthesized in the presynaptic neuron from acetyl CoA (A) and choline (Ch). It is then transported in a presynaptic vesicle to be released as ACh into the synaptic space. Here, it attaches to the postsynaptic membrane at a highly specific cholinergic receptor. Acetylchoinesterase is a membrane-bound enzyme that degrades acetylcholine into acetyl (A) and choline (Ch) groups. 138

FIGURES 6.2A–6.2C APP exists in various forms but in nervous tissue, it is usually 695 amino acids long. It is a transmembrane protein that is normally cleaved by α-secretase, which does not produce β-amyloid. β-amyloid formation requires the actions of both β- and γ-secretases. 144

FIGURE 6.3 The presenilins 1 and 2 share very similar structures. They are both membrane-spanning proteins. The diagram represents presenilin 1 (*PSEN1*), showing ten membrane-spanning domains (numbered boxes). Known mutations that cause early onset Alzheimer's disease (AD) lie mostly within the membrane or in a loop extending inside the cell (the cytoplasmic loop). 145

FIGURE 6.4 Schematic of the tau protein gene. There are six isoforms of tau in the nervous system (gray) obtained by splicing exons (dark gray). Mutations that cause frontotemporal dementia and related disorders (see text) are shown as solid stars. 148

FIGURE 6.5 The spread of neurofibrillary tangles in the cortex follows a consistent pattern. This diagram illustrates the progress of NFT deposition by six stages (I through VI) that typify the development of Alzheimer's disease. 149

FIGURE 6.6 Examples of theories of aging placed in broad classes. For clarity, only selected main theories are shown. 152

FIGURE 6.7 In the absence of disease, dendritic spine numbers change over the life course. In autism, increased numbers of spines are maintained throughout life. Individuals with schizophrenia have a greater than normal pruning of dendritic spines. Individuals with AD have greater loss of spines in late adulthood and old age. 160

FIGURE 7.1 The principal cognitive functions of cortical regions. 166

FIGURE 7.2 The organization of cognitive functions. Sensory information is presented on the left side of the diagram. Information from the sensory organs (eye, ear, etc.) enters awareness (the broken circle) through gates that can be opened and closed or hold sensory information briefly in temporary stores (rectangles). Selective attention is shown as downward arrows above the sensory input labels. Working memory is a temporary store of memories needed to organize responses required to meet immediate goals. Although shown as a box below the central executive, working memory involves the central executive, the articulatory loop, and the visuospatial sketchpad. The central executive exerts voluntary control over learning and retrieval of memories, planning, and voluntary attention. In everyday language, the central executive has top-level control over behavior. It can organize all the mind's resources to plan and achieve goal-directed behaviors while remaining flexible and adaptable. The broken circle represents a mental area of consciousness. 168

FIGURE 7.3 The word list and the Audrey Brown story explained. A model of the effects of age on relationships between two types of verbal memory test: recall of a list and recall of a prose story. The important difference is that speed probably does not predict recall of a prose story. This recall is much more affected by working memory having access to automatic processes that aid understanding of the story. Age has major negative effects on speed and working memory, and until very old age, it has many positive effects on vocabulary. At the foot of the diagram, the double-headed arrow shows that performances on list and prose recall are interconnected so that people who do well on one tend to do well on the other.   169

FIGURE 7.4 This figure shows how attention is directed by the central executive (the "stage manager" shown holding the spotlight) on a player selected by the stage manager to be shown in consciousness to the audience, including the brain regions concerned with self-awareness and discrimination among different types of experience arising from internal (body and mind) and external worlds (the environment). Logical continuity between successive experiences is maintained by a narrator (or "inner voice") who explains experiences to the self. A chorus is shown off–center stage, not at this moment, in the spotlight. The chorus represents the emotions associated with immediate or recent experiences or as recalled from emotional memory of experiences relevant to the moment.   178

FIGURE 7.5A A diagram showing a neural network made of neurons (circles). This is highly oversimplified and does not show the high density of connections made between neurons. Some connections are inhibitory and some are excitatory.   183

FIGURE 7.5B In one pathway, successful transmission of signals across the network is shown by a succession of open circles. When a link is broken, the signal takes a detour and thus takes longer to complete the route that increases from eleven neurons to twelve. On another pathway there are four breaks so that detours increase the route from eleven neurons to sixteen and so substantially increases the time to cross the network.   183

FIGURE 7.6 The associations between age and crystallized and fluid intelligence. Note that crystallized intelligence (Gc) is maintained after reaching its maximum in middle adulthood and, in some people, might continue to improve until the ninth decade of life. Fluid intelligence begins to decline around the age of forty-three and is significantly lower by about the age of sixty-seven.   187

FIGURE 7.7 Plots of longitudinal data from 791 participants in the Aberdeen 1921 and 1936 Birth Cohorts Study. Data are cognitive scores on between one and five occasions from the ages of sixty-three to eighty-two. The lines connect data from one person over each occasion. These scores show the slight negative effects of brain aging and the stronger effects of repeated practice.   188

FIGURE 7.8 Schematic showing how Piaget visualized stages of cognitive development. Note his primary focus is on the early stages of cognitive development and how late adulthood is relatively neglected.   192

FIGURE 8.1 Schematic summary of the origins of maladaptive behavior using psycho-dynamic terms. Mental life is divided by the vertical dashed line into the "unconscious" and the "conscious." The ego lies on the boundary between the two so that some aspects remain in conscious awareness; others are hidden in the unconscious. The integrity of the ego is protected by defense mechanisms. These serve the ego by protecting it from conflict between the id (source of primal drives such as libido and self-preservation) and the superego (the repository of rules and behavioral controls). Overuse of defense mechanisms (like denial) impairs the efficiency of goal-directed behavior and allows maladaptive behavior to emerge.   224

FIGURES 8.2A AND 8.2B This is a type of sociogram overlaid on a social landscape. Sociograms were invented by Jacob Moreno to study how groups establish preferences about group members. This figure does not distinguish between social isolates and those making "Mutual Choices." Some individuals make choices that go unreciprocated, the "One Way Choice." Cliques are groups of three or more people within a larger group who all choose each other.   228

FIGURE 9.1 AD is not a simple concept. After the age of eighty years, most people living with dementia have increasing evidence of both Alzheimer-type brain changes and brain–blood vessel disease. The figure shows this as the larger area of "mixed" brain–blood vessel disease and amyloid with less common dementia types of relatively "pure" disease lying alongside. Life course factors contributing to dementia are shown as cumulative age-related contributions to dementia susceptibility. As for many types of blood vessel disease, these factors do not directly cause dementia but typically involve, for example, a predisposition to midlife high blood pressure linked to prenatal exposures or to diabetes linked to maternal nutrition.   235

FIGURE 9.2 Three scenarios are presented in schematic form. In the top row, there are three dementia entities with clear-cut boundaries ("areas of discontinuity") between each. In the middle row, the three dementia subtypes overlap, and in the lower row, all three overlap to the extent that they form a continuum.   241

FIGURE 9.3 This shows the boundaries between three phases of a substance that passes from solid to liquid to gas as temperature rises. The point to stress is the clear-cut nature of the boundaries between each phase when temperature is plotted against density. This diagram shows nothing unusual: It simply shows a physical property of matter when the methods of measurement are precise.   241

FIGURE 9.4 This shows the same substance passing through the same three phases (solid, liquid, and gas), but now the methods of measurement of both temperature and density are prone to error so the true relationship might lie anywhere within each range of error. Observations obtained in this are more likely to fail to detect boundaries and presume that passing from one phase to another occurs over a range of temperatures and densities. In short, it might be erroneously concluded that there is a continuum of phases in which mixtures of gas–liquid and liquid–solid frequently occur.   242

FIGURE 11.1 Candidate pathways from anesthesia to dementia: Preoperative pathophysiology, surgical trauma, and aging biology interact synergistically with anesthetic agents to trigger cascades of inflammatory and prothrombotic molecular events to induce perturbations of cerebral activity at multiple levels. These include disrupted blood-brain-barrier function, proinflammatory microglial activation, and release of potentially neurotoxic excitatory amino acids. These events can precipitate the formation of cerebral amyloid and localized disturbances of cortical microvasculature that lead to selective death of cortical neurons in brain structures critical to memory, attention, and comprehension.   311

FIGURE 11.2 Dietary fish oils are rich in omega-3 fatty acids. Omega-3 acids provide precursors (EPA and DHA) of anti-inflammatory prostaglandins. Their effects are partly balanced by proinflammatory prostaglandins derived from the omega-6 family. Before the industrial revolution, human diets contained roughly equal amounts of omega-3 and omega-6 fatty acids. In the modern era, diets contain much more omega-6 fatty acids and are deplete in omega-3.   318

FIGURE 12.1 Major influences acting during the life course that increase the risk of dementia. There are major pathways from childhood IQ toward educational attainments and also to the acquisition of lifestyle habits of a balanced diet, exercise, moderate alcohol, and not smoking. Other childhood influences on IQ are too complex to be illustrated here; these include early fetal nutrition, childhood illnesses, "positive parenting," and other components of early education. Important hazardous influences on the risk of dementia include an acquired predisposition to develop diabetes, heart disease, and stroke and the influence of family factors and childhood IQ on job success and the lifelong accumulation of social and material capital. Protection against dementia is inferred as influences that oppose these hazards. Interventions that improve early childhood education, enable higher linguistic abilities, and help establish adult patterns of healthy eating, exercise, tobacco avoidance, and reduced alcohol and illicit drug use are largely matters of public policy and individual judgment. Management of health promotion to reduce risks of vascular disease is one of the great medical successes of the late-twentieth century and where gains seem likely to be maintained.   335

FIGURE 12.2 The schematic sets out interconnections between the neurobiology of compromised neural health that impairs neural connectivity and efficiency of information processing and places demands on cortical reorganization. The features of cognitive aging and the clinical dementia syndromes emerge alongside the forces imposed by age-related physical illness and sensory deficits. Emergent behaviors are modified by concurrent social processes specific to the aging condition and individual capacity to make good any deficits or initiate appropriate adaptive behaviors.

The complete range of possible outcomes is shown as a spectrum that ranges from success to cognitive failure. 336

FIGURE 12.3 Homocysteine is a naturally occurring compound that is converted to methionine, and methionine has a vital role in the expression of DNA. Deficiencies of folate and vitamins B6 and B12 cause homocysteine to build up to toxic levels in tissues where it promotes damage particularly to neurons and to the linings of blood vessels. 339

FIGURE 12.4 The risk of dementia was increased by about three times in fifty-five people with the highest homocysteine concentrations. All 199 participants were born in Aberdeen in 1921, were without dementia at the age of seventy-eight and followed up for eight years. The incidence curves are adjusted for education, childhood IQ, socioeconomic status, and blood antioxidant concentrations. 340

FIGURE 12.5 Major influences on current cognitive performance in late adulthood in 248 adults age sixty-eight years and without dementia. Negative influences include (1) brain shrinkage attributed in part to presymptomatic Alzheimer's disease; (2) smaller brain size in early adulthood; and white matter hyperintensities attributed to brain blood vessel disease and exacerbated by high blood pressure and abnormal glucose metabolism. Positive influences include (1) higher original intelligence from childhood and (2) longer duration of formal education and more complex occupations often with supervisory responsibilities. These relationships are shown as bold arrows, and combined, they account for about 50 percent of influences on current cognition. 347

FIGURE 12.6 The effect on dementia onset of interventions that delay dementia onset by five years. These produce an overall reduction of dementia prevalence by 50 percent. At an individual level, the opportunity to postpone dementia onset optimizes quality of life in old age and discourages a fatalistic view that dementia is always inevitable. 351

PROLOGUE

In the sixties, it was popular to quote from publications on "sociobiology" and to attribute behavior as being "all in our genes." Experts made confident predictions of "genes for success," "a gene for intelligence," and even "blue genes" to cause depression. These trends diminished at the start of the twenty-first century, so much so that it became crude to use these expressions. Much more complex ideas have since taken shape, offering the prospect of a coherent framework that could replace once fashionable but now tired "one gene for every trait, one gene for every disease" predictions. In the popular imagination, however, these ideas about the primacy of our genes are taken to mean that something as common as Alzheimer's disease (AD) should be solved by now. After all, you can hear people ask, "Aren't these just disorders of something physical? Don't biological tissues obey physical laws? The brain is just another lump of tissue, isn't it, like the heart or liver?"

So what is the current state of play? Have we reached consensus on where we are on the road to finding answers to the problems of the aging brain? Consider the following two newspaper reports.

In August 2010, the *New York Times* reported a conversation that took place in 2003 between Neil Buckholtz, chief of the Dementias of Aging Branch at the National Institute of Aging, and his old friend William Potter, a neuroscientist at Eli Lilly who had just left the National Institutes of Health (NIH).[1] Potter had been thinking about how to speed up the glacial progress of Alzheimer's drug research. The two men agreed that what was holding up most drug research in AD was a lack of reliable biological data that showed how people living with AD differed from healthy individuals and how these differences changed as AD progressed. This type of difference is called a "biomarker," which is shorthand for a biological measurement that is a reliable feature ("marker") of every case that is always found in people who will develop the disease. Importantly, a biomarker can be present before disease onset. It can identify people who are most likely to get the disease.

Almost simultaneously, the two scientists realized that no single company or institution had the resources for a breakthrough in drug research. Their solution was to persuade the government (which contributed $65 million) and big drug companies (which contributed $27 million) to help fund a huge collaboration that would provide the needed biological data. The result was the AD Neuro-imaging Initiative (ADNI). "It was unbelievable," said Dr. John Q. Trojanowski, from the University of Pennsylvania, "It's not science the way most of us have practiced it in our careers. But we all realized that we would never get biomarkers unless all of us parked our egos and intellectual-property noses outside the door and agreed that all of our data would be public immediately."

That was in 2003. Speeding forward to 2010, another piece was published in the *New York Times*.[2] This article reported a meeting that had convened in the National Institutes of Health in August 2008[3] and announced the "NIH Consensus Development Conference Statement on Preventing Alzheimer's Disease and Cognitive Decline." The shared conclusions of the participants were that on the current evidence, they could not suppose that "any modifiable risk factor" was associated with cognitive decline or AD. The panel made some suggestions about the direction of future research, leaving a lasting impression that their hard work and diligence were unrewarded.

Picking over the document and wondering how such a sense of disappointment could prevail, there were occasional pointers. Considering nutritional and dietary factors, the panel found either no consistent association or "very limited evidence" to support a protective role for any foodstuff or single nutrient. This finding later was debated by nutritional scientists who agreed that evidence from association studies was thin, and without any positive clinical trials, this conclusion held up.

Later, the panel acknowledged that "much of the available evidence derives from studies that were originally designed and conducted to investigate other conditions, such as cardiovascular disease and cancer." The panel primarily was composed of experts in the fields of cardiovascular disease and cancer, and the panel was not balanced by the presence of clinical neuroscientists with expert knowledge of the problems and pitfalls of preventive studies. But who else would have an opinion worth inclusion? Who conducts clinical trials to prevent neurological disease?

It is difficult to think of clinical trials that would be relevant to dementia prevention. At the top of any list likely would be those trials that teams have organized to prevent stroke and, of great relevance, to prevent dementia following stroke. Nevertheless, relevant trials have been conducted to prevent developmental neurological defects. Here, the best-known result from such a trial is the success of vitamin supplementation to prevent congenital neural tube defects (spina bifida and anencephaly).

These pioneering studies by Richard Smithells and Elizabeth Hibbard in Leeds (United Kingdom)[4] persuaded expert advisory committees to recommend fortification of some foodstuffs with folate. In the nineties, action was taken on the advice of Godfrey Oakley at the Centers for Disease Control and Prevention and the support of the wider medical community. These actions resulted in 1998 in the United States being the first country to introduce mandatory fortification of flour and other grains with folic acid to help prevent neural tube defects. Sadly, as in the rest of Europe, the United Kingdom has procrastinated on fortification. This is especially galling given that the pivotal preventive studies were completed in the United Kingdom.

The 2008 NIH panel recommended that long-term studies in dementia prevention should be supported but did not recommend what their aims should be, what could be measured, or even at what age they should start. With hindsight, any long-term study would benefit from knowing which measures should have been included from the beginning. This is rarely possible, however, because technical advances are so difficult to predict over time intervals even as short as twenty years. Much as the vitamin supplementation study supports starting folate as soon as pregnancy was planned, is it possible that preventive measures for dementia should start so early? These studies would take a lifetime to complete. Is this even feasible?

More optimistic views about dementia prevention are now well rooted in the scientific agenda and seem likely to bear fruit. At least three U.S.–based studies started in 2013 involve people from Australia, Columbia, the United Kingdom, and the United States. One study focuses on those at increased risk because participants carry a genetic mutation that inevitably will cause dementia, another identifies people with a buildup of abnormal brain protein (amyloid) associated with dementia, and the third recruits people who carry two copies of a gene (APOEε4) that substantially increases the risk of dementia. All studies began when participants were without clinical features of dementia (i.e., these are prevention not treatment studies) and will include sensitive mental tests and brain imaging measures. The interventions to be tested mostly are aimed at reducing the risk of dementia by reducing the toxic effects of amyloid. These U.S. studies differ importantly from the European approach to late-onset dementias, which are viewed as complex multifactorial conditions. Effective prevention must address simultaneously as many causal factors as possible, if these are to produce results.

Dementia prevention is an urgent priority in health research with huge potential economic benefits. Given the wide-ranging, albeit scientifically sound, proposals to prevent dementia, pulling together what is known about the risk of dementia and then forming a plan is a demanding task. International meetings of dementia researchers have given dementia prevention the highest priority, and most contributors have accepted the need for collaborative efforts that should span continents. An increasing sense of urgency is encouraged by the universal awareness that a "dementia epidemic" poses a major threat to the economic and social well-being of modern societies. Experts no longer anticipate an antidementia drug—a "magic bullet"—that will prove effective against the most frequent causes of age-related dementia syndromes. Instead, much more complex ideas envisage how drug development could address genetic and environmental hastening of dementia onset, but do not sit comfortably with gene-centered preconceptions about the causes of the dementias. These views are mirrored in historical differences between European and U.S. evolutionary biologists in understanding how genes act at the population level and not in how processes that control genetic expression influence the pace of evolution.[5]

The life course approach provides the framework for this book. It is used to explain the causes and courses of the dementias. Although written unashamedly for the general reader, the approach may be unfamiliar to neuroscientists working in their own sometimes-narrow field of dementia research. It is hoped that their work and that of others caring for people living with dementia will benefit from reading this book.

ACKNOWLEDGMENTS

My interest in the life course approach to brain aging and dementia draws heavily on my friends and former teachers in psychiatry. I had a wonderful start in Edinburgh working for Ian Oswald and Henry Walton, learning research methods from Norman Kreitman, Ralph Mac-Guire, and George Fink. My Edinburgh contemporaries Ivy-Marie Blackburn, Douglas Blackwood, and W. John Livesley provided the essential "grit in the oyster." During those early years, a life course approach to mental health care provided what seemed to us to be an obvious and practical framework for our clinical histories, social assessments, causal explanations, and long-term care plans.

For almost thirty years, I have enjoyed having two superb research collaborators in Ian Deary and John Starr. Together, building on our backgrounds in psychology, geriatric medicine, and psychiatry, we developed a program of life course research in brain aging and the dementias. We followed great Scottish traditions of clinical, social, and educational research to fulfill the intentions of those pioneers who aimed to change Scottish life for the better through improvements in health care, social resources, and education. Many colleagues and friends have helped. Rob Wrate, Mario de Parra, and Gernot Riedel made important contributions. Tom Russ read the completed text and made many appreciated comments—and more specifically, David St Clair helped with valuable comments on neurogenetics. Susan Duthie, Andrew Collins, and Allan Wright helped explore molecular contributions to genetics, nutrition, aging, and dementia. I owe much to the anonymous reviewers and to my editors Patrick Fitzgerald, Kathryn Schell, and Bridget Flannery-McCoy at Columbia University Press without whom this book would be less relevant and less well structured.

In particular, I am indebted to my copy editor, Maureen O'Driscoll, whose attention to detail and consistency throughout has contributed hugely to the accuracy and readability of this book. Finally, I wish to thank the Wellcome Trust whose generous support of my Clinical Research Fellowship (2001–2006) allowed my research career to flourish. Their thoughtful and considerate direction of my intellectual development contributed enormously to the ideas presented here.

UNDERSTANDING
BRAIN AGING
AND DEMENTIA

1

INTRODUCTION

I. LIFE HISTORIES AND THE LIFE COURSE APPROACH

This book explains how and why individuals differ in rates of brain aging and vulnerability to the dementias of late life. The overarching viewpoint derives from more than twenty years spent on the study of brain aging and health in Edinburgh and on two birth cohorts living in Aberdeen, Scotland. When first recruited at the age of sixty-four or seventy-seven, these cohorts agreed to a follow-up study of brain aging and mental health using well-tried and robust methods. Within a year or so, the study was extended to include sequential brain imaging, molecular genetic data, and the results of nutritional surveys. The intentions remained the same throughout the study: How does normal (i.e., nonpathological) aging of mental abilities differ from the early stages of decline into dementia? Why do some people remain vibrant and mentally able well into their ninth or even tenth decades? Eventually, topics that would be vital in understanding our follow-up data were translated into research reports. Advice and lessons learned from editors and reviewers helped shape the next stages of research, and ultimately, these lessons became the key areas covered in this text. This book ambitiously provides an overall grasp of brain aging and shows the importance of genetics and environmental influences on aging mental abilities and emotions. An understanding of the science behind the promise of delaying or even preventing dementia onset also is presented.

This book is written for the nonspecialist. The interested reader may accept a broad view of a typical life history and should be able to identify major stages in development and aging and, critically, to recognize important influences that determine individual differences in rates of brain and cognitive aging.

The "life course way of thinking" provides the framework around which a great deal of knowledge about brain aging and dementia can be organized. This life course approach is a scientific discipline based on the human life history, and it extends beyond a simple narrative to include influences from biology, psychology, and the social sciences in a single integrated causal structure.

II. GREATER NUMBERS OF THE ELDERLY

Our world will be transformed by population aging. Increasing life expectancy and falling birth rates will alter the age structures of every country. Exact predictions are difficult to make, but by 2050, more than 22 percent of the world population could be older than sixty years, of whom about one in five will be older than eighty years. The good news is that substantial numbers will have remained healthier with shorter and later periods of illness. Obvious economic changes will arise: Older adults have different needs and capacities. It is likely that worldwide economic conditions will change; people will remain in work longer and draw down their savings later. At national levels, changes in individual behavior of older people will affect how governments manage macroeconomic systems (including health care) and how governments will be influenced strongly by the political behavior of older people whose voting intentions will address their interests and needs in ways that differ from the young.

This book explores the consequences of brain aging at three levels. First, what are the facts needed to understand why people differ in rates of brain aging? Second, can the sources of differences in rates of brain aging be translated into practical and effective countermeasures to slow or even prevent dementia? And third, can a useful distinction be drawn between factors that determine individual susceptibility to dementia and differences in rates of brain aging?

Effective antidementia interventions present two main challenges. The first challenge concerns the health of brain cells.[1] We understand clearly the importance of good nutrition and the control of harmful effects of unwanted inflammation and failures to repair networks of brain cells. In parallel with these important examples, a second challenge appears. This challenge concerns the work performed by the brain to self-organize; to develop new networks and devise new ways of working to improve performance; to adjust to changing environments; and, potentially, to make good any age-related damage to brain cells. Importantly, some experts regard brain aging as an overall consequence of failure to maintain and repair brain cells. Mental activity is a major determinant of the patterns and density of connections between brain cells and, not surprisingly, the beneficial effects of effortful mental exercise on the aging brain are promoted widely.

III. THE ROLE OF EPIGENETICS
IN BRAIN AGING AND DEMENTIA

In step with a wider acceptance of the importance of early development for adult health, molecular biology began to probe questions that had puzzled scientists for more than a century. Whenever two genetically identical individuals (identical twins) are not identical in their medical histories or when a change in the environment (as with acute famine) has a biological effect that lasts long after the event, the science of epigenetics aims to provide answers. When scientists talk about epigenetics, they are trying to explain observations in which changes in gene expression cannot be explained by an alteration in the underlying genetic code. Advances in epigenetics were applied to biological programming, and a better understanding of pathways from impaired early development to adult disease began to emerge.

Important advances in epigenetics took place as several long-term studies from birth to midlife started to bear fruit. These findings linked childhood disadvantages to the onset of the common chronic diseases of adulthood. The British National Survey of Health and Development was among the first of these studies, starting with children born in one week in March 1946. By around thirty-six years of age, these children had begun to suffer the first symptoms of chronic disease. Early results showed, for example, that the risk of lung disease at the age of thirty-six could be linked to the accumulation of adversities in early life, including an overcrowded home, a neighborhood with high atmospheric pollution, and tobacco smoking from an early age. Although these types of harmful exposure already were implicated among the causes of disease, the long-term follow-up of a single birth cohort showed for the first time how these effects could accumulate across the life course. Sometimes the harmful effects were greater than the simple sum of separate adversities.

IV. SOCIAL CLASS AND THE LIFE COURSE

The idea that social class could explain why some individuals were at greater risk of chronic adult diseases has become accepted in the twentieth century. By the twenty-first century, however, a more nuanced view was taken. Social class differences in disease mortality appeared so exact—and persisted in the face of overall improvements in material standards of living—that a more substantial explanation was needed. Social scientists acknowledged the important role played by social structures in determining the provision of health literacy and education, access to and effective use of health services, and the use of social supportive networks to mitigate the effects of disease.

Life course studies showed how the common illnesses of childhood, parental social class, and education could predict poor dietary habits, obesity, and low occupational status. Complex links were found that showed how, for example, manual workers were affected more significantly by childhood adversities than those in white-collar jobs. Higher educational attainment was not always beneficial for health. For example, better-educated women more often delayed marriage, gave birth at a later age, and therefore were understood to experience a greater risk of breast cancer. Brain science has now reached a point at which understanding the interplay of genes and environment is strong enough to explore the brain mechanisms of development and aging.

We must navigate many pitfalls, however, before introducing the results of long-term studies of aging into a comprehensive picture of differences among people in rates of aging and disease. For example, a sample of older adults—say, those around the age of seventy-five—will be made up of those who have survived to that age. In the case of those who have been disadvantaged throughout life, survival alone is noteworthy and sometimes exceptional. Often these "hardy survivors" have employed many ways to resist deprivation and hardship, and they even may possess genetic advantages useful in disease resistance. Although important, these are not the only beneficial attributes of "survivorship." Lacking biological factors to protect against disease, some survivors may have been supported throughout their life by others, such as social groups who consider it virtuous to provide assistance to the less fortunate. Such concerns as these strongly influence the design of studies of aging of mental abilities. Timothy Salthouse of the University of Virginia has shown how to improve aging brain research, and many sections of this book have been influenced by his critiques.[2]

V. THE STUDY OF INDIVIDUAL DIFFERENCES

To understand how and why individuals differ in rates of brain aging and vulnerability to the dementias of late life, it is useful to grasp the relevant principles of brain development and organization. The physical changes that accompany brain aging are just as pertinent to the problem of the dementias, and so these problems are also set out in this manuscript. Likewise, important advances recently have been made in understanding the evolution of aging biological systems—of which the brain is the supremely complex example—and in understanding the key roles played through the regulation of genes. Although the topic of evolution may not appear to be of direct relevance to dementia, its introduction in this manuscript will open the door to the discussion of pathways that determine brain development and, by implication, their potential modification as strategies in neural regeneration. These implications are summarized before two key chapters on the *disconnected mind* (Chapter 7) and *emotional aging* (Chapter 8) are introduced. These chapters are influenced by the work of my colleague Ian Deary from the University of Edinburgh.[3] The intention is to explain the major biological, psychological, and social influences on the aging mind. It is also necessary, however, to strike a balance between acceptance that some but not all mental abilities decline in late life and the solid observation that not all abilities are affected; indeed, some will improve. This is especially true of emotional life in which, contrary to popular belief, emotional health tends to improve with age.

VI. RESILIENCE AND SOCIAL CAPITAL

A life course includes a succession of defined events and roles played by an individual over time. Timing of events rarely can be fixed precisely, but many major events (e.g., age at marriage) have "sociocultural norms." Opportunities to take on a particular role (e.g., starting a first job) will vary by economic circumstances and training or education requirements, so the timing of these events also will vary. Events and roles make up an accumulation of experiences that may be positive or negative. These experiences determine later successes or failures with growing physical and mental maturity. Some types of experience, such as the death of a parent, may lead to different long-term outcomes that vary, depending on the support available and the age of the child at which parental death occurred. Poor outcomes following death may "sensitize" some children to the breakdown of other close attachments or, for others, may provide an opportunity to learn how to manage stress successfully. In early development, the role of the family is of paramount importance.

We experience and accumulate memories of our world from infancy to old age. Many experiences influence our personal resilience in the face of aging and disease. From conception onward for almost a hundred years, lifetime experiences shape our brain and provide the resources necessary to ensure good health. When disease arises, our resilience will buffer its effects and preserve the functions of mind and body. We will question what is happening to us: "Is this normal? Am I just growing old? Could I be ill? Can I hope that from experience I will acquire many positive, helpful attributes?"

What are we to call this reservoir of experiences, family supports, and social networks that integrate us into our communities and shape our social brains in such mutually helpful ways? In modern times, the term "social capital" is used widely to convey these personal resources, but its use suffers from never having been defined satisfactorily. Robert Putnam (quoted by David Halpern) defined social capital as "features of social organization, such as trust, norms, and networks that improve the efficiency of society by facilitating coordinated actions."[4] Seen in this way, social capital is particularly fluid. In contrast, other forms of capital seem quite solid. We can speak easily of "financial capital" or "commodity capital"[5] and understand how fixed these can be, while retaining their value. Social capital is energized by the people making up its matrix of relationships, dreams, desires, and acts of generosity. To make sense of social capital, some social scientists focus on the capabilities of individuals to obtain benefits from their social networks in much the same way that a leading actor might advance a stage career by working with a company that "gets the best out of every performance."

Difficulties with definitions, especially concepts that rely on context to be explained, pose considerable problems for measurement of social capital. In Chapter 8, Emotional Aging, Susan Charles from the University of California–Los Angeles and Laura Carstensen from Stanford University address relevant issues.[6] Social networks, emotional support, and opportunities to share confidences are shown to be relevant to two important corollaries of brain aging: cognitive and emotional aging. Both of these corollaries are affected by their social context so that cognitive aging affects the individual's immediate social network and, at the same time, the individual's social network affects his or her cognition.

Social capital also can be understood from the perspective of the political structures that govern how we live. The idea that the state should meet all our needs "from cradle to grave" is an aspiration of some political activists. Claimed health improvements in the former Soviet Union, contemporary Cuba, and Israel have been attributed to comprehensive social networks that improved quality of life from near-destitute serfdom to the expectancy of a longer life. When these social networks collapse—as in the former Soviet Union—major inequalities emerge with greatly divergent health outcomes for the population.

VII. DEVELOPMENTAL ORIGINS OF ADULT DISEASE

Writing in the *Journal of the American Medical Association* in 2009, Jack Shonkoff, Thomas Boyce, and Bruce McEwen[7] described a consensus among scientists that the origins of many common adult diseases occur in early life. The immediate impact of this level of agreement led to a reinvigorated health policy. A less obvious outcome was to promote scientific studies that aim to understand the biology of the mechanisms that transform childhood adversity into increased risk of disease. What is clear by now is that the brain plays a central role in modifying risk of disease. For these reasons, it is necessary to consider how exposures to stressful events might harm the brain and increase the risk of dementia.

Given the sheer volume of research on the brain and the wide attention given to it by the popular press, it can be difficult to remain levelheaded. Because this book centers on the life course approach to brain aging, most relevant studies are descriptive or observational and do not depend on any sort of human experiment. They most often are nonexperimental; in a natural way, observations of people are made in one setting and compared with observations made in other settings. Life course research follows the trajectory of human development over time. Observations made on entry into a longitudinal design detect subsequent changes, and these changes then are related to variations in the environment (e.g., in quality of parenting, an adequate diet, or exposure to toxic substances).

VIII. LIMITATIONS OF ASSOCIATION STUDIES
AND THE PERILS OF REDUCTIONISM

Association studies support relationships between differences in the environment and different types of long-term outcome. Problems with these types of studies include events that rarely arise independently—that is, one type of event may predispose another (the so-called chain reaction). For example, recurrent lung infections may be linked to overcrowding at home, school absences, and inadequate health care. Any one of these might be linked plausibly to poor lung function in old age. Sometimes, the early effects of the disease cause the supposed effect of a factor believed to increase the risk of disease. This is a type of "reverse causality."[8] A good example is the poor diet habitually consumed by old people who later are thought to be living with dementia. Closer investigation reveals, in some instances, that poor diet was determined by the early symptoms of dementia that impaired planning of meals and their preparation. Therefore, associations do not establish causes and, by themselves, do not provide a framework for understanding brain aging; associations are observations that require explanation. Critically, association studies do not provide a sound basis for advice about interventions as in, for example, the recommendation to increase consumption of specific foodstuffs to "prevent dementia."

Ideas about associations, their strength, and direction, nevertheless, do lead to testable propositions concerning possible biological mechanisms that could explain an association. Under the principle of "biological plausibility," a comprehensive grasp of the underlying biology is essential. In the case of the brain—more than any other biological system—the biology is often extremely complex, and so many steps can be missing in seemingly intricate pathways that what is left looks hardly better than conjecture. This point takes us to the last item on the list of criteria to be added to the checklist of problems with association studies—that is, be wary of "biological reductionism."

Reductionism is a process that shows how a larger number of items (e.g., properties or facts) can be dispensed with in favor of a smaller number of connected ideas or theories. There are many examples of valid reductionism (e.g., how the laws of optics are reducible to the principles of electromagnetic theory), and it is certain that for science to progress, a process of reduction invariably is required. By itself, reductionism presents no problems. A scientific argument begins with the separation of useful information from what is irrelevant or unreliable. The information that is left requires some degree of ordering with an emphasis on economy of explanation. This is not the "ugly reductionism" damned as "biological determinism" by existentialists but rather what must remain at the core of scientific thinking. Determinism can be equally misleading, especially when a gene is reported as being linked to a trait described as "innate."[9] When robust correlations are found, for example, between the presence of a gene and a specific behavioral trait, it is tempting to state that the trait is "innate" among those people who carry that gene. The concept of innateness is attractive but never precisely defined. Does it mean that the potential for a particular behavior is present from birth? Or, in addition to "innateness," must the environment contribute another component before the behavior is manifest?

What is known about the biology of the brain limits how theories can be developed. Speculation about the true nature of mental function requires critical knowledge of the underlying biology; without it, the intrepid reductionist will be mocked for making plans without a suitable foundation. In *Lifelines: Biology, Freedom, Determinism*, Steven Rose[10] has constructed a framework from which to view the life course. He stresses the capacity of individuals to maintain themselves and to self-organize. His framework aims to mirror the complex interactions between experiences and biology across a lifetime. Rose is certainly opinionated (and very entertaining) but not necessarily right. He does, however, counter the careless acceptance of genetic determinism that can pervade medical thinking.

IX. SYSTEMS BIOLOGY, REDUCTIONISM, AND THE VALUE OF VISUALIZATION

Being lost in a familiar place is a common cause of anxiety and sometimes annoyance. Like most animals, we rely on reliable ways of establishing "Where am I?" "Where am I going?" "What is that place over there?" "Where will I find food?" Reliability is sustained by establishing the means to process information repeatedly without error, each time achieving the same result. Navigation is an example of how systems develop and learn how to do this accurately. A navigation system must show how one place relates to another and then store this information in a way that is easily retrievable. Comparisons between animal and human behavior provide potentially useful insights. For example, Luca Tommasi and colleagues[11] examined the suggestion that animals have evolved systems for navigating their way with increasing success so that primates would be better than, for example, rats and mice. In similar fashion, adults would be superior to children.

Over the course of evolution, the vertebrate brain has conserved systems and their component structures devoted to processing spatial information. To survive, these structures must be linked efficiently to the capacity to integrate various types of visual, topographical, tactile, and olfactory data with planning movement and coordinating many behavioral routines. The accurate mental representation of space introduces the possibility of adding abstract reasoning about the geometry of spatial relationships. Possession of this ability allows new ways of imagining the spaces we inhabit. Perhaps this is where humans differ most from other primates: We share many basic abilities (like spatial navigation) but are unique in our capacity to elaborate on this basic ability and to use our most recently evolved brain structures to do this. Without evolutionary understanding, a great deal of human biology—including human behavior—is extraordinarily difficult to study in health and disease.

The best-known current example of the perils of biological reductionism stems from the human genome project. A massive international research effort was made "to crack the human genetic code." By 2000, newspaper headlines claimed that "the book of life" now could be read and predicted a revolution in medicine. The first indication that something was not quite right came when it was realized that humans had fewer than 25,000 genes in total—about 25 percent of the number expected by many biologists. If the human genome contained too few genes to account for what makes us human, scientists began to take more seriously all the other noncoding DNA previously dismissed as "junk DNA" (or more exactly as "enigmatic DNA") because it did not function by providing a blueprint for proteins. From this unlikely position, geneticists started to explore how this noncoding DNA might act by regulating the activity of coding DNA. Quite quickly, ideas were advanced about how the environment (especially diet) could alter the molecular structure of this regulatory DNA. Interest in the actions of noncoding DNA currently goes beyond the proposition that there is "a gene for each disease." Eva Jablonka and Marion Lamb have challenged the philosophical and empirical foundations of these reductionist ideas comprehensively.[12]

Unavoidably complex ideas are at first difficult to grasp. One way to start thinking about these biological systems is to look at a map of an urban underground electric railway that serves a major city. At first, these maps were based on the precise geographic relationships between stations and the routes followed between stations. As systems expanded, these maps became confusing and a stroke of imagination was needed. Harry Beck, who made a comprehensible plan of the London underground, provided a breakthrough. The public took immediately to his novel design, and it soon was adapted around the world. When we imagine biological systems, we can reach similar levels of understanding. Taking our lead from Harry Beck, we can reduce the "clutter," get down to basic relationships, and spot where we can switch lines. These are regulatory switches present in biological systems that direct the flow of biological processes, switching "on" or "off" as needed.

Biologists sometimes insist on a "right way" to analyze living things. They argue that we must give priority to large regulatory biomolecules that are highly conserved in evolution and that subordinate all other data (at levels from whole populations to cellular systems) to the discipline of molecular biology. Many other scientific disciplines are available for the study of biological systems and all merit consideration. Some incorporate advances in mathematics and physics and were not added merely as concessions to those who want biological research to seem more "scientific." They were made to improve the independence of analysis of precise data and to compare the results of different and often-competing analytical methods.

These deliberations surface from an ocean of scientific data that could overwhelm our capacity to make sense of it all. What emerges is a scientific approach to understanding our own biology that promises a great deal. By adding the suffix "-omics" to a word in science, an attempt is made to convey a sense of comprehensiveness.[13] Many widely used similar words (like genomics, proteomics, nutrigenomics, metabolomics, and pharmacogenomics) are now of recognized value to medicine. For example, proteomics refers to all types of proteins in a specific tissue, and metabolomics includes all types of molecules in metabolic pathways. If all the "-omics" could be pulled together, their collected aggregation would represent the whole biological system of an organism; dividing the "-omics" by using prefixes makes it easier to apply these terms effectively. This division is important in a complex disease like Alzheimer's disease (AD) in which relevant molecular pathways are becoming much clearer, although as yet not fully understood. So far, molecular measurements have not supplanted clinical judgment. They can, however, support a clinical decision. In time, knowledge of molecular pathology may become be indispensible when choosing a drug to treat diseases like AD and to monitor treatment effectiveness.

Rapid expansion of already vast quantities of data tests threatens to overwhelm our capacity to understand biological systems, particularly when these data are connected with other nonbiomedical databases like genealogy, economic, or social media–generated information. The present solution is to try to build data maps that allow for the automated implementation of rules that alter structures of molecular components and that show how molecules form functional units that need not be close together. These data maps pose complex problems when attempting to explore the expression of specific genes in defined intracellular locations. The preferred method to solve this type of problem is to visualize which genes are "up-" or "down-regulated" in specific experimental conditions, sometimes using color gradients to show these variations and then animate the results. These approaches confirm the adage that a "picture is worth a thousand words" and appear well worth the current huge investments in developing visual analytical methods to investigate biological processes.[14] These maps represent a quantum leap in understanding the connectedness of biological systems with obvious parallels in the solution devised by Harry Beck to guide travelers through the London Tube. Using this analogy, the maps produced in systems biology are expanded to include the surface topography of central London with color variations showing passenger flow around critical choices in journey planning.

An immediate contribution is to underscore the importance of reaching an ultimate goal of "personalized and preventive medicine" that makes available the results of all possible technological advances and captures all that is knowable about an individual's biology. Such a detailed and complex data set will not only predict individual disease risks but also help prevent disease, make exact diagnoses, and anticipate favorable or unfavorable responses to drug treatments. So far, these methods have provided limited means to stratify, within diagnostic categories, groups of patients sharing similar features. This is much more like "stratified medicine" than the hoped-for "personalized medicine" promised by the analysis of "-omic" databases. So far, disease prevention strategies based on approaches to the personal classification of disease risk outlined in this and the preceding paragraph have not challenged the established benefits of public health policies or revealed novel approaches to drug design.[15,16]

Faced by rapid progress in biological research in brain science and aging, clinical scientists consciously avoid attempts to reduce a vast array of new information into a simplified theoretical model. Instead, acceptance is increasing of the need for an approach to understanding diseases as disturbances of biological systems. In its initial form, this approach incorporated data from scientific disciplines as different as physics and molecular genetics. At first, it seemed that once sufficient knowledge about molecular genetics became available, the full complexity of the individual would unravel. Better-considered views are now widely discussed. A full understanding of interacting biological systems will include an analysis of how these systems function at different developmental stages. But what exactly is the science behind such a daunting task? The term "systems biology" is used to describe this type of science and is defined as follows:

> Systems biology is a groundbreaking scientific approach that seeks to understand how all the individual components of a biological system interact in time and space to determine the functioning of the system. It allows insight into the large amounts of data from molecular biology and genomic research, integrated with an understanding of physiology to model the complex function of cells, organs and whole organisms, bringing with it the potential to improve our knowledge of health and disease.[17]

The inclusiveness of systems biology means that a huge amount of data is needed to model systems and—given the speed of technological advances in data collection—to devise computational models that incorporate results of high-throughput and high-content experiments and contemporary advances in underlying theory. The science of brain aging and dementia now faces many problems of this type, sometimes regarded as "data-rich but hypothesis-poor."[18]

X. THE BIOLOGY OF THE AGING MIND AND BRAIN

When trying to make sense of all the genes, neurons, and brain regions that support mental activity, it is tempting to assign functions to parts. This is what nineteenth-century scientists did when assigning a function to a single brain area. A gene functions in a definable way and makes just a small contribution to higher mental functions. The same gene can contribute to several—if not many—higher functions. A mutant gene, for example, is spoken of as if it "causes" a brain disease. It does no such thing. From analysis of genes, AD proves not to be a single genetic disorder but several. Molecular pathways toward AD involve multiple genes. Some of these pathways are involved with the abnormal processing of a large cell surface molecule and others are involved with insulin signaling and inflammation.

A mutant gene codes for an abnormal protein that leads to a sequence of molecular events that lead eventually to a disease. This is not the same as "causing" a disease. As Kenneth Kosik at Harvard Medical School has written,

> Databases from which integrative neuroscience can be built are appearing—the genomes, transcriptomes, proteomes and interactomes of all model systems that will imperfectly map onto imaging and physiology databases. No individual component of these intertwined data sets can be ascribed an independent function. Species comparisons will be as essential for defining physiological function as proteomic signatures are for identifying functional domains. Indeed, conservation in the organization of dendritic architecture is not obvious in that homologous neurons identified in two organisms with virtually identical genomes exhibit detailed branch differences. But other information such as receptor distribution or synaptic patterning might show conserved features that serve the same function across species.[19]

This explanation makes very good sense. It strengthens arguments against classification of AD as a single entity. Instead, it points the way to using knowledge about genetic mutations as landmarks that identify which of many pathways lead to AD and where along those various pathways there are points at which disease may be slowed or even halted.

The brain is made up of layers that are organized to perform tasks. Information flows between brain areas each contributing to the successful completion of a task. In an important way, it is misleading to assign functions to specific brain parts when the parts are all so closely interlinked. How these links are made, especially the fine connections between brain cells, the bundles of nerve fibers that make up the nerve tracts of the white matter, present one of the greatest challenges in brain science. The connections are so complex that the task of mapping them in their entirety is many times greater than completing the human genome. The shorthand term for all of these data on brain connections is called the "connectome," and its completion is a priority in aging brain research.

Two methods of studying brain systems in aging and the dementias are the hottest topics in understanding the biology of mental functions: molecular genetics and functional brain imaging. Like brain development, the science of molecular genetics is easier to understand from an evolutionary perspective, and this is where we will go for a comprehensive overview. Much as a map of an underground electric railway can give a newcomer a near-instant route to plan a journey, functional brain imaging provides pictures of the brain at work in health and disease that would require many words of text to explain.

The cortex makes up about three-quarters of the brain. Diseases of the cortex are major causes of dementia and premature death in children, adults, and old people. Well-known examples include epilepsy in childhood, schizophrenia in adolescence, and AD in late life. Although animal studies can provide a great deal of what is known about the human cortex, very large (and obvious) differences exist among humans, chimpanzees, and rats or mice. These differences are more than just the size of the cortex. The human cortex is vastly more complex in structure and connections. Human cortical cells are much more diverse and make more intricate local networks of connections. These differences between humans and other animals are achieved through longer development in the womb. The basic cellular structure of the rat cortex is established within six days, whereas the human cortex needs about seventy days to develop. Thereafter, these cells build networks of connections locally within and between cortical regions and with more distant brain structure lying below the cortex.

One of principal applications of this type of research was to explore how knowledge of the development of the cortex could be applied to the treatment of AD. Is it ever realistic to invest efforts in generating sources of new brain cells to treat AD? If these new cells have the potential to become one of the many different types of cortical cells, can we expect them to make exact connections with other brain cells to make good lost mental abilities? This type of work is included under the heading "regenerative neurology" and is linked closely to the role of synaptic plasticity in the developing nervous system and to its roles in brain maintenance and repair in response to brain injury or age-related neurodegenerative disease.

The biology of the aging brain provides a platform for the "information processing model" of brainwork. This model depends on the acceptance of three key assumptions: first, that people are actively engaged in processing both the amounts and types of information; second, that measurable aspects of performance reveal differences in information processing efficiency; and, third, that information is passed through a hypothetical series of stages or stores. An example of this is provided by the powers of attention, which are divided into actions that direct attention, select matters requiring thought, and at times divide attention so that multiple tasks can be tackled simultaneously. Attention is affected by brain aging and the consequences are seen in the slowing of mental performance and in managing complex tasks—like driving an automobile.

The information processing model of brain aging raises questions about changes in memory with aging. Are age differences between young and old people in short-term, long-term, and autobiographical memory explained by differences in information stores? In routine tasks during daily life, many different types of memory are used—for example, for places when finding a route between familiar landmarks (topographical memory)—and these types of memory are affected by aging. The search for the biological components of information processing systems is expected to be fruitful.[7] It would be helpful if this search yielded systems that are compact enough to explain succinctly how in youth we acquire the powers of memory, storing, organizing, and retrieving information often with great efficiency. More complex information processing systems are emerging, however. With aging, this mature complexity begins to unravel, causing changes in brain structures and their connections.

XI. SOCIAL COGNITION, AGING, AND DEMENTIA

The concept of "cognitive reserve" is useful shorthand for what is available to the brain to buffer the effects of injury and disease. Although the concept of reserve connects diverse data about dementia, it is not a unified functional entity with a well-understood substructure. It is a composite of many different aspects of brain function, social adjustment, and mental life. The concept of cognitive reserve includes the social resources available to the aging individual and overlaps, to an uncertain extent, with social capital.

From a life course perspective on brain aging, these "social components" have critical roles in the integration of many positive ("protective") influences on individual resistance to the presence of dementia. The contributions of childhood education, occupational training and complexity, a socially active and engaged lifestyle, and mentally effortful recreational pursuits contribute to cognitive reserve. Whether these additions are made passively through brain structural compensatory pathways or through active reorganization of brain information processing networks are incompletely understood.

Nutritional factors play important roles in brain development and aging. These factors will be easier to understand later in this book when looking at poor nutrition and responses to brain injury and neurodegenerative disease. In later life, the key contributions of specific micronutrients (e.g., antioxidants) in the effectiveness of cellular ("intrinsic") defenses against age-related damage become more important. The same applies to foodstuffs that have anti-inflammatory actions, as do some essential fatty acids found in fish oils. The importance of dietary nutrients in brain development will be summarized in Chapter 3, The Well-Connected Brain. Dietary habits have major effects on the risk of heart disease, stroke, high blood pressure, and late-onset diabetes. Obesity increases the risk of these diseases and, in turn, the same diseases are linked to the risk of dementia. These issues are of considerable public health importance and are sufficient to jeopardize many of the health gains made in the prevention of heart disease and stroke.

Individual differences in rates of brain aging and vulnerability to the dementias of late life lie at the core of this book. A complete understanding of the sources of these differences requires knowledge derived from physical studies of aging brain structures and functions, the role of genes, nutrition and mechanisms of aging, and the causes of dementia. Together these present a complex evolving landscape of personal growth tempered by mental impairments, loss of family and friends, and progressive disengagement in the face of depleted personal and social resources. The life course approach provides a unifying framework upon which these age-related developments can be placed. This approach also establishes a platform for personal action to anticipate brain aging and dementia and to implement measures to lessen the impact of adversities whenever these are encountered across the life course.

2

THE LIFE COURSE APPROACH

I. INFLUENTIAL IDEAS

Many older people take easily to the idea that their lives have progressed by stages with periods of stability followed by periods of change. For them, this is "normal aging." Many of these individuals can recall events that were critical for them at specific times and motivated major changes. Examples are found in decisions to change jobs or to divorce. These ideas have helped inform their understanding of aging processes. From a personal perspective, these events shape ideas about the influence of critical phases or stages of adult development. An individual, looking back over their life, may question whether these influences were as important as they seemed at the time. Did things really happen that way? Were events really so important? Generally, such self-doubt seems rare. In conversations about a personal "life story," it appears to be easier to relate what happened when the most important events occurred, typically when these concerned the actual or threatened loss of a loved one or independence. This type of personal understanding of a lifetime of experiences provides a framework in which to place the findings from many different scientific disciplines. This life course framework forms the basis of this book.

Many life course factors link poor health and premature death. A strong relationship exists among adults between poverty and mortality. In addition, many survey data support links between hardship in childhood and impaired adult mental and emotional development.[1-5] One explanation for this link is that those with less access to material resources and services always will be disadvantaged when trying to maintain health. Although it seems obvious, this idea is contradicted by the fact that in the face of substantial improvements over the past century in personal wealth—even among the poorest people—those who are less well off remain at a health disadvantage compared with those who are better off.

Could some other factor or group of factors cause health inequalities to persist in the face of material betterment? This apparent paradox has prompted a search for alternative explanations of the connections between poverty and poor health.

One plausible alternative explanation comes from health promotion research. Many studies show that exposure to severe or chronic social stress in childhood leads to unhealthy behaviors like substance abuse, risk taking, and failure to follow healthy lifestyles. These all contribute to early death. Careful analysis of these surveys suggests other routes from childhood to poor health in adulthood. Adults with low self esteem, depressive symptoms, and chronic hostility tend to have poor health. The origins of these psychological traits have links to childhood homes disrupted by divorce or separation, and boys appear to be more susceptible to these traits than girls. Links among poor health, poverty, and acquisition of unhealthy behaviors may be explained by poor people having few opportunities to manage stress effectively.[6]

Figure 2.1 shows a schematic of possible pathways from socioeconomic stress in childhood, from greater exposure to stress, and through specific brain structures to some of the deficits recognized more often in adults who have experienced hardships and stress in early development. Notice that four intermediate boxes appear along these pathways. These intermediate boxes represent brain structures with established links to the psychological and emotional functions shown in the left-hand boxes. Exposure to stress can affect these structures to their lasting detriment. Understanding how these ill effects arise and why they can persist for many years—even decades—is a difficult task. Answering these questions lies at the heart of this book, which will draw on the findings of many different scientific disciplines—for example, biomedical scientists concerned with developmental origins of healthy well-being and disease have studied the life course approach intently.[7] Later in this chapter, a biological paradigm is set out as a response to cues identified in a poor intrauterine environment that predict hardship in early postuterine life. When expectation and reality are mismatched (as when a disadvantaged intrauterine environment precedes a fortunate postuterine life), the greater the mismatch, and the greater the hazard posed for adult health.[8]

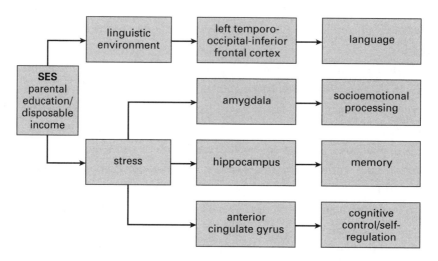

FIGURE 2.1 Proposed pathways from socioeconomic status (SES) in childhood that can affect linguistic development and language skills and from SES through greater exposure to stress and changes in socioemotional processing and cognitive control. Intervening brain structures are identified on these pathways and explained more fully later in the book.

THE TERMAN LIFE-CYCLE STUDY

The history of life course research begins in the early twentieth century. From this point, it is possible to pick out two historical strands. The first strand is concerned with improving the overall well-being of the population at large (as in the Terman study discussed in the next paragraph). The second strand explores the individual life course (as charted by Adolf Meyer from about 1915).[9]

Lewis M. Terman was a pioneer of psychological testing. He began a study in 1921 of the personal characteristics of Californian children born around 1910, with high ability, and aged about eleven years.[10] He recruited 1,528 children from the top 1 percent of the population. These children were described as "gifted children" and sometimes more colloquially as "Termites." Data collection continued as long as participants survived. There were 720 survivors in 1991. Terman successfully dispelled myths that intellectually superior children were sickly and emotionally weak by documenting their life histories. Follow-up over seventy years was conducted by questionnaires and personal interviews. The Terman Life-Cycle Study eventually provided uniquely detailed life histories. Records included health, physical, and emotional development;

school histories; recreational activities; and home life. The study also tracked information about family background as well as educational, vocational, and marital histories. Income, emotional stability, and sociopolitical attitudes also were documented.

The Terman study investigated contributions of childhood factors to long-term outcomes that include health and mortality. In a subsample of 1,285 Terman children, 560 had died by 1991. Predictors of shorter survival included male sex and family break-up before age twenty-one. Childhood personality traits of "lack of cheerfulness" and "conscientiousness" predicted a longer life. Socioeconomic factors and low birth weight among this predominantly middle-class sample did not contribute to the prediction of time to death.

The findings from the Terman study influenced the design of many later investigations. These findings emphasize the importance of supporting intensive data collection over the entire life course. Although sometimes criticized because their baseline data measurements seem quite out of date some ninety years later, they are often the most reliable information available on the influence of childhood on long-term adult health.

Terman led the way for other psychologists to explore life courses followed by children of different levels of ability. Among them was Godfrey Thompson, a Scottish psychologist who surveyed the mental ability of all 89,498 children born in 1921 and attending school in Scotland in June 1932. His data archive was investigated and as well as the links between mortality and various health outcomes, including dementia.[11] These surveys are unique because they included children of both sexes drawn from all ability levels.

ADOLF MEYER

Using life charts, Adolf Meyer (1866–1950) developed a personal vision of a holistic integration of physiological responsiveness with an individual's lifetime cumulative experience of advantage and adversity. The key feature of Meyer's Life Chart is that these are pragmatic, commonsense tools to show over time how quite complex and diverse factors are associated with consistent variations in adaptive success. By these means, Meyer hoped to discover how biological structures were related to mental functions. He avoided attempts to identify through introspection the elements of mental life (e.g., consciousness). Eventually, he hoped that his life history methods would show how physiology and psychology were organized in ways that produced the uniqueness of each individual. His methods were neglected by his successors who found Meyer's Life Charts to be unsatisfactory and time consuming. More telling, perhaps, was Meyer's disparagement of Freudian language saying that he preferred "to turn the verbal into the visual"—not, as psychoanalysts had done, turn the visual (dreams) into the verbal.

To modern eyes, Meyer's Life Chart makes only superficial sense: Calendar dates are listed vertically down the right and chronological age down the left. Specific events or episodes of illness are noted, as are their durations. When Meyer represents physical growth by weight (e.g., brain weight) down the central vertical column, it suggests that his measures were acting as proxies for other more fundamental but as yet unknown biological substrates of psychological function. This is a most unsatisfactory aspect of Meyer's Life Chart methods. Nevertheless, Meyer was making an important point: Life charts can illustrate relationships between personal growth and development, positive and negative influences and the onset of symptoms. Adolf Meyer was strongly opposed to the tortuous explanations of Freud and his followers. He believed his Life Charts were simple illustrations to reveal complex associations between life events and psychological outcomes in a way that can be grasped quickly by patient and clinician alike. His aim was always to contribute an understanding of the part played by experience in explaining personal development.

A systems biology approach combines the high data throughput available to drive models of aging with new hypotheses tested in experiments. Molecular genetic studies provide examples of large data sets (Chapter 1 introduced the suffix "-omics" to illustrate this point). The contest is open to link that which is observable in the mental and physical changes of aging with the overabundance of "-omic" data now available. Optimists believe these links eventually will relate to the fine intricate structure of neural networks that underpin how behavior changes with age. Some theoreticians of aging now argue that these connections will be understood in terms of the evolutionary biology of aging and longevity.

The phrase "resilience in the face of adversity" sometimes is used to convey the effects of adversity over the life course. Charting these complex issues is by no means easy. Acute adversity, such as the death of a sibling, differs from the chronic accumulation of adversities over long time periods. The severity of an experience differs by subjects' ages so that, for example, a teenage girl might react quite differently to the death of her mother had the death occurred when she was much younger. Adverse experiences are not exclusively those events perceived by most outside observers as likely to be stressful. Some types of experience may be stressful because the individual is uniquely sensitized to their recurrence and cannot be generalized to others. Additionally, some types of stressful events are so closely related that if one occurs, another is likely to follow.

Figures 2.2, 2.3, and 2.4 show how the life course can be charted much as Adolf Meyer suggested, but these figures use data that are easier to analyze from a statistical viewpoint. Figure 2.2 shows a schematic representation of a woman's

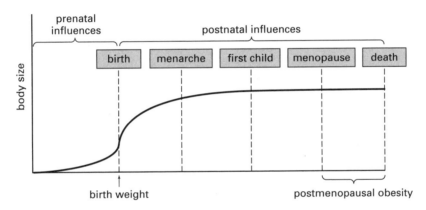

FIGURE 2.2 Growth in body weight over a woman's life course in relation to changes in reproductive status.

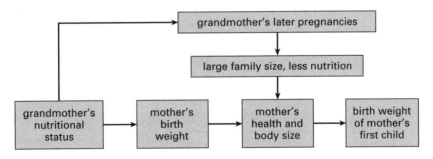

FIGURE 2.3 How grandmother's nutritional status and family size may affect the growth and health of her daughter and later the birth weight of her grandchild.

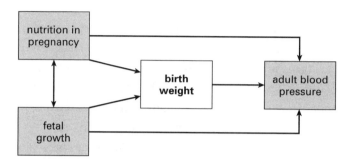

FIGURE 2.4 This figure illustrates the links between what happened during growth in the womb with birth weight. A possible link between birth weight and the regulation of blood pressure in adulthood is shown as a direct line between birth weight and blood pressure. Two other possible links are shown. These links suggest that in addition to a direct link between birth weight and adult blood pressure, the major influence is between mother's nutrition and the adult blood pressure of her child. A third possibility suggests that fetal growth can affect adult blood pressure. The diagram would be much more complicated if we added factors present during pregnancy that persisted into early life and beyond. One such factor might be overcrowding, which can be stressful. If this persisted from pregnancy into maturity, then it might influence blood pressure.

life course. Powerful influences on female body size are placed across the life course, where they can influence overall body size. Figure 2.3 adds a further layer of complex influences on a woman's body size by showing links among three generations of a family (grandmother, mother, and child) and by illustrating how large family size affects availability of food, influences access to material resources, and dilutes parental care.

Another layer is added when we think about how high blood pressure could pass from one generation to another (shown in Figure 2.4). Here, grandmother's nutritional status affects the growth of her child in pregnancy and early life. Her child's growth is less than optimal and her poor diet persists. This might increase her risk of increased blood pressure and that of her children, too.

II. DEVELOPMENTAL STAGES

The life course is popularly understood as a series of stages. Shakespeare likened these stages to the "seven ages of man." When thinking about how best to study the life course, should we try to place findings within the context of each stage?

Or, will findings be easier to understand (and relate to one another) if we accept from the outset that these stages will be continuous without any breakpoints? In everyday speech, we use terms like infancy, childhood, adolescence, adulthood, and old age to place people within age-specific categories. In psychological theories of emotional development, the stages of emotional development in infancy ("oral," "anal," or "phallic"), which were systematized by Freud and his followers, have entered popular speech. Later concepts of lifetime emotional development also have relied on the identification of stages that extend from birth to old age. Erikson has provided influential descriptions of these stages (Freud and Erikson's contributions are discussed in Chapter 8, Emotional Aging).

These widely used stages help us talk about aging, and we will continue to use similar terms throughout this book. It is helpful, however, to qualify the results of many studies of aging processes. These studies provide observations about aging as continuous data, best visualized as smooth changes over time and few abrupt breaks in rate of change.

Take, for example, the proposal that there is a consistent change in midlife that is so sharp and sudden that it is called a "crisis." Among life-changing critical stages, the *midlife crisis* is perhaps the most frequently commented on stage, but it is surprisingly difficult to show that it actually occurs. A midlife crisis is thought to be triggered by an individual facing their mortality. An individual in crisis typically is described as a man who conforms to a stereotype and takes up with a younger woman. Evidence to support the occurrence of crises, especially among women, fails to convince either gender that such crises even occur among women; in fact, they are as frequent among men and as they are among women, making them near universal.

When asked, most middle-age adults do not talk about crises but rather identify many positive aspects of aging. Men talk about job satisfaction, and women talk about self-esteem and becoming better at handling stress. These self-reflections are much more typical of middle age than are assertions of crisis. Many middle-age people become better able to describe their inner feelings, pains, and pleasures.

Successful aging builds on these advances in self-realization by acknowledging both gains and losses. With age comes acceptance that the price of becoming more competent—that is, taking control of matters that are important to oneself—will be offset by loss of physical attractiveness, increasing frailty, and risk of illness and disability.

It is easy to find evidence of transient and accelerated rates of change in physiological studies of human development from conception through adolescence and into maturity. It is much more difficult to find similar crucial (or critical) stages in late adulthood and old age. Rapid growth comes at a price. There are well-recognized periods in early development when children are especially sensitive to their environments. When these environments include harmful agents, the ill effects may persist throughout later development and even carry through to late life. The recognition of such sensitive periods is a fundamental part of health care of children. Identifying these periods and timing interventions are essential to protect children from poor long-term health outcomes. The concern in an account of brain aging and dementia is not with child health but with the possibility that adverse effects on health encountered in childhood may not be detectable until late life.

At the most fundamental level, the relevance of childhood disadvantage is explained in biological terms. During pregnancy, development occurs at great speed, and this high rate of growth makes the child more vulnerable to the effects of harmful environmental exposures than at any time in later life. In general terms, the harmful effects of exposures during pregnancy can be divided into early (before sixteen weeks of gestation), when major structural abnormalities arise, and late (sixteen to forty weeks). Harm in later pregnancy disrupts development and leads to impaired growth (low birth weight), inefficient regulation of body functions (including defenses), and obstetric problems like prematurity and complicated delivery.

Important stages of development are not confined to life in the womb. The brain and nervous system continue to develop throughout childhood and adolescence. Brain cells (neurons) continue to migrate to their correct positions in the brain, new brain cells continue proliferating, and new connections (synapses) form between neurons. Insulation of nerve fibers (myelination) continues throughout early childhood to about ten years of age. Later, in adolescence, the number of connections between nerve cells (synapses) is reduced—a process that probably continues throughout adulthood. Outside the brain, other body systems are established to meet a changing, sometimes hostile, environment. The immune system gradually becomes more competent in the recognition of foreign substances and is linked to the mature body's capacity to deal effectively with stress.

III. CRITICAL PERIODS

Educationalists have long known that opportunities are much greater to achieve excellence as a virtuoso musician the earlier learning and practice begin. The same is true of acquiring a second language (bilingualism). Early learning makes the task easier if the opportunity is available. The benefits may extend beyond childhood. Practicing a musical instrument may improve not only auditory discrimination but also fine motor skills, vocabulary, and nonverbal reasoning.[12] But what of experiences that a child would want to forget? Is it feasible to think that the windows of opportunity opened during critical periods could be reinstated, allowing traumatic experiences to be erased and forgotten? This supposition lies in the realms of science fiction, but what of the damage caused by stroke or disorders of memory like age-related cognitive decline or dementia? Could an understanding of these critical periods prove to be relevant to repair brain tissue that has been damaged by aging or dementia?[13]

In studies that compare members of the same nonhuman species, evidence suggests that these "windows" are tightly controlled. In commercial husbandry, for example, precise knowledge of the timing of critical periods in maternal attachment behavior can be vital. Livestock farmers know that poor maternal care and abandonment of offspring often persists for several generations with consequent increased losses and greater costs of fostering.

The biological basis of memory has several competing theories. Most of these theories assume that learning depends on strengthening the connections between specific groups of neurons in the brain. Each theory has argued for a different way such strengthening might be achieved, including sprouting new nerve endings, increasing connectivity between nerve cells, and increasing sensitivity of nerve cells to "read" the messages from connecting cells.

It is reasonable to assume that learning and memory make some sort of mark on the brain. We know that memory disorders appear with normal aging and, to a catastrophic extent, in dementia. But where in the brain are these marks made, and what is their nature? Gabriel Horn from the University of Cambridge (United Kingdom) took on these questions when he investigated the mechanisms of imprinting in the chick of a domestic hen.[14] After exposure to a stimulus, a day-old chick not only prefers to follow that stimulus but also can recognize it and coordinate its movements appropriately. Such learning is linked in the chick to increased transcription of DNA-coded proteins in regions of chick brain already known to be associated with learning. These regions integrate visual and auditory information, which eventually translates into changes in the local connections made in the nervous system. Horn thought these changes in nerve cell structure were evidence of remodeling of cell-to-cell connections in the chick brain. He thought that the molecules that held nerve cells together (cell surface adhesion molecules) were important components of remodeling. These molecules have important roles in the self-organization of the brain's nervous circuitry and in the molecular biology of Alzheimer's disease.

A critical period in development is an exactly controlled window of time during which specific environmental stimuli trigger biological changes that are essential (i.e., "critical") for normal development. Figure 2.5 shows how four brain systems that underpin competencies in the management of sensory information, understanding and use of language, thinking strategies, and high mental performance are arranged in logical sequence. Eventual maturation of all these competencies is based on successful passage through each stage of development and in the correct order. As an acute observer of infant development, mothers recognize how important these maturational steps are for a child, how these overlap, and how they can depend on each other. These steps were first

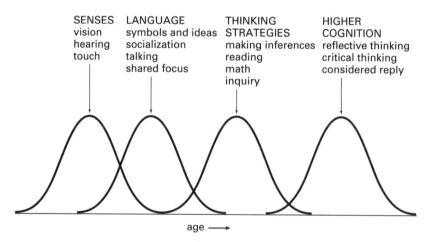

FIGURE 2.5 This illustration shows in diagrammatic form multiple successive critical periods in the acquisition of sensory functions, the use of language, thinking strategies, and the development of higher mental abilities. These are just four systems that could be represented in this way and are overlapping systems without abrupt onsets and offsets.

recognized as "imprinting" in goslings by Konrad Lorenz.[15] Starting in the thirties, Lorenz showed that if he took on the role of a mother goose within a few hours after goslings hatched, the baby geese would follow him as though he were their mother until they were fully grown. Pending major advances in molecular genetics, it was impossible for Lorenz to unlock the molecular machinery that controlled the timing ("onset" and "offset") of these critical periods.

Torsten Weisel and David Hubel in 1963 at Harvard Medical School described how the anatomical structures in brain regions that processed visual information (the visual cortex) were modeled or "sculpted" by incoming visual stimuli.[16] They showed that within the visual cortex, there are cells that respond to one eye and not the other. When one eye was closed permanently, cells that previously had responded to that eye began to respond to the other. The term "plasticity" was introduced to indicate that the brain passed through a period when learning was enhanced quickly, presumably because the brain structures involved were more malleable or "plastic" than at other times. It generally was assumed that the triggers for onset of a critical period would be excitatory.

As new tools were invented to analyze the underlying molecular events in the timing of critical periods, it became clear which pathways were important in controlling their onset and offset. Rather counterintuitively, these pathways turned out to be inhibitory rather than excitatory and, once discovered, suggested ways they could be manipulated.[13,17] Among the first clinical disorders likely to benefit from these discoveries was a type of blindness caused by a congenital squint (amblyopia). Attempts to improve memory and learning after strokes seem likely to follow before major mental disorders are investigated. Pharmacological interventions that "reopen" the windows of opportunity uncovered in these types of studies likely will be combined with training programs to restore neurological function when damaged by stroke, brain injury, and, perhaps, even Alzheimer's disease. Advances of this sort are expected to form parts of programs in restorative or regenerative neurology.

Underlying critical periods are important biological processes (explained in some detail in later chapters), and the growth and proliferation of cells that drives all development is perhaps the most obvious of these processes. When cell numbers expand rapidly, they place great demands on the body's capacity to secure sufficient nourishment. Rapid growth must be controlled, and among more than 210 different types of cell, the body is required to regulate cell replication, which can vary from a few hours to many weeks to complete just one cycle of cell division. Surprisingly, a process as complex as the regulation of cell division is achieved with fewer than ten different regulatory proteins.

Cell division is not the only process that regulates cell numbers. A process that causes cells to die (some say commit "cell suicide") called apoptosis or "programmed cell death" is essential. This process allows unwanted cells to be removed. Apoptosis is critically important in the developing brain.

The DNA that writes the genetic code is translated into proteins that make up the structure of every cell. DNA controls all cell division and mechanisms of cell death. Throughout development, DNA is exceptionally busy, being "switched on" or "switched off" as needed. Gene expression or suppression is regulated by DNA that does not code for proteins ("noncoding DNA"). This type of DNA can be modified by the nutrients present in the womb to make overexpressed or underexpressed DNA, which in turn provides one pathway along which maternal nutrition can affect child development and, possibly, influence vulnerability to disease when that child reaches adulthood. The biology of critical periods is closely involved with the timing of changes in gene expression.

To ensure accurate expression of DNA, it is essential that when errors arise in DNA, they are corrected. Even in simple organisms like bacteria, DNA repair systems are present, and in something as complex as a human brain, it is unsurprising that the DNA repair process is proportionately more intricate and refined. Environmental toxins will interfere with DNA repair. Predictably, several categories of diseases are caused by defects in DNA repair. Some are genetic diseases in their own right caused by inherited errors in the structure of DNA repair genes, while others are the result of direct DNA damage caused by substances as diverse as alcohol (giving rise to fetal alcohol syndrome) and certain types of atmospheric pollution.

The developing child does not have a complete set of mature responses to harmful environmental agents or stressors. Some types of harm can overwhelm a child's ability to manage or compensate for damage. What may be commonly encountered without ill effect in an adult can be transformed to a potentially fatal hazard in the developing child.

These considerations prompted many laboratory studies on the effects on animal development of exposure to possibly harmful environmental agents, but these studies offer only clues as to what is happening in the real world. Studies on whole populations of people are needed to determine whether a possible harmful agent poses a definite danger to child development and, when sufficient numbers of people are studied, to estimate the relative importance of the threat. The outcome of these different scientific disciplines has been a marriage of epidemiology with developmental biology, each working to complement the other to advance our understanding of the roles of heredity and the environment on human development.

Key questions remain about the timing of critical biological processes underpinning development and whether abnormal development is more likely when a fetus or child is stressed during those sensitive times. Answers to most of these questions are substantially incomplete.

IV. PROGRAMMED DEVELOPMENT AND AGING

Developmental processes require precise conditions to be met to ensure that essential body functions mature usefully and efficiently. Although it is easily understood that exposure during childhood to extreme privations like famine will delay development, it is less obvious how the experiences of a baby can influence the effectiveness of the stress responses of a mature adult who was exposed repeatedly to stressful circumstances in early life. The proposition that a genetic code can be read ("transcribed" then "translated") into proteins soon followed the discovery of DNA. The idea that a genetic code can be "programmed" to respond in a predictable way to the environment derives from the first use of punch cards in the control of machine tools. In control engineering, a program of instructions follows a precise sequence of steps to complete a complex task. This entered popular speech in the use of programmable machine tools, of office machinery, and later in military cryptography. "Programming" was applied to learning in the fifties, so it was not unusual to hear that people could be programmed to respond with a specific pattern of behavior in certain circumstances.

Used in developmental biology, the term "program" is closer to the way the term is used in computing. For example, by the thirties, punch cards could enter an external series of instructions to a machine that would perform as designed. In biology, the instructions of the program are coded internally (as DNA), and gene expression is controlled internally with opportunities for the environment to influence (or better, as in the terminology of Eva Jablonka and Marion Lamb,[18] to make an educated guess about the future environment) a set of innate instructions that will determine when each developmental gene is switched on or repressed. Successful optimum development requires that switches operate in exact sequence. One way of looking at a series of switches is suggested by maps of urban metro systems. Points at which passengers can switch from one line to another are shown as diamonds on the underground map. An exact sequence of switches is needed to ensure that a passenger makes the shortest logical journey to a destination. Likewise, developmental processes require a similar exact sequence of genetic switches to achieve optimum results.

Some people are more outgoing, whereas others are more tense or anxious. People are different, too, in how they respond to stress and how resilient they appear in the face of a threat. Some people find stressful experiences exciting, even exhilarating. We often explain these individual differences in terms of a previous experience of adversity or danger. It also seems to be true that many people who once endured unpleasant effects of stress take care to avoid future stress, now fearful of the consequences. If we were able to follow a person over an entire lifetime, to tie in illnesses with early experiences like poor parenting, poverty, and economic insecurity, we might discover links between early life adversities and some specific long-term health outcomes. This type of investigation is rarely practical largely because records are never as comprehensive as needed. Critics argue that bigger and better survey data would be too expensive to undertake and health gains too small to justify the effort.

Nevertheless, from pioneering studies in Europe and the United States, evidence has accumulated steadily to link early life circumstances with a range of adult diseases. These are mostly heart, lungs, and blood vessel diseases, but they also include strong links with increases in overall mortality. Many possible pathways could link adversity or hardship in childhood to adult disease. The relative contribution of each of these pathways is as yet not fully understood and is a current priority in life course research.

Interest among scientists in connections between circumstances experienced early in life and health in later life dates from the work of David Barker in Southampton in the seventies.[19] Barker first proposed the idea that a fetus is "programmed" to respond to the expected environment in quite specific ways. Because of his pioneering studies, conditions in the womb and in the first few years of childhood were linked to specific causes of death and longevity. The idea is new, however, that different pathways to aging and the chronic diseases of late life can be traced to early living conditions and that this pathway would be explained by some sort of programming error. Certainly, many people now accept that the origins of vulnerability to aging and age-related diseases are traceable to early life, but it is not yet accepted that programming is responsible. The most likely outcome seems to be that the term "programming" will be replaced with a more precise lexicon of terms that distinguish among the different regulatory pathways.

Consider, for example, the problem of stroke. In the United States, the incidence of stroke is greater among African Americans than among those of European ancestry. Geographic variation in stroke incidence in the United States was first recognized in the sixties, when higher stroke mortality was reported in a cluster of states in the southeast (later known as the "stroke belt"). At first, higher rates of smoking, poorly treated elevated blood pressure, obesity, diabetes, and alcohol abuse were blamed for the increased rate of strokes. Later explanations implicated some sort of local environmental toxin (like lead or cadmium) or unidentified genetic factors or ineffective delivery of health care, any of which seemed potentially responsible. By the early twenty-first century, however, the problem of increased stroke mortality in the U.S. "stroke belt" was found in both adults and children. Detailed studies showed that neither adult environment nor ethnicity could adequately explain the existence of this so-called stroke belt. The problem affected only U.S. citizens born in the stroke belt and not those who moved there. The numerous studies on this question cannot yet provide an exact explanation of the association between place of birth and risk of stroke.

The original studies by David Barker emphasized the importance of maternal health and how successfully a child grew in the womb as critical factors that could increase the risk of heart disease and stroke in late life.[20] His explanation was that when only a fetus, the developing child was programmed to respond to stress, thereby ensuring resistance to disease and greater longevity. Biological mechanisms now have been sought to explain how such programming could work. Could an inappropriate reaction to stress explain the increased incidence of stroke?

David Barker emphasized the role of dietary influences on growth and maturation of physiologic responses as relevant, even critically important. Eileen Crimmins and Caleb Finch from the University of Southern California proposed a more complex model that includes diet but extends to involve the role of infections.[21] Like Barker, they made use of historic records of adults, but they considered a much wider range of birth epochs from 1751 to 1899. Heart disease and stroke begin with harmful changes in the walls of blood vessels (atherosclerosis), where inflammation has a key role in processes established in the womb. In the survey data, mortality fell among adults born into the same birth cohort that also had experienced a sharp decline in childhood mortality. In other cohorts from the same survey, adult height was lower when people were born during epochs when childhood mortality was greater.

Childhood infections can leave traces of damage throughout the body. Heart, lungs, and blood vessels can show the effects of childhood infections in late adulthood and old age. Crimmins and Finch strengthened the view that early life adversities that include poor diet and childhood infections can jeopardize the chances of good health in later life. In addition, they supported strategies to improve the health of mothers and children, thereby enhancing the prospects of a healthy old age. It seems plausible that comparable effects lay at the root cause of increased incidence of strokes in the U.S. southeast.

How the brain responds to environmental factors, acting early in life, will prove critical to the developing child. The effects of programming on the brain produce lasting organizational changes that are detected most readily in the brain's responses to adversity. The developing brain's regulation of responses to stress has consequences for the mature brain that extend beyond the control of stress-responsive hormones to learning and memory. Cortisol is the most important stress-responsive hormone. It is released from the pituitary gland, a pea-size structure located on the midline at the base of the brain. The corticotrophin-releasing hormone controls the cortisol release system (shown in Figure 2.6), and elements of this system are all potential targets for programming. Like all control systems, the cortisol system must be preset to operate within stipulated limits.

Retarded growth in the womb is a questionable direct cause of an increased risk of adult disease. Genes that affect growth and the risk of later disease are implicated. A good example is shown by the system that controls blood sugar levels. Stress during pregnancy affects fetal brain structure and (in animals) will reduce the number of hippocampal receptors for cortisol (or its equivalent). Among older animals, those who were stressed during their time in the womb learn less well than those who were not stressed. Reduced cortisol receptor numbers could account for this effect on learning.

Currently, two principles explain how programming works. The first involves cortisol, some important neurotransmitters, and naturally occurring opiate-like compounds. The second focuses on alterations in the number of receptors

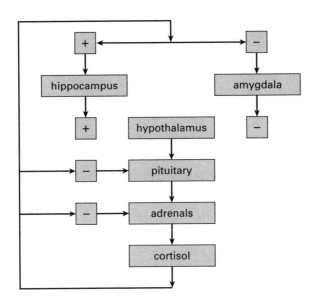

FIGURE 2.6 Cortisol is a stress hormone released from the adrenal cortex. Its release is stimulated by adrenocorticotrophic hormone (ACTH) released from the anterior part of the pituitary gland lying on the base of the brain. Above the pituitary, the hypothalamus integrates many stress responses, including the release of corticotrophin releasing factor (CRF), which stimulates ACTH release. The hypothalamus senses increased cortisol release, and this "feedback" inhibits further release of CRF.

for cortisol found in the hippocampus and amygdala. Either of these pathways could be sufficient to explain either alone or in combination the "thrifty phenotype" hypothesis. This hypothesis proposes that undernutrition of the mother or fetus during pregnancy predisposes the fetus to expect poor nutrition during development and maturity. As an adult, a fetus programmed in this way will aim to retain all available nutrients by laying down stores of energy as body fat. The reasoning here is that fetal physiological control systems are reprogrammed ("preset") during pregnancy to expect a short supply of nutrients. Faced by an environment with an abundance of food, however, this leads to diseases linked to obesity, high blood pressure, and diabetes.

By themselves, these two principles do not provide a satisfactory explanation of how programming takes place. Genes involved in the regulation of fetal growth also play an important role. Evidence for this role is found in the large differences between different strains of closely related animals that respond differently to fetal malnutrition. In humans, stress during pregnancy is linked to birth weight and to the way cortisol production is regulated throughout the life course.[22,23]

Maternal cortisol and maternal psychological stress influence the programming of the fetus. Both maternal cortisol and maternal psychosocial stress (pregnancy-specific anxiety) are associated with the course of infant development. These factors contribute to mental functions by one year of age among healthy, full-term infants. These effects on cognition seem likely to persist throughout the life course. The precise effects of maternal stress are affected by the exact timing during pregnancy of exposure of the fetus to stress.

V. THREE LIVES

Anne was irritated by John. He was forever asking her to repeat something she had said only minutes ago. He needed a list to run even the easiest errand. He had lost all interest in his garden, once his pride and joy. Last winter he stayed in by choice, "too tired" to go out fishing with an old friend, saying he was not "interested in that sort of thing anymore." Anne asked, "Is this normal for a man who is only seventy-two? He still looks young, he gets around, seems as fit as he was ten years ago. We have no money worries, nothing has happened to upset him. Should I get him tested for dementia?"

A life course approach concerns the biological changes of aging, examines the growth and decline of higher mental abilities, and explains how friendships change. Anne's complaints about John can be judged soundly when there is good information about John's previous behavior. Was he always like this or has this happened only recently? What do these symptoms mean for John? Do they stop him from doing things he used to do? What were his jobs, hobbies, and pastimes? Only when comparisons can be made between the man he has become and the "old John" everyone knew could we gauge the nature and severity of his change.

When we place John's life in its historical context, we first identify key points on his life chart, his date and place of birth, his father's job, his early education, military service, marriage, and jobs. John was born into poverty in 1935 in the Midwest after his father left home looking for work. His first years were difficult. His younger brother and sister had died in infancy, and his father failed to find work, never to come home again. Nevertheless, he graduated from high school and worked his way up in a company that prospered in the decades after World War II. After his own marriage and the birth of his children, he brought his frail mother to live with them in her final years. Although his childhood was difficult, he was grateful for the opportunity to care for his mother.

An observer looking over John's shoulder at his whole life so far, the trials and tribulations endured over seventy-two years, would see his strengths and failings. Standing a little farther back would be the points at which some negative influences had once acted, elsewhere would be the sources of his positive attributes and how he coped with adversity so far. At times, out of adversity, he had emerged stronger, empowered by his success. At other times, adversity had defeated him when he had expected to overcome whatever obstacle was put in his way. To the observer it would sometimes seem that "timing is everything." In his maturity, John had coped very well, where as a child he had felt incompetent. Later in his old age, he began to fail but in ways that differed from the dogged determination of his childhood.

Consider two more people, just like John but born at different times. Pauline was born in New York in 1915. Her parents prospered; she completed college and went to work. By the outbreak of war in 1941, she was employed in a large insurance office hoping for a promotion when, as men volunteered for military service, her promotion became a company necessity. Her life seemed "charmed"—marriage, children, and a secure home came in quick succession so that by age forty she had returned to a senior position and by age fifty she was a senior vice president in her organization. When memory problems began at age seventy-five, her family found a well-resourced clinic that gave her an early diagnosis of dementia and, importantly, time and help to make important decisions about her later care, should she need it, and the disposal of her assets ("advance directives").

Johan was born in Northern Holland in 1945. At this time, the retreating German army prevented much essential food and fuel supplies from reaching local people, and through the first months of her pregnancy, Johan's mother starved, finding fewer than 600 calories per day. His birth weight (4.4 pounds) was low for the length of his mother's pregnancy, and although his appearance was unremarkable, he always seemed to be a bit behind in school. He spent his adult life as a manual laborer; his mother died and he never married. He was unemployed when he was found wandering late at night in Groningen, no sign of alcohol or drugs but uncertain of his whereabouts or how he had gotten there. By this time, he was just fifty-nine years old, out of touch with his family, and taking poor care of himself. Careful examination showed that he had many features of dementia of the Alzheimer type. His medical records showed he had been rejected for military service because his reading age placed him at the same level as a child of about nine years. In contrast, his comprehension, memory, and manual dexterity seemed relatively unimpaired. More complex demands that needed abstract reasoning were certainly difficult for him, and his overall speed of thought was slower than expected for his age. Although Johan had a similar performance profile on mental testing as had Pauline, his condition was much more disabling. The care team realized he would need support to live at home, access to regular meals, and a place to go during the day.

VI. THE DEVELOPMENTAL ORIGINS OF AGING AND ADULT DISEASE

The life course approach encourages appreciation of what has happened over a life lived so far, tracing back to the earliest influences in childhood, even back to time spent in the womb. It is now commonplace to consider the childhood origins of many adult diseases—like heart disease or some cancers—but rarely to think that differences between individuals in how quickly they age might originate in childhood.

The idea that adult diseases may originate in early life circumstances was first recognized by physicians working in Europe in the early nineteenth century. In antiquity, a link sometimes was made between the living conditions of poor families and impaired growth of their children, but poor nutrition was not considered the most likely reason. Chronic infections, inadequate hygiene, and maternal care were all thought to be more likely causes. It was not until the twentieth century and the discovery of vitamins that the importance of diet and sunshine was established as essential for the physical development of children.

An association between nutrition in childhood, including time spent in the womb, and adult health was established by follow-up studies of children born during times of great economic hardship. An early example is a study of children born after the famine in North Holland (the 1944 *Hongerwinter* when Johan was born). To punish the Dutch for their unwillingness to help the Nazi war effort,

German forces blockaded Western Holland cutting off not only food but also fuel supplies as well. Food stocks were soon so depleted that people in affected areas were barely surviving on about 600 calories per day. More than 18,000 Dutch people died in a famine made worse by an unusually harsh winter.

Children growing up in adverse conditions tend as adults to remain in similar circumstances. It is difficult to be sure that hardship in childhood determines their poor health and greater mortality or whether the effects of their hard times as adults were more important. Researchers in Glasgow (Scotland, United Kingdom) followed up more than 5,000 men (ages thirty-five to sixty-four years old) and showed that after adjustment for age and adult socioeconomic status (SES), adverse childhood SES was associated with greater mortality, especially deaths caused by stomach cancer or stroke.[24] Other causes of death that affect the heart or lungs were linked to adversity in adulthood. Specific links also have been found between childhood respiratory illnesses (e.g., whooping cough[25]) and reduced adult lung function, suggesting that residual lung capacity available to compensate for adult lung disease is lessened. Evidence also is strong that early life stress can cause mental disorders in adolescence and adulthood. These connections are attributed to lasting changes in the hormonal regulation of stress response and by alterations in brain structure. There are associations between extreme childhood adversity and reduced size of brain regions that are associated with memory (the hippocampus) and regulation of emotions and attention (the caudate nucleus and the anterior cingulate cortex).[26]

The Dutch famine birth cohort study was undertaken in Amsterdam (Holland) and Southampton (United Kingdom). This study followed up the children of pregnant women who were exposed to the Dutch famine and showed that these children were more likely as adults to suffer from diabetes, high blood pressure, and heart disease. Poor nutrition in the womb seemed to have led to health problems in adult life (the fetal origins of adult disease hypothesis). Opportunities to test this proposal are relatively rare and did not arise again until the Chinese famine (1958–1962). Compared with children born immediately before or after the famine, children of pregnant mothers exposed to the Chinese famine were more likely to suffer the same types of health problems as had affected the children of pregnant mothers exposed to the Dutch famine. These problems include diabetes, stroke, high blood pressure, poor respiratory health, impaired mental development, and schizophrenia (a neurodevelopmental mental disorder affecting young adults). More recent follow-up studies have examined children about eleven years old whose mothers were provided with food supplements in feeding programs in the Gambia (West Africa) while they were pregnant or breastfeeding. These African studies have not yet shown whether these children face the same health problems as did the much older survivors of the Dutch and Chinese famines.

Important criticisms have been made of the proposed link between adverse conditions in the womb and adult disease. These criticisms arose because the first reports seemed to be just "fishing expeditions" in which risk factors were teased out of large data sets without good reasons for suspecting their role in disease. The original findings are now so frequently replicated in different countries, however, that the association is regarded as robust. A more important problem is the strong association between poor maternal care in pregnancy, hardship in childhood, and lasting poverty and social disadvantage that persists through adulthood. In the U.S. Nurses' Health Study cohort, the associations between low birth weight, adult heart disease, and stroke were found after adjustment for the effects of adult smoking, poor diet, and lack of exercise in adulthood. The fetal origins of adult disease hypothesis is now accepted widely.[27,28]

Economic records, vital statistics of births and deaths, and notifications of epidemics are all used to examine a link between childhood circumstances and adult health. Using records for the complete business cycles in Holland from 1812 to 2000, an association was found between life expectancy and economic activity. These showed that adults born during times of economic recession would live significantly shorter lives than those born during more prosperous times. A study from France provided a similar result. In wine-growing regions, adults born during the phylloxera outbreak (which destroyed the vines and caused great hardship) were significantly shorter than adults born in the same regions at other times.

The U.S. Adverse Child Experience Study identified more than 17,000 adults and interviewed them about current health and adverse circumstances in early childhood. This study found that a greater number of adverse childhood experiences reported was associated with a greater risk of premature death and worse health problems in later life. Like the famine studies, increased mortality could be attributed to heart disease and chronic respiratory disorders as well as to the nature of the adjustment made by individuals to the society in which they lived. Long-term prospective studies are not, however, without possible sources of major error, and some circumspection is essential when weighing their value.[29]

VII. BRINGING IT ALL TOGETHER

Strong links exist between early life experiences and cognitive function in later life, with some evidence that these differences persist through to old age. Although it is clear that general mental ability in adulthood is strongly determined by inheritance—probably by a large number of genes each of small effect—the effects of extreme poverty, early death of either parent, deprivation, and mental or physical cruelty are all associated with lower than expected mental performance in later life. These adverse experiences act alongside or possibly in tandem with genetic factors. Orphaned children raised in institutions when examined in late life have below-average function of brain regions that are critical to memory and learning. These findings are most marked when compared with same-age adults who never were institutionalized. When institutionalized infants are removed to foster homes, their capacities for mental development are much improved. A lot depends, however, on the timing of such a move. If this move is made before age two years, the benefits are made consistently and less so if the infant is moved later. There is, therefore, a strong possibility that a "critical period" exists for the development of brain processes that will support cognitive function for much of life.

Early life experiences can affect brain regions that support learning and memory. The hippocampus is one of these regions, and it is known to be affected by chronic severe stress in childhood. For example, in some but not all studies, adults who have been abused as children possess a smaller hippocampus than adult who were not abused. This type of research points to the need to investigate the biology of mechanisms that transform adversity over the life course into disadvantaged cognitive ability. The importance of this type of work should not be underestimated given that more than half the children in the world experience chronic severe stresses with effects on their mental development and that so few opportunities exist to prevent or mitigate these effects.

The scientific ideas discussed in this chapter originate in the principles of evolutionary biology. From the time of conception until early infancy, offspring can decipher critical characteristics of their environment. Without secure means to keep warm and well nourished, children gradually will refine a system that anticipates an external world that might be unsafe. The developing child has programs that prepare for a range of possible life circumstances, including many that are highly stressful and unpredictable. It is to the advantage of the maturing child if their biology retains innate programs to anticipate stress and the capacity to transform a potentially adverse experience into one that strengthens their resilience. Later, we will consider the possibility that some biological determinants that enhance coping with stress in early life also improve reproductive success. The same biology may be disadvantageous in postreproductive life and increase the risk of disease. This is termed "antagonistic pleiotropy" (pleiotropy means that a gene can effect more than one character of an offspring; thus, antagonistic pleiotropy means that the different effects of the gene oppose each other).

Since 2000, the United States has invested heavily in home visiting to improve the living conditions and opportunities for young children. The drive behind these programs came from a thorough evaluation of the evidence that early life adversity harmed the development of children to their considerable long-term educational, occupational, and emotional disadvantage. Critical to the acceptance and funding of these programs is the evidence that interventions of this type have long-term benefits for disadvantaged children. These positive returns are found in higher school grades, greater employment prospects when joining a skilled labor force, and, potentially, becoming U.S. citizens better prepared to play a full and active role in the community. It is at this intersection where the study of the life course integrates the findings from many scientific disciplines and provides a substantial foundation for future progress.

APPENDIX A: TERMINOLOGY

The array of terms used in studies of development and aging can be confusing. These terms sometimes are used interchangeably, and frequently their precise usage is not defined. The following definitions clarify the terminology used in this book:

Life span is defined as the maximum length of life observed in any one individual from a group. The *life span perspective* takes a holistic overview of all psychological and social factors acting on individuals at any stage of development or aging. The *life span construct* is used to summarize how an individual develops a coherent and unified sense of their past, present, and future. *Life span psychology* aims to integrate biological with cultural perspectives on the attainment of intellectual developmental goals.

Life expectancy is a statistical concept that is used widely in studies of populations. It estimates the number of years of life remaining at any given age. Life expectancy is expressed as the average complete number of subsequent years of life that an individual can expect.

Life cycle encompasses all stages of development and aging that remain consistent between and within successive generations. The life cycle includes stages of reproduction, embryonic, and fetal life and continues through to old age.

The *life story* idea is based on the proposal that beginning in adolescence, youngsters create a narrative with a beginning, middle, and end. During adulthood and into old age, narratives are revised continually to accommodate life changes that confirm personal continuity and to support representations of oneself often in an idealized fashion (e.g., as "reliable," "hard working," or "dutiful").

Life history identifies the expected timing and duration of key events in biological, social, and psychological development. The concept is used in evolutionary biology to explore how natural selection acts at many points in the life history to increase the number of offspring. Evolutionary forces are known to act on the timing of sexual maturity, mate selection, maternal behaviors, parenting, and many illness behaviors, including help seeking and altruism. It is unclear how evolution has affected senescence, but this seems likely to be influenced by wider options for child care (as provided by the postmenopausal grandmother) and stronger emotional ties between offspring and grandparents.

The *life course* perspective examines an individual *life history* to determine how early events influence subsequent life choices (like marital partner), specific behaviors (like crime), or disease incidence (like a heart attack). Life course theory is not yet adequately defined. At present, life course theory addresses connections between sequences of historical events (like famine, war, economic crises); the social roles enacted by individuals; and diverse social, psychological, and health outcomes. The life course approach includes a multidisciplinary scheme to study individual life histories and includes observations from psychology, sociology, social history, economics, demography, developmental biology, and aging. Therefore, the life course perspective identifies the contributions of social and historical contexts, underlying developmental processes, and the attributions of meanings to experiences to human development and aging across the life course. These changes take place throughout life and are not confined to particular chronological age ranges.

A *life chart* is used in medicine to assemble all past and current data on the course of a specific illness recorded by the patient or clinician. Typically, a horizontal line across the middle of the chart represents well-being, and movements above and below the line show positive and negative variations in overall health. Significant complications arising over the course of an illness—like hospitalizations—are identified, as are interventions like treatments. Chronic, fluctuating conditions like multiple sclerosis, diabetes, and manic–depressive illness (bipolar disorder) are charted in this way. Chronic progressive illnesses, such as late-onset dementia, also can be life charted successfully.

APPENDIX B: FURTHER READING

Several competent introductory accounts of the life course approach in medicine are available. Di Kuh and Yoav Ben-Shlomo[30,31] have provided one such account in their 2004 text, *A Life Course Approach to Chronic Disease Epidemiology.* A helpful undergraduate account of developmental psychology can be found in *Adult Development and Aging* by John Cavanaugh and Fredda Blachard Fields.[32] *Adult Development and Aging* is well complemented by Rudolph Schaffer's *Introducing Child Psychology.*[33] Together, these three books are helpful when aiming to see the bigger picture. The references to this chapter include many relevant sources to extend and explain the science behind the life course approach.

3

THE WELL-CONNECTED BRAIN

I. A BRIEF INTRODUCTION TO THE NERVOUS SYSTEM

Interested readers will find more than sufficient detail of brain structure in many excellent student texts.[1,2] This brief introduction emphasizes the outermost surfaces of the brain called the cerebral cortex. The two aims are (1) to describe a healthy mature brain without the effects of dementia or aging and (2) to introduce some of the science behind recent discoveries of how major brain regions are connected.

Faced with a brain for the first time, medieval anatomists wondered whether it was just one machine with individual parts contributing to the whole or if the different parts of the brain completed different tasks. This was a reasonable question: Are mental functions located in specific brain structures? By the nineteenth century, the locationist view—that the brain is organized in a very structured way—was widely held. By the early twentieth century, the basic cellular structure of the brain and the nature of cell-to-cell connections (synapses) had been well described although incompletely understood. Initially, the locationist view seemed to hold up quite well, but with many technical advances, mental functions were found to be widely dispersed around cortical areas.[3,4]

WHAT YOU SEE WITH THE NAKED EYE

Neuroanatomy is the study of the structure of the nervous system. It describes how the various parts are connected. The gross neuroanatomy can be seen easily with the naked eye by removing the brain from the skull and placing it in a container of preservative (formalin) until it hardens. After removal of the tough membrane around the brain (the dura mater), the external surface brain (or cerebrum) is revealed to be made up of two halves–the cerebral hemispheres. The most obvious features of the brain surface are the gyri (protruded rounded surfaces) separated by hollows or creases (sulci). When the cerebral hemispheres are divided along the midline, this reveals their internal structure. Viewed from midline, the brain appears layered so that the external surface of the cerebral hemispheres is made up of sulci and gyri arranged as a continuous folded sheet. Figure 3.1 shows how these features are imaged within the living brain by MRI.

The brain is best imagined as the greatly enlarged outer surface of the top end of a vertical tube running up the spine into the skull (see Figure 3.2). The cavity inside this hollow tube develops into the ventricles of the brain. These are

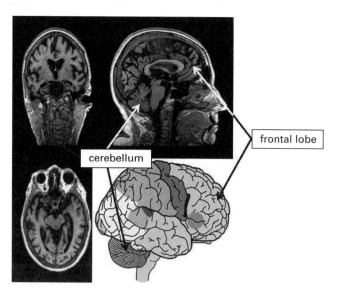

FIGURE 3.1 This figure shows aspects of the external and internal structure of the brain. A diagram in the lower right panel shows how the outer surface of the brain is made up of a deeply folded layer of brain tissue (the cortex) with prominent gyri separated by clefts (sulci). The upper right panel shows an MRI scan of the head with structures that lie along the midline (through the tip of the nose backward between the eyes) and points to two structures: the frontal lobes and the cerebellum. The left-hand panels show how the MRI scan can slice through the head through planes chosen to reveal specific brain structures. The upper left panel shows the brain along a slice through the top of the skull and both ears to reveal fluid-filled spaces in the middle of the brain. The lower left panel slices through a horizontal plane at the level of the eyeballs to expose relationships between the cortex and deeper brain structures.

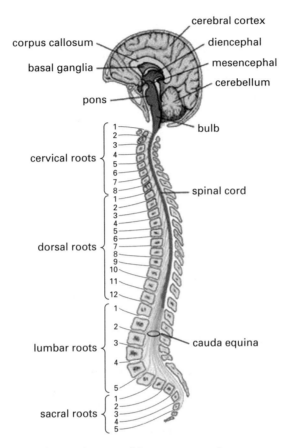

FIGURE 3.2 This figure shows a diagram of the mature central nervous system. The neural tube which formed from the neural groove in the embryo has become the tube-like structure of the spinal cord. The highest (anterior) part of this tube has developed into a recognizable brain now seen as a greatly enlarged swelling encased in the skull.

fluid-filled spaces contained within the brain and connected to the fluid that bathes the brain and spinal cord. Parts of the base of the brain and the brain stem contain dense groups of brain cells that are richly interconnected. These groups are called nuclei and are connected along pathways (called tracts) of nerve fibers to other nuclei, the cerebral cortex, and the spinal cord. The cortex covers the outer surface of the brain and is rich in nerve cell bodies. Like the central part of the spinal cord, the cortex is referred to as the gray matter.

This folding of the cortical surface allows much more cortex to be packed inside the skull. In fact, if the cortical surface were smooth, the skull would

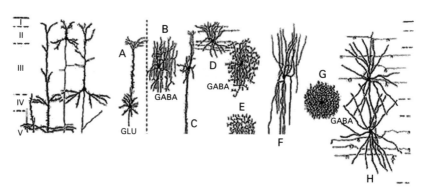

FIGURE 3.3 Examples of different types of neuron.

need to be about 35 percent larger. Folding of the cortex also means that the cells (neurons) of the cortex are brought much closer together, ensuring that the cortico-cortical connections can be shorter and faster.

WHAT YOU SEE DOWN A MICROSCOPE

Nerve fibers that extend from the cell body of individual neurons make connections between neurons. Figure 3.3 shows the types of neurons, and Figure 3.4 is a diagram of a single neuron with a transmitting neuron connecting through a dendrite with an axon making more connections. As their name suggests, dendrites look like the branches of a tree extending just a short way from the cell body. Axons are much longer nerve fibers that connect with other neurons sometimes at the cell body but more often on other dendrites. Axons are made up into bundles (a bit like telephone wires) and are wrapped in a fatty substance (myelin) that not only insulates each axon but also makes information transfer more efficient and quicker; thus, they are called "white matter pathways" or more often just "white matter."

The thickness of the cortex varies between 1.5 and 4.5 millimeters (mm) and contains the cell bodies of neurons, their dendrites, and some of their axons. The gray matter of the brain forms a continuous sheet as an outer layer that encloses inner white matter. Although called "gray matter," this is something of a misnomer caused by preservation. In life, the gray matter appears more pink than gray because of the rich supply of blood vessels to the cortex. The gray matter of the cortex is made up of brain cells called neurons or glia. White matter tracts are

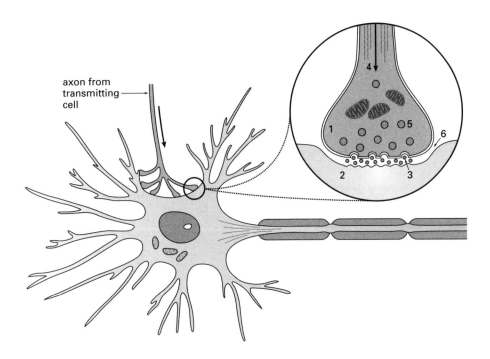

FIGURE 3.4 Schematic diagram of the components of the neuron. Incoming electrical information arrives through dendrites to pass through synapses as chemical messages. The neuron responds to the sum total of incoming information and generates an outgoing electrical message through its axon.

extensions from neurons covered by fatty material (myelin) so that these appear milky white to the naked eye. The brain and spinal cord are supported and protected by the bones of the skull and the vertebral column.

The cortex and large groups of subcortical neurons (the "subcortical nuclei") are the most important information processing parts of the brain. The human cortex is termed the "neocortex" because it is the most recently evolved portion of the cerebral cortex and supports higher mental functions for humans. The cortex contains some 100 billion cells, each with 1,000 to 10,000 synapses (connections), and has roughly 100 million meters of wiring. If spread out flat from its normal folded state, it would appear like a formal dinner napkin ranging between 1.5 and 4.5 mm in thickness. The cells in the cortex are arranged in six layers that differ between cortical regions (Figure 3.5). Different cortical regions support vision, hearing, touch, the sense of balance, movement,

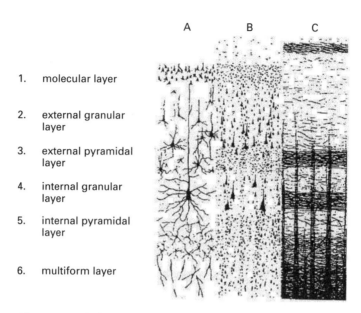

1. molecular layer

2. external granular layer

3. external pyramidal layer

4. internal granular layer

5. internal pyramidal layer

6. multiform layer

FIGURE 3.5 The cortex includes six layers each containing neurons that are typical of each layer. The relative numbers of cell types vary between cortical regions.

emotional responses, and other feats of cognition. The subcortical nuclei are groups of neurons that relay information between different parts of the body and the cortex and between cortical areas. The cortex and subcortical nuclei both contain the cell bodies of neurons.

The fine detail of the cortex has been studied extensively. Once the principle of localization of specific mental functions to precise cortical areas of the cortex seemed acceptable, anatomists began to look for cellular differences between cortical areas with different functions. The first major step was to find a way to visualize individual neurons, and this was achieved first by an Italian, Camillo Golgi (1843–1926), who found a way to stain neurons using silver. Golgi's breakthrough was taken up by a Spanish anatomist, Santiago Ramón y Cajal (1852–1934), who showed that neurons were individual structures and not one very large cell with a myriad of extensions. Later, Cajal demonstrated that electrical information was transmitted in just one direction along axons from dendrites to the axonal tip. The discoveries of these two great men were celebrated by their sharing the 1906 Nobel Prize in medicine and physiology.

Figure 3.5 shows a cortical neuron and identifies the points of connection between neurons as synapses.

NEURONS AND GLIAL CELLS

The human brain contains about 100 billion neurons, which make up about 40 percent of all the cells in the brain. Glial cells are much more numerous than neurons, and the functions of neurons and glia are quite different. *Glial cells* play three major roles: in brain inflammation, as brain structural supports, and in the guidance and regulation of the movement of molecules across the blood-brain barrier. *Neurons* are the information processing cells of the brain and the connections between neurons (synapses) are the functional element of information transfer between neurons. The number of synapses greatly exceeds the number of neurons: Because neurons can make up to about 10,000 synapses, the brain has at least 60 trillion synapses. Neurons make functional connections with other neurons to set up "neuronal networks" that support all types of mental activity.

Connections between neurons make up highly specialized neuronal circuits or networks. Different networks fulfill different functional roles, and these networks now can be identified using a combination of methods. Brain areas that control complex behaviors are interconnected to form systems such as those that regulate, for example, the movement of limbs or attention. Until recently, most people (including many neuroscientists) believed that the mature adult brain produced no new neurons. In fact, some brain areas continue to produce new neurons (neurogenesis) throughout life, and this finding has encouraged the idea that if neurogenesis could be stimulated in brain areas damaged by aging or disease, then some brain functions might be preserved that otherwise would be lost.

Interest in neurogenesis dates from the 1990s, although birth and differentiation of new neurons was detected in the 1960s.[5,6] The pace of research quickened once it was realized that neurogenesis occurred in adult humans. Did this realization open the door to novel therapies that might restore function (restorative or regenerative neurology) in patients with neurological damage from trauma, stroke, or disease? So far, it is too early to conclude that neurogenesis has a place in medicine. Certainly, interest remains in areas as diverse as the mode of action of some antidepressant drugs and some types of memory loss following trauma or stroke.

SANTIAGO RAMÓN Y CAJAL, NEURAL PLASTICITY, AND CRITICAL PERIODS

The scientific study of brain connections dates from the amazing work of Santiago Ramón y Cajal. Using limited equipment, Cajal described the nature of neuronal connections, their growth, and some aspects of their electrical activity. Although he did not use the term "plasticity" to describe how neuronal connections were shaped and remodeled into specific patterns, he certainly made full use of the term. He applied "plasticity" to understanding how the environment influenced brain development and function.[7]

Early experiences in the life course significantly affect brain development and later behaviors. Sensory systems typically are more plastic than pathways between brain regions that are not close together and, once established, would be quite a challenge to modify. For example, once the brain had set up pathways to process stereoscopic vision, it would cause mayhem if it needed to be rerouted. So, just as molecular mechanisms control plasticity, comparable molecular systems stabilize neural pathways during and after brain development.

Plasticity exists at many levels in brain circuits, and a great deal of effort is being made to understand these changes at the cellular level of connections between neurons (synapses) and in the patterns of connections made by the dendrites and axons of neurons. Discovery of the biological basis of brain plasticity has many potential applications not only in restorative neurology but also in education and in the slowing of brain aging and delay of dementia onset. Eventually, an exciting possibility will be realized when factors are identified that control the expression of genes involved in plasticity to reveal new ways to enhance memory in aging and the dementias.[8]

For a relatively brief period during brain development, the brain remains highly plastic. When injured, a child's brain has a much greater capacity to regenerate than an adult's brain. Even when nerve cells can be stimulated to grow again in an adult, newly sprouting nerve fibers fail to make functional connections. This is not true of all vertebrate animals. For instance, when the spinal cord is completely severed in certain fish species, the cord will regenerate completely following an injury that would have left a human permanently paralyzed. Growing new nerve fibers and making these fibers connect to form functional networks is the aim of many research programs in restorative neurology.

Some workers anticipate success will be achieved only once they have learned to recapitulate development.[9]

Immature neural networks require careful pruning to function properly. This type of plasticity is highly dependent on experience that strengthens some connections between neurons while weakening others. Much learning during development is based on these processes and remains fundamental to the acquisition of efficient behavioral and motor responses to the environment. A key element of this type of learning is the capacity of the developing offspring to respond to different types of environment. This development is possible only during critical periods when the animal retains plasticity to remodel neural circuits in response to sensory experiences.

Critical periods are precisely defined developmental periods during which time neural networks have the greatest plasticity. In a complex self-organizing system made up of overlapping and intermeshed networks that process information and regulate responses to the environment, critical periods are essential. This is true throughout the animal world. Among zebra finches, song learning occurs during a critical period from twenty-five to sixty days old. Neurons in the visual cortex respond to stimuli from one eye more than the other. This is called ocular dominance. In the cat, ocular dominance takes place between 28 and 120 days. In the human, ocular dominance is established between twelve and thirty-six months.

II. FUNCTIONAL ORGANIZATION

The basic structure of the brain locates many brain functions to specific regions. These are not haphazard arrangements but rather conform to a basic plan. Imagine the brain as a tall office block with perhaps forty levels (floors), and the top floors are home to the top executives and their support staff. The basements and lobby will house a good many service personnel, many with critical roles in building maintenance, security, heating, lighting, and ventilation and so on. This type of multilevel organization is found in the brain where functions at the base of the brain are critical to the survival of the individual. These include the control of respiration, blood pressure, and important networks that determine the level of arousal. Much as the senior executives do not keep a constant check on the functions of the heating boilers in the basement, top levels of the brain are not consciously aware of what is happening in the base of the brain. Relay stations ("junior staff") at intermediate levels act as gatekeepers who filter the flow of information to the executives working higher up in the brain.

The cognitive systems rely on both cortical and subcortical components. This means that cortical networks that support a cognitive function (like verbal memory) extend across several brain areas and do not map exclusively on to one cortical region.

THE CEREBRAL CORTEX

The cerebral cortex is about the size of a large dish cloth that is folded into pleats to cover the brain yet remain within the skull. The cortex accounts for most of the differences in brain size between humans and other species and is made up of six layers of cells as shown in Figure 3.2. Neuronal patterns of these layers differ between cortical regions just as the functions of these areas also differ. For example, separate cortical areas process visual data from the eyes in the occipital lobe. Auditory data from the ears are processed in the temporal lobe, and like other cortical sensory areas, when an area is damaged, the ability to process sensory data specific to that area is lost. Other cortical areas are connected to the sensory areas and are able to link different types of sensory data. Likewise, other cortical areas control voluntary movement (motor areas).

Frontal lobes have two major subdivisions, the motor cortex and the prefrontal cortex. Figure 3.6 shows the distribution of functions across the cortex. The motor cortex communicates with the spinal cord along axons that can extend from cell bodies in the cortex for several feet before connecting with motor cells in the spinal cord. The anterior regions of the frontal lobe are concerned with complex high-level cognitive tasks. These tasks involve decisions that require several types of data to be combined. Such data can be based firmly in perceptions of the external world but can include some data that are entirely abstract.[10] Examples of the complexity of tasks undertaken by the prefrontal cortex include moral judgments and forecasting the behavior of others. These tasks are among the most complex cognitive tasks and are linked to other higher cognitive abilities. Prefrontal lobe functions are particularly involved with planning, decision making, and balancing risk as well as some types of abstract reasoning, social behaviors, and sexual behaviors. If you wanted a place for the senior executives working on the top floor of the office block, look no further than the prefrontal lobes.

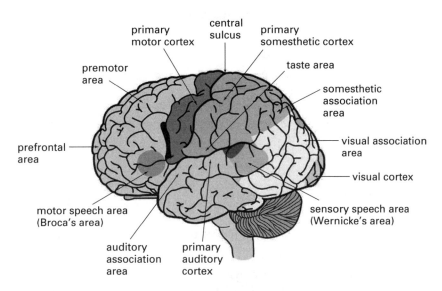

FIGURE 3.6 Schematic diagram of the divisions of the cortex into lobes with areas linked to specific brain functions. The brain is made up of two symmetrical halves, and the diagram shows the outer surface of the left half. The brain faces to the left where the anterior parts are the frontal lobes. The temporal lobes lie under the part of the skull between the eyes and ears (the temples). At the posterior part of the brain lies the occipital lobes, and the parietal lobes occupy the space between the frontal and occipital lobes.

The dominant cerebral hemisphere (the left side in right-handed people) in humans is highly specialized in the processing of auditory data (understanding speech), verbal memory, and many language functions, including comprehension of the spoken word. It is tempting to think that these sophisticated cognitive abilities would be located in cortical regions that are not present in other great apes. Although it is certainly true that mammals that are not primates do not have the equivalent types of cortical complexity, little in the human cortex cannot be found in other higher primates. The structural complexity of the cortex can be related to age-related decline in mental abilities. A narrow age-range sample showed that greater cortical complexity was associated with superior retention of mental abilities from eleven to sixty-eight years of age.[11]

The layers of the cortex are highly interconnected. When all the connections between brain cells are examined, 75 percent lie entirely within the cortex. The remaining 25 percent connect brain structures with the cortex and each other. When compared with other animals, the human neocortex accounts for more than 80 percent of the total brain weight, whereas the cortex of the chimpanzee is only at around 70 percent and in some small mammals it is as low as 15 percent. The neocortex is the most recently evolved part of the cortex.

HIGHER COGNITIVE FUNCTIONS

We are an intelligent species and the use of our intelligence quite properly gives us pleasure. In this respect the brain is like a muscle: when it is in use we feel very good. Understanding is a kind of ecstasy.[12]

Levels between the base of the brain and the cerebral cortex contain major relay stations that pass information from body to brain and from the external world via the sensory organs (ear, eye, mouth, and nose) to specialized relay stations and, after some sorting, onward to specific regions of the cortex. The convention is to rank human abilities (like language, planning, and judgment) as "higher cognitive functions" and to place these in the cortex. These cognitive functions traditionally were labeled higher than other human cognitive functions, and by implication, they are higher than those possessed by other nonhuman animals. Although humans share some of these "higher cognitive functions" with other great apes, we have expanded our abilities beyond the capabilities of all other animals.

Uniquely, humans can reason about things we cannot see, finding explanations of effects that are largely unobservable. We recognize some of these capacities as advanced types of "abstract reasoning" and use terms like "intuition" or "insight" for others. Alongside these terms, we can identify the character of other individuals. In our relationships, we can acquire and apply skills in "mental time travel" to recall previous events and "mind reading" to show empathy and to act altruistically. Many humans can add talents for creativity in language, music, and the visual arts, sometimes coupling these with a facility for profound analytical thinking. These superior cognitive abilities rely in turn on intact higher cognitive functions, like language comprehension.

Empathy is disrupted in some types of dementia that affect the frontal lobes and can have profound implications for the care and safety of someone living with these types of dementia. A new branch of neuroscience (social cognitive neuroscience) has investigated empathy and has shown that at least three types of mental processing are involved. These are (1) the suppression of one's own viewpoint, (2) activation of those brain areas that would be stimulated if one were experiencing the emotion shown by the person empathized with, and (3) the capacity to assign agency to a stimulus that is not affecting oneself. These are each major tasks and the work needed to integrate all three is a highly complex undertaking. When advanced brain imaging techniques were applied to this problem, reductions were found in the size of specific cortical regions. These regions included the right pole of the temporal lobe and parts of the frontal lobe, and they point to a key principle in understanding higher cognitive functions. This principle is that brain networks are required to work together to support a specific cognitive ability; it usually is not possible to assign a cognitive ability to a specific cortical area or brain structure.

We can be confident that higher mental abilities rely on brain complexity, and we suspect more than a grain of truth in Lyall Watson's aphorism that "if the brain were so simple we could understand it, we would be so simple we couldn't." Although it seems relatively easy to appreciate the main large brain structures, and with some extra effort, place smaller but essential brain structures in exact relation to the larger, it is probably beyond our comprehension to understand fully how neurons are connected. These are not trivial points. To understand how the exact pattern of connections between neurons account for cognitive functions, three types of information are needed. First, a detailed map is needed of the connections of neurons that make up each neural circuit serving each cognitive function. Second, the flow of data through the circuit needs to be related to the cognitive function served by the circuit. Third, these types of observation need be put together in a way that explains how cognition can be understood reliably by the observations.

Progress has been made in meeting the first of these requirements. Using specific types of dye and tracers, short-distance connections between neurons can be revealed, but ultra-short and long connections cannot be visualized using these methods. Unfortunately, it is often impossible to distinguish the connections made by adjacent neurons. Long-distance connections can be studied only in animals as these techniques depend on making surgical lesions in pathways that then degenerate showing which brain areas are interconnected. In very small animals like the roundworm *Caenorhabditis elegans*, it is feasible to use electron microscopy to reconstruct all neuronal connections in a single animal, but the process is extremely laborious and rarely done.

A new solution is to use multiple, distinct labels that distinguish one neuron and its connections from another, much as a telephone engineer uses various colors to tell apart one cable from another.[13] To succeed with mapping neuronal circuits, more than 100 different color labels are needed (imaginatively called the "Brainbow" technique). But that is only the beginning of the problem. The sheer density of connections requires advanced computational methods to build even a partial image of a single neural network. Eventually, it is hoped that these anatomical techniques will be combined successfully to provide detailed maps of neural circuits. These maps, however, will not be applicable to studies of the living brain. To achieve this goal, noninvasive techniques are needed probably in conjunction with the anatomical methods described thus far. To date, these methods have proved incompatible and new techniques are needed.

Throughout the development of the nervous system, the brain builds untold numbers of neural networks that vary from large-scale extensive systems to quite localized arrangements. All are critical to brain functions, and many are relevant to aging and the dementias. These networks make up groups that perform related functions and when studied at rest show synchronous electrical activity that is consistent with what is known about the microanatomy of their connections. Later, the clinical features of five major categories of late-onset dementia will be described. Each of these categories is distinguishable by their characteristic clinical features. Using magnetic resonance imaging (MRI) techniques to image the brain, functional and structural networks can be delineated.[14]

Some dementia-specific findings agree with other investigations of intrinsic networks found in the healthy brain. Studies of this type aim to correlate the clinical symptoms of each type of dementia with brain areas already known to support specific brain functions. Imaging studies that build on these findings are likely to improve the early diagnosis of dementia. An important limitation, however, is that some individuals may retain a capacity to remodel networks affected by dementia and thus maintain brain functions that otherwise would have been impaired by the dementia. In life, it seems, not all brains will respond to the presence of dementia in exactly the same way.

Structural analyses of the aging brain include techniques to quantify the complexity of cortical layers. One technique exploits the observation that the cortex includes small units that are in many ways similar to the whole cortex. MRI is used to estimate the total area of the much-folded external surface of the cortex and also of the area of the boundary between gray and white matter that lies beneath it. In normal development and aging and in the absence of brain disease, the thickness of the cortex increases throughout life into middle age; gray matter gradually is lost, while white matter remains much the same. From about fifty-five years of age, the cortex begins to thin. Individuals who have spent longer pursuing full-time education and who perform at high levels on tests of general intelligence tend to have greater cortical thicknesses than those with less education and who perform less well.[15] These findings suggest that learning and education contribute to cortical thickness and, by inference, reveal a cortex of greater complexity.[16] Some differences exist between men and women in cortical structure: Men have larger brains with a thicker cortex. Women have smaller brain with a greater proportion of gray matter and more complex cortical folding.

The structure of the mature adult brain reflects the sum of diverse genetic and environmental influences. Influences on brain development that act early in life can cause lasting impairment of brain growth and may lower the threshold when the effects of aging produce deficits in learning and memory. Childhood adversity is linked to reduced sizes of brain structures involved in the regulation of emotions (e.g., the volumes of the anterior cingulate and caudate) and memory (the hippocampus). Childhood low economic status, poor parenting, exposure to toxins (e.g., atmospheric lead), malnourishment, and an impoverished understimulating environment have all been implicated at one time or another in restricted brain development.[17,18] Importantly, most of these potentially harmful influences do not restrict the growth of the whole brain, but rather they affect specific brain structures in ways that certainly reduce their size and probably reduce the richness of their internal neural networks.[19,20]

Among the greatest challenges in biological science—greater than the Human Genome Project—is to understand how these cortical networks are built, function, and are maintained. Early insights came from the application of a largely unknown field of mathematics ("graph theory") to neural circuits in the visual system. Later, brain MRI was aligned with graph theory to show how small neural networks could give rise to great functional complexity. An important step was to show how relatively simple networks communicated inside a "small world" of neurons using pathways that rarely exceeded six steps (synapses) between neurons. The wiring of neuronal circuits in *Caenorhabditis elegans* was seen to be made up of a mixture of fixed, regular features as well as many effects that appeared to be random. Theoretical studies followed, and it soon became possible to design experiments that combined graph theory with the results of functional MRI (fMRI). This technique can detect changes in blood flow through small volumes of brain tissue and link these changes to work performed by the brain while being examined using fMRI. This has led to the detection of brain modules that make up several neural circuits and function coherently to perform specific cognitive tasks. The technique provides valuable insights into the brain "at work" and can reveal, for example, how the brains of some older adults retain youthful functional efficiency, whereas others begin to fail.[21]

The brain is made up of neural networks each performing complex computational tasks that can often influence each other. Knowledge of the structural connections that make up individual circuits shows what tasks each circuit does (or does not) and the extent to which one circuit can influence another. So far, it appears that a structural map of the living architecture of connections in neural networks is not only achievable but also critical to understanding brain function in development and aging and in the study of brain diseases. In 2005, Olaf Sporns suggested the term "connectome" to describe such a map. As an aside, he first checked online to determine whether the term "connectome" had been used and found just one competitor who could claim priority (this was a dating site called "come-to-me"!).

By 2009, the National Institute of Health (NIH) had committed $40 million to a five-year project acquiring and sharing a Human Connectome Map that will use multiple neuroimaging techniques, electromagnetic recordings, and genetic and behavioral data. One of the project leaders, Washington University's Dr. David Van Essen, has been quoted as saying, "We're planning a concerted attack on one of the great scientific challenges of the 21st century" (http://www.humanconnectome.org/consortia/). "This study will have transformative impact, paving the way toward a detailed understanding of how our brain circuitry changes as we age and how it differs in psychiatric and neurologic illness." The U.S. research teams will study 1,200 healthy adults, including twin pairs and their siblings from 300 families. The brain connectivity maps will reveal how genetic and behavioral data relate to each map, showing how brain networks are organized in development and aging. The findings will be relevant to brain disorders as diverse as autism, schizophrenia, and the dementias of late life.

III. DIFFUSION TENSOR IMAGING

Brain imaging techniques are increasingly powerful. Over the past forty years, methods of mapping brain structures at work and at rest have provided important information about the structure of the living brain. Many of these techniques are termed "noninvasive" and are used safely in adults and children. Used in conjunction with well-established anatomical methods described earlier, they have contributed substantially to understanding the development and aging of the human brain.

Diffusion tensor imaging (DTI) is one of these new techniques. It is based on MRI, which takes only a few minutes to capture the complete anatomy of a single brain. Although MRI is a major advance in brain imaging, it has four important physical limitations. These arise because MRI is a technique trying to measure a structure bathed in moving water. An MRI slice can take about 10–100 milliseconds (ms) to acquire data during which water molecules can move about 10 μm (this is the random thermal motion of water or "Brownian motion"). The problem is just like taking a photograph of a racing car: The image will be blurred if the shutter speed is not fast enough. A second problem is the sheer amount of data obtained by MRI when the aim is to obtain the fine definition of brain structures. In turn, this causes a third problem: The amount of data obtained is currently beyond our capacity to analyze it efficiently. Finally, although MRI gives excellent images of anatomical structures, contrast between structures is less than satisfactory.

DTI aims to overcome these limitations by making use of water movement to figure out the fine anatomy. Susomi Mori and Jiangyang Zhang at Johns Hopkins have suggested the following analogy to explain how DTI does this: Imagine a drop of ink falling onto moist blotting paper. A blob will form showing how the ink (through Brownian motion) has tracked through the fibers of the blotting paper. The shape of the blot is telling us something about the structure of the blotting paper. If the blot is circular, this shows that the paper fibers are oriented equally in all directions. Technically, this is called "isotropic diffusion." If the blot is elongated in one direction but not others, this shows that the paper's fibers are oriented preferentially in that direction—causing "anisotropic diffusion." DTI measures the organization of nerve fibers in the brain by measuring anisotropy. Water molecules move more easily along the direction of the nerve fibers because the sheath of the fiber limits movement that is perpendicular to the direction of the nerve fiber.

When conventional MRI is compared with DTI results, MRI does not distinguish between nerve pathways, whereas DTI does so. The DTI images obtained in animal and human studies agree with what was obtained in the past using anatomical techniques. The analysis of DTI data in three-dimensional imaging of major nerve pathways is called "tractography" (a tract is a region of the body that extends within clear boundaries; e.g., digestive tract). When nerve tracts are identified and delineated in this way, they can be given arbitrary colors so we can make sense of what we see. When tracts are visualized, DTI shows accurately where they run and, importantly, pathways shown by DTI agree with postmortem studies. The same is true of the developing brain. In this case, DTI findings agree with the known anatomy and confirm the patterns of change in nerve tracts during gestation and early childhood and adolescence. In old age, DTI provides a unique method of examining the whole brain, increasing the chances of discovering changes in the aging brain and dementias that might be overlooked if only small brain regions were studied after death. Later, when the role of childhood hardship is considered alongside other risk factors for dementia in late life, DTI studies give us the opportunity to consider whether dementia arises typically in a normally developed brain or whether—for example, when nerve fiber bundles are found missing—this could be caused either by brain disease or faulty development.

IV. BRAIN STRUCTURES AND FUNCTIONS RELEVANT TO AGING

With aging, our highest mental functions change in ways that are both subtle and profound. There are many well-detailed deficits in memory, attention, and language. Against this backdrop stand some positives, not least the fact that in the face of these deficits detected on mental testing, much mental function remains uncompromised. There is also potential for growth of wisdom and the capacity to use personal judgments in ways that appear dispassionate and impartial. From an historical perspective, brain scientists first made substantial progress in locating specific higher cognitive functions to specific regions of the cortex by identifying nuclei and pathways that were essential to those functions.

As suggested by studies of brain structural connectivity, this view may have been too simplistic. Much current evidence points to the importance of networks of brain cells that support higher cognitive functions.

Brain structural changes with aging lead to our first questions about the distribution of age-related brain changes. Do these changes arise uniformly among all parts of the brain, or are specific brain parts—like particular pathways or regions of the cerebral cortex—more vulnerable to aging than others? When deficits occur in mental performance, and these are attributed to aging, can those deficits be explained by age-related changes in key brain structures? Or is disruption of critical pathways or networks more important than structural damage? The answers to these questions have come largely from studies of the cerebral cortex and, to a lesser extent, from studies of brain structures that lie beneath the cortex (i.e., the subcortical nuclei).

Some mental abilities distinguish humans from the great apes. These functions are managed primarily by the cerebral hemispheres. When this is done exclusively by one hemisphere and not the other, as mentioned previously, this hemisphere is referred to as being dominant over the other. The term dominant is linked to "lateralization" of brain functions, whereby certain functions are largely performed on one side of the brain and others are performed on the other side of the brain. In most right-handed people, the left hemisphere is dominant. The left hemisphere is the cleverer half of the brain. It has language (it speaks) and can calculate (it has mathematical ability). The right hemisphere is superior at spatial perception and musical ability, but is much inferior to the left in language and visual perception.

Lateralization of functions develops during the first few years after birth. Until then, if an infant suffers damage to one hemisphere, then the other hemisphere has some capacity to compensate and to develop the functions of the damaged hemisphere. This capacity to compensate is another example of brain "plasticity" and diminishes in later childhood. Plasticity of brain structures and their interconnections is an important concept in brain aging.

When aging impairs higher cognitive functions, are cognitive deficits found in all types of cognitive performance, or are there specific patterns of loss? To find answers, we need to know where to start and where to continue to look. First, we could study how the brain evolved to make us better able than animals to solve cognitive problems. Next, we could take a developmental perspective and ask whether these uniquely human abilities are present from birth, or are specific experiences required if they are to develop? These questions are explored in the following paragraphs. Next, we can ask: How are cognitive abilities maintained along a life course that sometimes exceeds ninety years? How are superior cognitive abilities acquired so that "mental time travel" can allow an individual to recall previous events (episodic memory) and to travel forward to imagine future needs (episodic prospection)?

ARE SIMILAR CAUSAL PATHWAYS IMPLICATED IN ALZHEIMER'S, PARKINSON'S, AND HUNTINGTON'S DISEASES?

Patterns of loss of brain cells are associated with each subtype of dementia other than those caused by disease of the blood supply to the brain. The dementias are mostly age-related brain diseases that largely affect the cerebral cortex. Certain types of dementia also affect subcortical structures. These dementias include the dementia associated with Parkinson's disease, Huntington's disease, and some forms of cerebrovascular disease. Among the cortical dementias, it remains unclear why each has its own specific pattern of brain cell loss. The best guesses have been that (1) brain cell death is caused by excessive "wear and tear" linked to heavy traffic in processing information, (2) some sort of toxin spreads from brain cell to brain cell along the connections between cells, (3) cells stay healthy only if they receive molecular signals from their neighbors (these signals are called "neurotrophins" and are lost when a brain cell dies causing its neighbor to die in turn), and (4) not all brain cells are the same. Some brain cells share genes or proteins that could confer greater vulnerability to aging. If the causes of specific patterns of loss could be found, then knowledge of why brain cells die in such specific patterns might be the basis of a successful treatment. For example, if it could be shown that loss of a neurotrophin was the cause in one type of dementia, could a treatment to replace that neurotrophin be successful? This treatment might include designing a drug that mimicked the actions of the neurotrophin or giving a synthetic version of the neurotrophin itself. Examples of this type of approach will be given when treatments of dementia are described later in this chapter.

Alzheimer's disease, Parkinson's disease, and Huntington's disease are progressive disorders that affect specific subtypes of neurons. These three disorders have large differences in the exact sites of pathology. The clinical signs and symptoms of the three overlap only a little, but they share many similarities in their neurobiology. These shared mechanisms raise the possibility of exploiting common therapeutic targets for drug development. One idea is that a defect or a loss of specific neurotrophins causes these diseases, with a different neurotrophin deficit causing a selective loss of brain cells typical of each condition. As Huntington's disease has a known genetic cause, it is now possible to identify accurately those individuals with the Huntington's disease genetic mutation but do not yet manifest symptoms. It is rarely possible to predict Alzheimer's disease in the same way. People who carry the Huntington's mutation could be a model for Alzheimer's disease to test whether selective use of neurotrophins can become a therapeutic intervention that slows down or even prevents progress to Alzheimer's or Parkinson's disease.

These proposals emphasize those aspects of brain biology that are relevant to understanding brain aging and some of the dementias and their treatments. In the *life course approach*, a different emphasis is placed on the need to avoid pitfalls posed by an impulse toward reductionism. The many influential factors described in brain biology reflect the complexity of interplay between body systems. Not least among these levels of complexity is the brain itself, as indicated in the account of brain structural connectivity. On the one hand, it seems possible to appreciate a level of complexity by naming the main structures and the major bundles of fiber pathways along which structures exchange information. On the other, it seems impossible to understand how all of this interconnectedness actually functions.

V. BLUEPRINTS FOR A WELL-CONNECTED BRAIN

A striking feature of the mature central nervous system is the precision of the synaptic circuitry. In contemplating mature circuitry, it is impossible to imagine how more than 20 billion neurons in the human brain become precisely connected through trillions of synapses. Remarkably, much of the final wiring can be established in the absence of neural activity or experience; so largely the genetic program must encode the algorithms that allow precise connectivity. This program, honed over 1 billion years of evolution, generates networks with the flexibility to respond to a wide range of physiological challenges.

Deanna Benson, David Colman, and George Huntley, Mount Sinai School of Medicine, 2001,
writing about how the complexity of brain circuitry is achieved in early development[22]

We suggest that one set of controlling genes provide general instructions for indi-
vidual and species-specific changes in the proto map. Another set of genes control[s]
cell proliferation in the units. At later developmental stages, each proliferative unit
becomes a polyclone that, mostly through asymmetrical division, produces cohorts
of post mitotic cells that migrate along common radial glial guides and stack up in
reverse order of arrival in the cortex. These stacks of neurons, called ontogenetic
columns, become basic processing units in the adult cortex. The surface area of
each cytoarchitectonic region during evolution and in each individual, therefore,
depends on the number of contributing proliferative units, while the thickness of
the cortex depends on the number of cell divisions within the units.

Pasko Rakic, Yale University, 1988, writing about how the outer surface of the brain
(the cortex) conforms to a development pattern with exact ordering of the migration
and guidance of neurons that will become the cortex[23]

These quotations from the work of distinguished neuroscientists establish from the outset that the human brain not only is extremely complex but also presents a huge challenge to understanding how it was put together, how it works, and how in old age, after almost a century of service, it begins to fail. Both quotations indicate the importance of genes in the determination of brain complexity and the role of evolutionary forces in the creation of a human brain. Translated into the challenges faced in regenerative neurology when the aim is to restore functions lost by damage or disease, the work of great neuroscientists like Pasko Rakic at Yale University (United States) will resonate for decades to come.

Historically, much emphasis was placed on the relatively large size of the human brain. By the mid-1960s, it was clear that the great advantage of human brains over those of higher primates like chimpanzees was that human brains became not only bigger but were substantially reorganized during evolution and retained the potential to continue self-organizing over the life course. Of course, these views are not universally accepted and some continue to place more importance on brain size rather than the richness of internal connections between brain areas. In broad terms, we can understand that sheer size does not increase mental capacity. The blue whale, for example, has a brain that is about five times larger than our own, but it does not appear to possess five times greater mental abilities than us. Human brains are around four times larger than other mammals of comparable body size. This does not necessarily make us four times cleverer than say a chimpanzee with a brain weighing about 400 grams, whereas ours weighs around 1,300 grams, but it certainly helps develop additional mental abilities. It is also relevant that during our recent evolutionary history the human brain reduced in size (compared with the Neanderthals) by around 10 percent (150 milliliters [mls]). Some genes specific to human head size have changed so rapidly during recent evolution that the term "accelerated evolution" is used to describe what has happened to them.

The brain develops from the third week of life in the womb and continues into late adolescence and possibly into late life.[24] Specific genes acting at a molecular level determine brain structure of the human embryo. By the seventh week after conception, neuron production has started and continues for another twelve weeks. From the eighth week after conception onward, the fetus develops elaborate brain structures including the basic anatomy of the major fiber pathways. During infancy and later child development, environmental input plays a major role particularly in the fine-tuning of connections between neurons. The overall impression is that neither genes nor input totally determines mature brain structure and that the adult brain continues to be shaped by a complex interplay of dynamic and adaptive processes that allow new connections to be made and maintained and, consequently, new behaviors to be established.

Brain development proceeds apace throughout childhood, increasing by about fourfold in volume by age five, so that by age six, it has reached about 90 percent of the adult volume. The fine structure of the brain (termed "cytoarchitecture" or "cytoarchitectonics") is extraordinarily rich and detailed in early infancy so that the level of connectivity between neurons is much greater by far than that of adults. The vigor and enthusiasm of neurons in childhood to make connections gradually is pruned back in adolescence through competitive processes that are influenced greatly by experience. By this stage of brain development, the brain has hugely increased its adaptive capacity from infancy to adolescence. The principle of brain "plasticity" underpins many of these acquisitions.

The brain of the developing embryo takes shape from a mass of cells that must move to positions at which they can form key brain structures and differentiate from cells with differing roles to play in the nervous system. These processes are controlled by molecular signals that act on undifferentiated precursor cells with the huge potential to perform any of many functions (i.e., they are *pluripotent stem cells*) at sites in the developing nervous system. At this early stage, the brain and spinal cord resemble a hollow straw-like tube that soon will have its first brain-like structures. These form at the end of the tube as pouch-like expansions that will become the brain. These are termed the forebrain ("prosencephalon"), the midbrain ("mesencephalon"), and hind brain ("rhombencephalon"), which soon will divide into smaller regions. This is the rudimentary organization of the brain and central nervous system.

Once the basic brain parts and their relative positions are established, the developing fetus can begin to establish patterns made up of different neuronal types specific to the requirements of each brain area. Here, patterns found in the cortex are described, but comparable processes take place throughout the brain. Cortical patterns result from different types of molecular signal known as Emx2 and Pax6. These two molecules are produced in opposite gradients along the anterior–posterior axis of the developing brain. Their effects are that the cells destined to become cortical neurons will be exposed to a gradient of relative concentrations of Emx2 and Pax6. Those cells exposed to a high concentration of Pax6 and a low concentration of Emx2 will become the neurons of the motor cortex, while the opposite holds true for cells of the visual cortex with intermediate concentrations producing neurons for the sensory cortex. Of course, when brain structures are highly specialized, additional methods of regulating neuronal subtype and position are required. These molecules seem to conform to the same gradient principle as the Emx2–Pax6 gradient and are found as pairs of molecules and add new levels of refinement to the control of cortical structure.

During the embryonic period up to the eighth week of pregnancy, the brain has grown to become a smooth tube-like structure with bulbous protuberances. Next, during fetal development, the brain will change into a structure more easily recognizable as the precursor of an adult brain. The gyri and sulci (Figure 3.1) of the cerebral cortex are formed in orderly sequence. The first sulci appear as grooves on the surface of the cortex. The first fissure appears around week 8, running anterior to posterior and dividing the two cerebral hemispheres; the cortical sulci are given names corresponding to the overlaying area of the skull (e.g., the temporal sulcus appears in week twenty under the temple where gray hairs are said to appear as the first sign of aging). Other sulci include the sylvian, cingulate, parieto-occipital, and calcarine appearing in weeks fourteen to sixteen.

The migration of neurons that will make up the cortex also takes place in an orderly fashion. The precursors of these cells are not yet differentiated into specific types of neuron. The first precursors to move into the cortex do so by extending a process to the outer cortical surface, which allows the cell nucleus to move closer to the outer cortex, which at this stage is very thin. Then as the brain becomes larger and the cortex thickens, this type of movement becomes ineffective and is replaced by guidance from a special type of glial cell. These guiding cells make a type of scaffolding along which neurons can migrate out into the cortex.

The movement of precursor cells to become the neurons of the cortex results in the characteristic six-layered appearance of the cortex. Those neurons that migrated first into the cortex form the deepest cortical layer, while those arriving last make up the outermost cortical layer. A molecule called Reelin—a critical part of the mechanism that signals neurons to stop migrating—provides an important control of movement of neurons into the cortex.

New ideas about brain repair following stroke or in chronic conditions like Alzheimer's are now better understood and accepted. One idea is that *pluripotent stem cells* can be used to replace and take over the functions of cortical cells damaged by disease. The success of such a treatment proposal requires some working knowledge about cell differentiation within the cortex. How is this process controlled? Do specific types of precursor cell produce specific subtypes of neuron? Alternatively, is a single type of precursor cell able to differentiate into any type of neuron? This differentiation was shown early in the development of the cortex when precursor cells demonstrated the ability to differentiate into any type of cortical neuron. This capacity ("pluripotency") of precursors is gradually lost. If brain repair using stem cells is to succeed, the exact conditions required for differentiating implanted stem cells into the type of neuron lost by injury or disease must be achieved. This same caveat applies to the control of processes that determine the connections between neurons that constitute the information processing networks impaired by injury or disease.

Once the layered arrangements of cells in the cerebral cortex were recognized, a search started to explain how these arrangements functioned together. Initially, the vertical positions of the cell bodies of neurons that make up the layers of the cortex attracted most interest, which was described by Vernon Mountcastle in 1957. He suggested that the neurons in these vertical columns (or cylinders) make up the functional unit of the cerebral cortex. He found that neurons in a single column could be activated only by a specific type of peripheral stimulus (e.g., touch or joint movement) on the opposite side of the body.

He concluded that his observations supported the principle "that the elementary pattern of organization in the cerebral cortex is a vertically oriented column or cylinder of cells capable of input-output functions of considerable complexity, independent of horizontal [intragriseal] spread of activity."

These columns are arranged first as small "microcolumns" containing about eleven neurons about 50 μm in diameter and then aggregated as "macrocolumns" (termed "barrels") about 600–900 μm in diameter. It is widely supposed that the neurons in each column are all derived from the same mother cell in brain development.

Columns form a series of units that are repeated across the surface of the cerebral cortex. Evidence is strong that these macrocolumns are repeated across the cortex but that their exact arrangements differ according to the functions served by specific cortical regions. The task of demonstrating how the cortical connections of these columns are made up is yet to be completed and probably remains quite some time away. Although the laboratory rat has provided much information about how these connections might be made, they are probably a whole world full of whiskers and smells away from the rich visual world of the higher primates, including humans. Deriving associations of meaning from cortico-cortical connections is the goal of much research in human neuroscience in health and disease. Whether or not the cortical microcolumns are the fundamental unit of cortical organization is an unanswered question, although it is accepted widely as the best idea we have and as good a place as any to start.

VI. THE BRAIN AND LANGUAGE

Many major challenges are ahead to understand how the cortex is organized. The examples of speech production and processing often are cited to illustrate how even after more than 130 years of research these are not yet fully understood. Many unanswered and difficult questions remain with few established facts. Attempts to draw brain maps to distinguish the production of speech from the auditory comprehension of language provide a case in point. What quickly became clear is that systems are able to map what is heard onto the meaning of the sounds. Sounds change over time and must be broken down ("parsed") into intelligible blocks. These are called "phonemes" and are the basic building blocks of speech. Phonemes combine to form the smallest units of speech that have meaning. Brain imaging studies do not support the idea of a single "speech center" in the brain. The fact is that production and comprehension of speech are distributed widely across many areas of the cortex.

The first stage of speech processing requires an analysis of changes over time in frequency and amplitude of sound ("spectrotemporal analysis"). These computations take place on both sides of the brain and divide auditory information into streams. Take for example the distinction between the words "pets" and "pest." These words differ by just a few milliseconds in the order of sounds heard but the words have totally different meanings. fMRI has reinforced the idea that listening to speech takes place on both sides of the brain not just in the left (dominant) side. When brain responses to speech are compared with responses to nonverbal sounds ("pseudo-words"), some preference emerges for dominant-sided activation.

Speech comprehension is helped when information from other senses is available. For instance, with training, many people quickly learn how to lip-read. This shows that by watching but not hearing someone speak, language can be understood. fMRI shows that this is achieved through auditory and visual areas of the cortex working together—that is, language comprehension can be improved when sensory systems are combined.

Words and ideas must be stored, and these stores sometimes are disturbed in dementia. A store of information about words and their meaning is called "semantic," and when disrupted in dementia, this syndrome is called "semantic dementia." A store of information about how words are put together is called "syntactic" (from the word "syntax" meaning the grammatical structure of sentences). Information about word forms (how they sound and are spelled) also is stored. Together, semantic, syntactic, and word form stores make up a central store (or lexicon) available to find meaning in speech. The existence of a central store is the most basic idea about how the brain organizes information about language.

CLINICAL STUDIES ON BRAIN AND LANGUAGE

Some patients have deficits in their use of language after brain injury or stroke and in some types of age-related progressive brain disease. These deficits are caused by problems in understanding or giving the correct meanings to language. When the posterior part of the left (dominant) region of the cortex is damaged, as in Wernicke's aphasia, mistakes are made in speech production. These mistakes are called "semantic paraphasias." Examples include using one word in place of another (e.g., onion when they mean ball). Early in their illness, patients with progressive *semantic dementia* show deficits in their use of words, substituting one word with another from the same conceptual category. For example, saying animal when they intend to say rabbit. The exact pattern of semantic disruption can be related closely in anatomical studies to the precise location of brain damage or disease.

VII. BRAIN MAPS

Everyone knows that mental abilities differ among people. Some of us are great with math, some excel at music, and others have a way with words. We accept, too, that it is not unusual for people to be much better than average in quite a few areas so when we know someone pretty well, knowing their particular strengths, we often can predict whether that person will become proficient fairly quickly in another area of mental performance. Until the invention of MRI of the brain, it was difficult to show whether these particular mental strengths (or weaknesses) were linked to a specific part of the brain. When an identifiable brain area is damaged, we are not surprised that some types of mental ability may suffer or even be lost altogether. When damage is located in a discrete brain region and subsequent loss of a specific mental ability is consistently lost when that region is damaged, then it is logical to link that brain region to that particular mental ability. This type of observation provided the basis for brain maps of mental functions collected over the past 150 years.[25]

The idea that separate mental functions are linked to separate brain areas was developed in the nineteenth century and led to the proposal that variations in size of specific brain areas could account for variation in mental abilities among individuals. These ideas were linked to the belief that larger brains worked more efficiently, to the notion that loss of brain tissue impaired mental performance, and to the recent idea that replacement of brain tissue could restore lost functions. This type of research is based on the ability to determine individual brain structures, their precise location, and their boundaries. These methods detail the exact structural anatomy of the brain, but they do not explain how the brain works.

In addition to the structure of each brain region, more information is needed about connections between regions to investigate how the brain determines human behaviors and mental life. Old approaches to the study of connections within the brain required examination of postmortem tissue often obtained after animal experiments. These invasive techniques have been available for many decades. New approaches followed the invention of brain MRI to study the structure and functions of the living human brain. These novel techniques are largely indirect—that is, they rely on inference to interpret and are prone to error. No other methods, however, allow the connections between brain areas to be studied noninvasively during life. The results are comparable between individuals and between brain sites. When differences in connections are detected, they can be related to differences in mental function and behaviors.

Our aim is to understand how and why our brains work less well in late adult life, what makes people different in their rates of aging of mental abilities, and why for some people old age be catastrophic for mental function. For some the capacity for independent life will be completely lost. We could think about the aging brain in one of several ways. The first, following from ideas about "brain maps," would identify brain regions and describe how each separate region is affected by aging. A second way is to think about the brain as a well-connected grouping of "hubs" for the processing of information. Seen in this way, a devastating disorder like Alzheimer's is viewed as a disorder in which the brain's most highly connected regions are attacked by the disease before other less well-connected brain areas. This approach to brain diseases suggests that each brain disorder could map onto a distinct pattern of damaged brain networks. Of course, this does not mean that a particular pattern of damaged networks causes the characteristic features of a type of dementia. It is possible that the symptoms of dementia disrupt the networks that support those same symptoms.

VIII. BRINGING IT ALL TOGETHER

Future studies on the aging brain and dementia will be affected by the way medicine adapts and incorporates new ways of biological thinking about the brain. Contemporary research teams are made up of neurobiologists, clinicians, and physical scientists with expertise in mathematics, brain imaging, and much more. The individual contributions of these disciplines will extend beyond providing novel ways of examining the brain. The capacity to visualize single molecules in complex biological systems is just the beginning of a new biology that aims to understand how these systems function and how much of the regulation of apparently diverse systems is shared among relatively few molecules that were highly conserved in evolution. The great success in decoding the human genome has laid a secure foundation for the next phase of scientific discovery in aging brain research. Genomic studies gave previously unimaginably large amounts of information about genes and their variability. The next phase will generate data sets that are larger than the human genome by more than a thousand-fold.

Advances on this scale will be achieved, for example, by the NIH-driven program to analyze fully the brain "connectome." Once established, the connectome will become the foundation of a new era in brain research that allies "connectomonics" with information science, life course research, and new data on the regulation of gene expression.

It is reasonable to ask what purposes this enterprise might serve. Will it just be for the sake of understanding or will it be translatable into "point-of-care medicine"? In the United States, the 2006 Senate passed the Genomics and Personalized Medicine Act. This proposed that the application of genomic and molecular data (proteomics) would make clinicians better able to target the delivery of health care, facilitate the discovery and clinical testing of new products, and help determine a person's predisposition to a particular disease or condition. The science of informatics will be at the forefront of future health systems. How investigation, diagnosis, and care are delivered will be transformed by this emerging discipline. It is already feasible to envisage the screening of large numbers of different molecular markers to identify disease or preclinical states. This seems likely to be achieved through molecular arrays and defining patterns of biomarkers to redraw the boundaries between diseases and to monitor the effects of treatment. Making sense of these data will be a huge challenge. How best to apply each personalized data set to each specific individual and thus not only guide diagnosis and treatment but also lead to prevention is as yet unknown. It is suspected that every life course health strategy will prove unique to one individual. More than any other organ, the complexity of the brain and its regulatory systems will prove the greatest challenge of all. Of the many excellent books published on developmental neurobiology, *Building Brains* is recommended for a more detailed account.[26]

4

EVOLUTION, AGING, AND DEMENTIA

I. INTRODUCTION

Scientists often have a naive faith that if only they could discover enough facts about a problem, they would somehow arrange themselves in a compelling and true solution. . . . Nothing in biology makes sense without evolution.

Theodosius Dobzhansky[1]

Some people will think about evolution as "just a theory" that cannot have much relevance to aging and dementia. Evolution for them appears to be all about "survival of the fittest," all about who does best at reproduction. Superficially, there seems no good reason to speculate how the laws of evolution should affect aging. Once opportunities for reproduction have diminished in late adulthood, genetic variation in old age should not influence reproductive fitness. This line of thinking is consistent with the view that providing care of the elderly works against evolution. Such reasoning ignores the roles played by mutual cooperation and altruism as driving forces behind the success of our species. Viewed from the perspective of human evolution in particular and not just evolutionary theory in general, a highly complex picture emerges that is relevant to this account of aging and dementia. Understanding brain aging and dementia is also relevant to the application of "Big Science." Without an overarching theory it is difficult to pull together all the known facts (the "facts" cited by Dobzhansky, above) into a single comprehensive whole.

As a species, humans have acquired in the course of evolution effective means to transmit information from generation to generation. These means can be summarized as four major pathways.[2] They can be coded (1) as physical instructions in the structure of genes; (2) as epigenetic modifications of noncoding genes that control the expression of coding genes; (3) as physical information about the external world, as exemplified by changes in the DNA of grandmother's egg, which was in her womb at a time of hardship and so influences the DNA of granddaughter; and (4) through cultural transmission of acquired adaptive behaviors that pass information from one generation to the next.

An important aspect of the transmission of information from one generation to the next concerns the increase in noncommunicable diseases among modern societies in transition from times of food insecurity to assured plenty. Humans have the capacity to adapt not only their own metabolism but also that of their offspring to a harsh environment that favors energy-conserving genes. When the offspring's environment is better provided for, so-called diseases of the Western culture increase markedly in numbers.[3,4] These are coronary heart disease, atherosclerosis, Alzheimer's disease (AD), osteoporosis, and stroke (known by the helpful acronym CHAAOS). Later in this book, reported decreases in the prevalence of AD will be linked to interventions to prevent heart disease and stroke.

Evolution is driven not only by mutations arising in coding genes but through the second and fourth pathways (epigenetic modifications and cultural transmissions) listed in the previous paragraph. Each of these pathways possesses characteristics that can be selected for (or against). By these means, evolution can occur at the level of the individual, the family, or the wider community. From the perspective of disease genetics, the loss of some genes by being selected against may be as important as the occurrence of genetic mutations. In some unknown way, loss of certain ancestral genes may convey advantages to health and reproductive fitness. Other genes may be advantageous in early adulthood during which time they improve reproductive fitness, but the same gene could increase susceptibility to age-related disease. This is an example of antagonistic pleiotropy whereby a gene possesses more than one action (pleiotropy), but these actions antagonize one another.

Grandmothers and their daughters function as single units in ensuring the reproductive success of their family as a group. Upon entering menopause, grandmothers increase their availability for care of their grandchildren, freeing the mother for other family-related tasks. A similar line of reasoning suggests how leadership roles in community groups can prove critical to their prosperity and even their survival. Surveys of societies exposed to great variation in hardship strengthen the view that the family unit provides valuable insights into understanding the origins of poor health in adulthood.[5] Thus far, most studies have focused on links from maternal health to child development and onto midlife diseases, as shown later in this book, because exposures to risk factors for vascular disease also are linked to an increased risk of late-life dementia.

Because few have connected evolution with aging and dementia, it may be a surprise to find that anatomists and molecular biologists began to think some years ago about dementia in terms of evolution and natural selection. It may be perhaps more surprising to find that scientists working on brain development also have begun to exchange ideas about evolution and dementia. Nutritional scientists also suspect that rapid changes in human diets over the past few centuries have been so speedy that evolution lags some distance behind. For some commentators, our adopted dietary habits are now a possible source of increased risk of dementia and even may affect rates of aging.

The contributions of early nutrition to brain development, educational success, and social adjustment have attracted much research interest. In part, this interest is motivated by widely expressed public views that contemporary diets are "nutrient poor" and "energy dense" in ways that cause obesity as well as some chronic health and behavioral problems. The scientific evidence that can be assembled to support these concerns is notoriously weak and not least because of the difficulties of studying dietary habits in children allowed free access to snacks with irregular meals.

Among the best-understood sources of dietary variation in child development is the contribution of human breast milk. For varying periods, babies receive their mother's milk, artificial feeds, or mother's milk and artificial feeds in combination. Human milk contains high amounts of several types of essential fatty acids (essential because they must be obtained from diet and cannot be made in the body). These are eicosapentaenoic acid (EPA) and docosahexaenoic acid (DHA). They are necessary in the building of white matter pathways, in the modeling of neural networks, and in modulating the release of pro- and anti-inflammatory chemicals in nervous tissues. The cognitive development of breast-milk-nourished babies is reported consistently to be better than babies who were fed artificial ("formula") milk. These advantages do not appear to be attributable to the higher social status or acumen of mothers who breast-feed. In later life, surveys have suggested that diets that are replete in DHA and EPA are associated with better retention of mental ability and, possibly, with lower rates of dementia. Not surprisingly, randomized controlled trials of DHA and EPA food supplements have been undertaken.

Scientists who investigate how the brain is constructed regularly refer to the theory of evolution. When asked to explain how or why the human brain has a particular feature, they often point to its origins in evolutionary history. By comparing human and animal brains, brain scientists can identify the rules that govern brain design, how the brain responds to injury or disease, and what resources are available to assist brain maintenance and repair.

An evolutionary perspective provides a useful introduction to key questions about AD. One of the enduring uncertainties about AD is its relationship with brain aging. If it can be shown that AD and brain aging have different origins, then this would strengthen the idea that AD is a distinct type of brain disease. If AD and brain aging share the same causes, then this would support the idea that AD and brain aging lie on the same continuum, differing only in the extent or severity of brain changes.

In the biology of the dementias, genetic causes of the rare familial early onset form of AD are later shown to involve three major genes (amyloid precursor protein [*APP*]; presenilin 1 [*PSEN1*]; and presenilin 2 [*PSEN1*]). These three genes have many hundreds of possible mutations. We reasonably can conclude that AD is not a single genetic disorder, but rather it is made up of different kinds of genetic abnormality each to a varying extent. If we start from the standpoint that there are many pathways toward AD, we could speculate that brain aging contributes to most, if not all, of these pathways and probably contributes most of all to late-onset AD after the age of seventy. Conversely, we might presume that if brain aging is important in late-onset AD, then it is possible that some premature element of brain aging processes can trigger the early onset of AD.

This is a difficult puzzle to solve when we suspect that up to around 80 percent of the causes of AD are genetic while also recognizing that fewer than 30 percent of people living with early onset dementia carry a known genetic mutation and less than 1 percent of people with late-onset AD carry a mutation. Add this suspicion to our sketchy ideas about genes that influence rates of brain aging and the size of the challenge becomes apparent. The precise role of amyloid deposition in late-onset dementias has been discussed widely in this context. Some laboratory scientists support what is known as the *amyloid hypothesis of AD*. They reason that because some early onset AD has mutations in the APP gene, because amyloid accumulates in the brain, and because many studies show that amyloid fragments are toxic to brain cells, the case is strong to investigate the role of amyloid in aging and dementia as completely as possible. In this model of AD, multiple causal pathways can lead to brain cell death and dementia, but all pathways connect steps that increase amyloid formation or impair its clearance from the brain. Not all experts agree with this hypothesis, and some see too many examples of amyloid accumulation in aged humans and animals to accept that amyloid plays a critical role in AD. What if amyloid is a bystander, generated in AD as part of a brain response to age-related brain cell disease? In this and later chapters, these questions will be addressed and alternatives to the amyloid hypothesis will be evaluated.

One helpful approach to the study of individual differences in human brain aging and dementia relies heavily on comparisons between humans and animals. The more closely humans are related genetically to a particular animal species, the better the chances of discovering differences that are critical for humans. This single principle leads to one of the mysteries of AD. Does AD only ever affect humans? Can animals develop dementia?

The answer to these questions will influence how we investigate the link between aging and dementia. If brain aging in humans and animals is similar in nature but only humans develop AD, does this mean that brain-aging processes are not relevant to dementia?

Many domestic animals have lived well beyond what they might have achieved in the wild, and reliable published accounts have identified behavioral changes in old cats and dogs that affect learning and memory. Are these accounts of the animal actually equivalent to AD? To be certain, we would need to establish rules for making an AD diagnosis in an animal. We would need to make sense of specific observed behavioral and brain structural changes in the animal that are found in the typical brain structure of humans with AD.

The human brain has evolved complex structures that support higher cognitive functions. These neural systems are not present in the brains of the higher primates with whom humans shared a common ancestor. Are these systems preferentially affected by AD? If this were the case, and if we could show that the same human systems contained genetic networks absent from the brains of higher primates, then we would be an important step closer to knowing which genes might be involved.

Comparative studies that rely on brain tissue from aged humans with and without AD and aged higher primates explore much more than just genes and their proteins. These studies extend to comparisons with younger animals and aim to understand what is happening in neurons affected by AD. Brain cells have a capacity for self-repair and reorganization, and when these mechanisms are active, they involve genetic networks that also are important in brain development. Much of the genetic machinery for early fetal development was assembled early in our evolutionary history so it is perhaps not surprising to find those same old (or "archaic") genes are reactivated in brain repair in response to injury, aging, and AD.

To sum up, we have three questions: (1) Is AD specific to humans? (2) Are the most recently evolved human brain networks more susceptible to AD? (3) Is reactivation of brain developmental programs in the course of AD incompatible with the working of the mature human brain?

II. IS AD SPECIFIC TO HUMANS?

Brain aging is almost universal throughout the animal world. Old people without dementia, aged nonhuman primates, and aged dogs share the same age-related brain changes. At first, we believed these changes were caused by loss of neurons from the cerebral cortex; however, with more advanced methods of counting neurons, this idea was rejected firmly. In humans, about 10 percent of gray matter volume is lost but the total number of neurons decreases by less. Most of the age-related reduction in brain size is attributed to decreased size of individual neurons and to loss of white matter. Structural studies using a variety of techniques show that about 30–40 percent of white matter is lost during aging. This loss probably is caused by the loss of the fatty myelin sheath that gives the white matter its distinctive color and almost certainly contributes to impaired cortico-cortical information transfer and thereby to cognitive aging.

Most studies show fine microscopic changes in old age of synaptic density in humans and nonhuman animals.[6] Studies in rats and monkeys show that memory functions associated with the medial temporal lobe are vulnerable to age-related decline, although this is not a consistent finding. When nonhuman primates are given set behavioral tests that involve the dorsolateral prefrontal cortex (DLPFC), older animals do less well than younger ones. Brain examination of these animals after death does not show a significant loss of neurons; however, examination does show a nearly 50 percent loss of dendritic processes, which is similar to the 50 percent loss in cortex found in old people without dementia. Neural circuits that are vulnerable to AD in humans also are vulnerable to age-related synaptic changes in animals. These changes are probably responsible for decline in learning and memory performance and do not appear to be species specific. A marked contrast is evident, however, between this type of synaptic loss and the damage caused by AD. In the absence of dementia, neural circuits remain relatively intact, although behavioral deficits in learning and memory are frequent. Early in the course of AD, synaptic loss is extensive and is followed later by extensive neuronal death and the destruction of cortical neural circuits that support higher cognitive functions. What is remarkable and mysterious is the fact that age-related synaptic loss in animals does not progress to AD, whereas in humans, AD develops in more than 10 percent of old people and may be as great as 40 percent in the oldest-old.

Textbooks of veterinary neuropathology[7] describe brain aging in old domestic dogs and emphasize that in the absence of any neurological disease during life, the ventricular spaces are nevertheless greatly enlarged with an extensive loss of surrounding myelin. Aggregations of abnormal proteins occur in AD, and these contain large distinctive amounts of amyloid (described more fully in Chapter 6, The Biology of the Dementias). Amyloid (literally "starch-like") is the name given to aggregates of short fragments of a larger precursor molecule. These fragments contain between thirty-seven and forty-nine amino acids, whereas their precursor (APP) contains up to 770 amino acids. Amyloid fragments are prone to form sheets of fibrils, known as β-pleated sheets of amyloid; the term is shortened to β-amyloid, a convention we shall use in this chapter.

β-amyloid plaques are not uncommon in old dogs and other nonhuman animals, but their density and distribution always differ greatly from old humans. The fine structure of neurons in humans and nonhuman animals alike, however, does change markedly with age. All species experience loss of dendrites, dendritic spines, and synapses from large cortical neurons as well as sparing of the hippocampus and entorhinal cortex.

When looking at our closest genetic relative, the chimpanzee, Chet Sherwood[8] at the George Washington University, Washington, D.C., and his colleagues elsewhere used magnetic resonance imaging (MRI)-derived data to compare aging cortex between eighty-seven humans and sixty-nine chimpanzees. The researchers were interested in two questions: Are human brain aging changes unique to humans and, if they are, could these changes be explained simply by humans living longer than chimpanzees? In their samples, which were large enough to detect these effects, no MRI-derived measure of regional brain volume "showed a significant effect of age" in the chimpanzees.

When Chet Sherwood reviewed the data in light of the age differences between humans and chimpanzees, he found that none of their chimpanzees (maximum age of forty-five) had lived long enough to exceed the age at which humans first show age-related brain volume decline (from about the age of forty-five). Studies elsewhere on chimpanzees living up to the age of fifty-nine had suggested some slight reductions. When taken together, however, the somewhat small number of aged chimpanzees studied so far does not support the view that they undergo age-related brain changes that are similar in nature or extent to humans.[9,10]

The evolutionary significance of these findings was not lost on Chet Sherwood and his colleagues. Sources of regional difference in brain energy usage between humans and chimpanzees, in their view, could make much higher energy-consuming neurons of the human cortex especially vulnerable to the effects of aging. This vulnerability may be sufficient to trigger the cascade of molecular events in AD.

The evolutionary "tree of life" is shown in Figure 4.1, which illustrates the relationships between modern humans and many living primates. It provides only a general overview, and some relationships remain disputed. This type of map commonly is used in comparative neurology.

Most AD changes differ in extent but not in type from brain aging without dementia. In AD, the loss of synapses is much more severe and β-amyloid plaques are distributed more widely. An important exception is the development

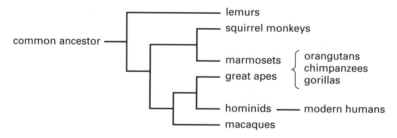

FIGURE 4.1 A branch from the "tree of life" that shows evolutionary relationships between modern humans and many living primates.

of neurofibrillary tangles (NFTs) that almost never are found in the absence of a clinical dementia syndrome. The occurrence of NFTs in nonhuman animals is also rare and mostly occurs in animals like sheep and goats that obviously are different from primates. Microscopic examination of NFTs found in nonhuman animals show that these are not quite the same as the NFTs found in humans. Although β-amyloid plaques similar to those found in human AD are detectable in the cortex of some aged nonhuman animals, it is almost unknown for both β-amyloid plaques and NFTs to occur in the same aged animal. We can conclude, therefore, that when strict rules for the pathological and behavioral diagnosis of AD are applied to old animals, in very few single cases, if any at all, can an AD diagnosis be confirmed. Scientists like Chet Sherwood, however, raise an important point: Even if a higher primate survives in captivity up to sixty years of age, outliving others in the wild perhaps by more than ten years, this remains below the peak incidence of dementia onset in humans, which doubles in incidence every five years from about seventy years old. Chimpanzees, in their view, simply never live long enough to develop AD. This criticism must be tempered by observations that frailty linked to senescence is frequent in chimpanzees from the age of twenty and by the fact that AD becomes symptomatic in humans only after a lag period perhaps as long as twenty years. This period overlaps sufficiently with the longest living chimpanzees to demonstrate that if AD occurred in chimpanzees, captive specimens in fact have lived long enough to show AD features when their brains are examined after death even though they remained asymptomatic.

This conclusion supports a strong suspicion that AD relates to the evolution of the human brain. If we accept this, we could proceed to make a case for the involvement of evolutionary changes in brain structure and function in the pathological processes leading to AD.

III. HUMAN EVOLUTION

Figure 4.1 shows that we are not descended from modern primates but share common ancestors with them whose offspring adapted to their environments in different ways. The great success of our species is founded on the way our brains developed to provide some individuals with the capacity to adapt to changes. Other members of our species could not adapt; their children were less successful and eventually their offspring died out. Meanwhile, those who could adapt prospered and had more surviving children who more effectively filled the niche first exploited by their ancestors some generations back. Looking at human evolution, we can see how at each step—when the environment changed, food became scarce, safety was jeopardized, and shelter was harder to find—some of our ancestors were more adaptable and so more successful. This sequence of events unfolded over many hundreds of generations. Among the many demanding environmental challenges our species has faced, the effects are well established of cold weather on food supply, causing human populations to move across a changing landscape as they coped with ever-changing seasons.[11]

Evolutionary biologists now speculate with some confidence that human evolution was spurred on by environmental change, but the reasons behind our success lie deep within us in the structure of our DNA. We are now at the point at which we can look in some detail at the molecular mechanisms underpinning the successful evolution of our brain, to see the molecules at work that make us distinctly human. The pace of genetic discovery is now so quick that we soon will be able to understand how our brains are constructed and how the brain can achieve all it does.

About 7 to 10 million years before the present (ybp), our distant ancestors had close relatives among the most advanced primates who would become the forebears of modern chimpanzees. Their ancestors remained in the tropical rain forests where their present-day descendants seem to have conserved many of the attributes that had adapted them so well to forest living: plentiful food year-round, better security in small social groups, and shared care of their children. Some early relatives of these chimpanzee ancestors moved into distant forests where a slightly smaller stature was advantageous, but others migrated for unknown reasons into the grasslands beyond the forest. Here, an upright posture provided better vision and allowed the skull to move back above the neck, easing the development of structures to allow better vocalization and relieving the chest wall of the burden of a four-limbed gait. Figure 4.2 summarizes these relationships between humans and higher primates and reveals the extent of genetic differences between species.

The greatest achievement of human evolution is the development of higher cognitive functions. This development was made possible by substantial increases in the size of the cortex, although this is not the whole story. When we observe our closest relatives, like the chimpanzees, we can see that our behavior is so

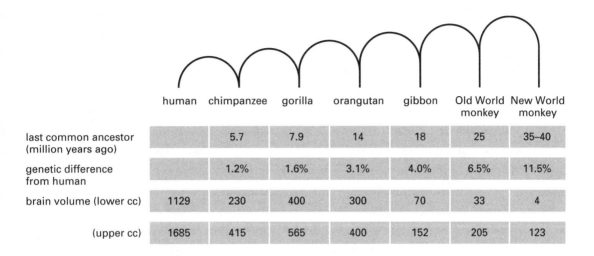

	human	chimpanzee	gorilla	orangutan	gibbon	Old World monkey	New World monkey
last common ancestor (million years ago)		5.7	7.9	14	18	25	35–40
genetic difference from human		1.2%	1.6%	3.1%	4.0%	6.5%	11.5%
brain volume (lower cc)	1129	230	400	300	70	33	4
(upper cc)	1685	415	565	400	152	205	123

FIGURE 4.2 This figure shows the relationships between modern humans and their most closely related primates in terms of time, as we shared a common ancestor (millions of years ago); percent of genes that differ between humans and primate species; and brain volume (uncorrected for body size).

much richer, much more flexible, and adaptive. If our adaptability is uniquely human, could this be relevant to understanding the biology of AD? A likely explanation concerns the architectural substructure of the cortex, not just its increased size. Our adaptability and flexibility probably relies on our capacity to organize specific neural networks to do this type of work. In AD, we already know from the work of scientists like Heiko and Eva Braak[12] that certain neurons become more vulnerable to AD with aging. Could it be that these neurons were those most recently evolved, and it is their absence from higher primates that greatly reduces their risk of AD?

CHANGING ENVIRONMENTS AND EVOLUTION

Evolution is about successful adaptation to changing environments. Our prehistory had many generations to adapt to environmental change. Each Ice Age, for example, spanned many thousands of years from slow beginnings to a maximum followed by a slow thaw. Humans are the only species to experience the effects of environmental changes that are entirely self-imposed. The most obvious changes concern the diets of factory workers during the Industrial Revolution. In our prehistory, we ate from a range of more than 3,000 plants and fruits, lean game, and many fishes. Within a few centuries, industrialized workers were eating less varied diets—around twenty types of foodstuffs—mostly grains and sugars in diets that became progressively more nutrient poor but energy rich. Notably, our consumption of marine fish oils fell sharply from a diet that was balanced in consumption of omega-3 and omega-6 essential fatty acids to one in which omega-6 predominates by about 20:1 over omega-3. Our consumption of animal fats and dairy produce increased as our consumption of essential fatty acids fell. We return to this topic in a later chapter on the prevention of dementia.

Women also changed their timing of reproduction. The timing of first menstruation (menarche) was much later in ancestral human societies than current and was followed much sooner by a first pregnancy and lengthy periods of (infertile) lactational amenorrhea. The onset of menopause allowed opportunities to share childcare with grandparents, improving parental care and children's chances of survival to reproductive age.

From an evolutionary viewpoint, in the face of rapid environmental change, we have had insufficient time to adapt successfully to radical transformations in diet and the hormonal life trajectories of women. We are aware that the public health consequences include increased risks of heart disease and many types of cancers. It is not known whether the risk of dementia also has increased.

IV. ARE THE MOST EVOLVED NEURAL NETWORKS MORE VULNERABLE TO AD?

The most obvious brain differences between higher primates and humans are our greater brain size and much greater lateralization of brain functions. Head size is determined largely by brain growth. At birth, infants' brains are contained within expandable plates that do not fuse completely for some years. With brain development, the skull expands but only as much as required by the developing brain within. In the nineties, some genes were found to play a major part in head size. This seems a good place to start a search for biological processes that supported the evolution of the human brain and possibly increased our susceptibility to AD. A mutation of one of these genes will impair brain growth and cause not only a small brain but also a small skull ("microencephaly") and profound learning difficulties. Although associations between genes and pathological conditions that impair mental development suggest that the same genes could have influenced the normal range of overall human ability ("general intelligence" or "g"), this did not happen.

GENES AND HEAD SIZE

Genes that are known to control head size include microencephalin and the abnormal spindle-like microencephaly-associated gene (*ASPM*), named after the genetic disorder microencephaly, which led to their discovery.[13] These genes have undergone accelerated modifications in the course of human evolution and raise the question of how they might change as we continue to evolve. They are linked to head size but not to intelligence. The possibility remains, however, that it is those genes that have undergone an accelerated evolution, and that the functional constraints on these genes are weaker in humans than in other nonprimate animals.[14,15]

So far, brain size differences between humans and higher primates cannot be explained by genetic differences associated with microencephaly (a pathological condition).[16] Likewise, as yet, no grounds support the idea that differences between people in mental ability—like mental speed, better use of language, or a more reliable memory—can be explained by differences in genes that influence head size. So, in answer to the question, no evidence supports a positive contribution of increased head size to risk of dementia. In fact, the evidence more often has pointed in the opposite direction. Smaller head size can be linked to increased risk of dementia in some but not all studies.

Lateralization of cortical functions could provide a clue to human vulnerability to AD. Even after more than 140 years since Broca's first reports, however, the biology of cortical lateralization remains poorly understood. Advances in the study of bird brains suggest why lateralization of perceptual functions might have evolved and provided cogent reasons why lateralization can be so advantageous. Set against these lines of evolutionary reasoning are the disadvantages of predictability of lateralized behavior when attacked by predators. Some biologists have concluded that almost all social vertebrates are lateralized at a population level, whereas solitary species are less often lateralized.

An important advantage of cortical lateralization is that functions that once were distributed equally in either cortex can remain in one or other cortex, allowing the other to develop new capacities. This type of cortical adaptive process allows human cognition to appear limitless in its creativity and adaptability. Whereas other apes differ little from one another in feeding, sexual behavior, and so on, humans are dramatically different from apes in terms of complex language, music, math, art, philosophy, and commerce. Many higher functions rely either on the development of spoken language or the use of symbols to express concepts.

Use of language is a higher cognitive function that is largely lateralized to the dominant side (cerebral hemisphere). Some linguists like Denis Bouchard at the University of Quebec at Montreal (Canada)[17] have argued that the potential for language comes from three distinct capacities: (1) physical properties of the brain that must have been present before language developed, (2) the development of a conceptual system that links ideas to symbols within a grammatical framework, and (3) an articulatory system that allows the use of language to communicate.

The speech center in humans is located in the left hemisphere. This hemisphere controls movement on the right side of the body. Broca's and Wernicke's areas are the two main centers on the left side of the brain and are activated when we speak. In the premotor cortex, an area is activated when humans perform an action involving their hand or mouth. The neurons in this area also are activated when a human or other higher primate *sees* another individual perform the same action so they were called "mirror neurons." This capacity to recognize the actions of other individuals is thought to be present in Broca's area and to be a key precursor of the development of language. Social neuroscientists have identified a role for mirror neurons in humans in predicting the behavior of others during social interactions and have suggested that they provide the basis for the development of a theory of mind (described in Chapter 8, Emotional Aging).

Once the firing of mirror neurons is brought under voluntary control, a basic means of communication is possible along the lines of "You know that I know what you are doing." The questions remain: How do we express that knowledge? Through a hand gesture or a facial grimace? How long does it take for this to become "You know that I know what you are thinking"? Evolutionary theorists are not sure how long this takes, but they speculate that because movements of mouth and face are used by modern primates to communicate and—allied with body gestures—can convey much more information than a single sound or drumming, human evolution refined these methods to develop speech. Because it is physically impossible for us to make more than one intelligible sound at a time, speech became a series of different sounds subject to rules about their construction and order and thus made sense to the listener.

The neural circuits involved in language production and comprehension were first studied in patients who were victims of localized brain injury or stroke. Modern techniques include positron emission tomography (PET), functional magnetic resonance imaging (fMRI), stimulation of the cortex during neurosurgery, and recording electrical activity on the surface of the cortex during language-related tasks. Currently, the neural circuits for language comprehension involve the "classical" areas of Broca and Wernicke, but they extend to the memory retrieval areas in the temporal cortex and to frontal areas that integrate information to make an understandable "whole message." Poorly understood neural circuits add control systems to complete this complex circuitry. These systems regulate motor response to language and the timing of responses (e.g., knowing when to respond in a conversation).

How does this circuitry compare with the succession of areas affected by NFT formation described by Braak and Braak? Certainly, Braak and Braak[18] interpreted the involvement of temporal and frontal association areas as evidence that NFT formation occurred preferentially in the most evolved of the brain's neural circuits; thus far, no opposing view has gained support.

David Neill[19] of Newcastle University (United Kingdom) and Stephen Rapoport[20] of the National Institute of Mental Health (United States) have emphasized that neurons affected by NFTs are those that differ most from primates and are more evolved than neurons elsewhere in the brain. Cortical association areas affected by NFTs show evidence of greater evolutionary progression than other brain areas. In addition, neurons affected by NFTs are late to differentiate during brain development. Late maturation of myelin in white matter corticocortical connections was speculated by Braak and Braak[21] to follow the sequence of formation of NFTs. A consequence of delayed myelination is that these neurons retain greater synaptic plasticity than other brain areas. Vulnerability to the formation of NFT formation may be linked to greater plasticity. Figure 4.3 shows the myelination of major brain pathways from life in the womb through adolescence up to age about 30 years.

Synaptic remodeling is an important component of brain compensatory responses to injury and AD pathology. These processes are best understood as the brain's response to repair damage and restore function. Some brain scientists

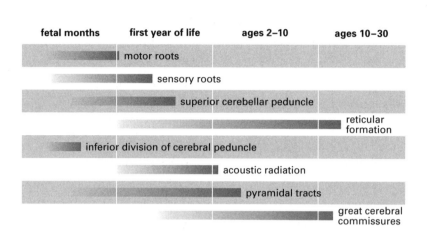

FIGURE 4.3 The cognitive development of children and young adults closely follows the pattern of progressive myelination of cortical axons. Regions myelinated first include the spinal cord and brainstem. Myelination continues toward the frontal cortex, with sensory pathways myelinating before motor pathways, and downstream projection pathways myelinating before cortical association pathways. The prefrontal cortex myelinates last. Although initiated prenatally in humans, most tracts and regions are myelinated completely during the first year of life. In humans, some regions do not complete myelination until the second and third decades of life.

have thought that in certain circumstances brain efforts at compensatory remodeling may go wrong. Is it possible that these responses contribute to AD? Tom Arendt is perhaps best known for placing abnormal compensatory remodeling among the primary causes of AD. Seen in these terms by David Neill[22] and Tom Arendt,[23] vulnerability to NFT formation was believed to be related to enhanced synaptic plasticity in the most evolved brain neurons.

These ideas may, however, find support in the normal functions of APP in the healthy brain. Although the complete range of APP functions remains to be determined, APP is indicated to be involved in synaptic plasticity and repair and in brain iron metabolism. This evidence shows that APP is located in or close to synapses, and its maximum expression occurs during brain development (neurogenesis). In health, α-secretase is involved in these functions. This involvement raises an intriguing possibility—running counter to the amyloid hypothesis of AD—that APP is promoting synaptic connectivity to make good synaptic loss in AD. Although the presenilins also are expressed in the developing brain, no clear-cut role has been established in synaptic plasticity. Presenilins, however, may be involved in pathways that determine the numbers of neurons that make up the mature brain.

V. IS REACTIVATION OF DEVELOPMENTAL PROGRAMS IN THE COURSE OF AD INCOMPATIBLE WITH THE MATURE BRAIN?

For almost two hundred years, biologists have remarked on structural similarities between the developing human embryo and the embryos of other vertebrates. These similarities are most evident in the first few months of pregnancy, and it is not until later fetal development that the human fetus becomes obviously different from other primates. This similarity suggested that human development recapitulated human evolutionary history, a proposal that remained poorly founded until the tools of modern molecular biology were applied to the comparison. The recapitulation of evolution in the embryology of vertebrates is illustrated in Figure 4.4.

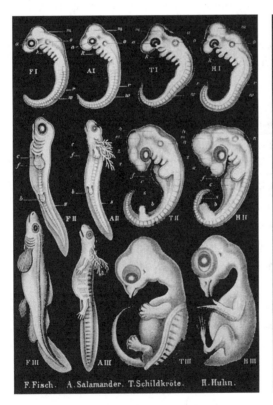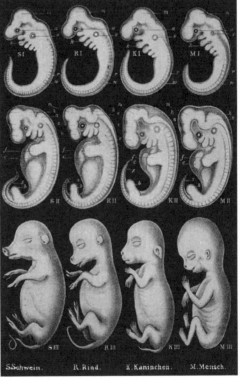

FIGURE 4.4 Embryos of selected vertebrate species. The striking similarities between species are most marked during early development. As development progresses, species begin to show major differences.

The first part of this chapter looked for differences in susceptibility to AD between species before considering the possibility that the most evolved neural circuits were more vulnerable to AD. We recognize that evolution tends to conserve basic blueprints for living things and adds features that increase opportunities to pass on parental genes into successive generations. Therefore, when we locate precisely the sources of differences between healthy human brains and our most closely related higher primates, we likely will find clues to the origins of higher mental functions and, just as possibly, to AD as well.

Over the past thirty years, Linda Partridge at University College, London University (United Kingdom), has investigated the basic molecular blueprints for aging and has shown, by exploring physiology and anatomy, the mechanisms that allowed aging to evolve could be discovered gradually.[24] Studying model organisms like the fruit fly, Partridge has helped unravel the biology of aging by applying evolutionary theory at critical steps as her thinking progressed.[25,26] A new scientific discipline was envisaged combining evolutionary biology with gerontology in a way that she described in short form as "evo-gero." This discipline now occupies a key position among explanations of aging and their relationships to developmental programs. In fact, they are relevant to a biological understanding of programming in general.

What emerges from years of patient application and thoughtful experiments is an intricate system that has layer upon layer of complex pathways that determine aging processes. With colleague David Gems, Partridge wrote,

> Homo sapiens and chimpanzees walked the Earth only some 5.4 million years ago, yet our maximum lifespan is twice that of our closest living relative (around 110 years versus around 59 years). Do the genes and processes that have been the focus of [work on a] model organism . . . specify species differences in ageing? Do they also control the remarkable phenotypic plasticity of lifespan seen in, for instance, social insects? Answering these questions will require an approach analogous to that used in understanding the evolution of differences in development that lead to differences in anatomy (i.e., evolutionary developmental biology, or evo-devo). One might naturally refer to such an approach as evolutionary gerontology (or evo-gero).[27]

In Figure 4.5, Partridge and Gems provided a scheme that shows how differences in rates of molecular aging could influence molecular changes that

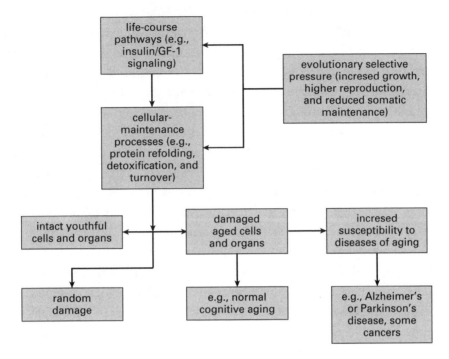

FIGURE 4.5 Life course pathways control improved cellular maintenance, which leads to slowed aging (e.g., slowed normal cognitive ageing) and protection against diseases of aging (e.g., neurodegenerative diseases of aging, such as Alzheimer's and Parkinson's disease, and cancer). Aging can evolve via selection to reduce investment in energetically costly somatic maintenance processes and instead to increase early fitness traits, such as growth and reproduction.

Adapted with permission from Partridge and Gems, "Beyond the Evolutionary Theory of Ageing," *Trends in Ecology and Evolution* 21 (2006): 334.

predispose humans to age-related disorders like AD and Parkinson's disease as well as to certain cancers.

Molecular mechanisms of aging and development could not be thought of in this way until the effects of genes on development were discovered. Until these breakthroughs were made, there just seemed to be piles of detailed information accumulating in the work of evolutionary developmental biologists (evo-devo in the previous jargon) that provided neither explanations of what was controlling aging and development or predictions about where these controls might go wrong.

Sean Carroll[28] from the University of Wisconsin at Madison proposed a general theory to explain how the forms of animal species have evolved and how species like chimpanzees and humans that share so many genes (around 99 percent) can differ so much in anatomy. He has gathered together much relevant data to provide a modern synthesis of developmental biology. Like others, Carroll has suggested that anatomical evolution depends much more on changes in the regulation of gene expression than on fundamental changes in protein structures through genetic mutation.

Carroll noted an intriguing observation among the molecular discoveries made at the beginning of the revolution in molecular biology around 1984. At that time, it was first realized that despite totally different species having anatomical forms that had evolved from branches of the evolutionary tree (as shown in Figures 4.1, 4.2, and 4.4) that split apart millions of years ago, these forms were governed by similar sets of genes. Previously, it was thought that each species had developed its own unique genetic program to control development and that this uniqueness was responsible for evolutionary progress.

The scale of technical achievements in molecular biology is appreciated throughout the life and social sciences with repercussions in philosophy and religious thought.[29] The first advance is that we are able to read sequences of genes in a way that allows us to read whole genomes. This technique reveals how genetic sequences differ between and within species. It was not surprising to find that so many of the genetic differences between humans and other primates involved genes that control brain development. These differences were found in gene structure and in the regulation of gene expression.

Discoveries of this magnitude generated many new hypotheses about how genes were selected positively during human evolution; why some genes were deleted and others duplicated; and why evolution should be associated with noncoding regions of the genome, especially among genes that change gene expression.

In Chapter 2, The Life Course Approach, we examined some scientific ideas that encourage and support a life course approach to brain aging and dementia. These ideas are highly interrelated and complex, not just affecting development and aging in the absence of disease (including dementia) but also showing strong associations between childhood disadvantage and late-onset adult disease.

The principles of evolutionary biology are relevant to these associations. They are major determinants not only of physical growth and development but also of an infant's anticipation of the external world. Earlier, the idea was fashioned that every baby is born with an innate repertoire of developmental programs that could be reprogrammed by the need to adjust to a changing environment. Every life is a versatile variation of a highly evolved template. In Chapter 3, The Well-Connected Brain, brain structures and their connections were described and related to higher mental functions that are shared among all humans but are absent from all other animals. We have seen how these brain structures and their neural networks are damaged selectively in the course of AD. These specific observations provide the best evidence that what makes us human makes us vulnerable to AD. The search for the origins of AD can be extended to examine the extraordinary and rapid evolution of the human brain and possible links with accelerated evolution of specific genes. These changes appear to be greatest in those genes involved in human brain development from conception to maturity and, potentially, these are the genes implicated in diseases of the aging nervous system.[30]

We must keep sight of our aim to understand how the brain develops and then begins to fail in old age, and how along the way it tries to make good any faults caused by wear and tear. These faults certainly accumulate with age and, given time, will make thinking inefficient and error-prone. If the human brain were a new product expected to last a hundred years or so, the project development engineers would have quite a job on their hands. Most likely, they would look around the animal world for what already works and then imagine what they would need to add to make a brain uniquely human. This is the best way to think about brain design.

In this account of our search for the origins of AD in human evolution, we can now state two overriding principles of brain design and identify three basic requirements.[31] The *first* principle is highly conservative. If evolution took many millions of years to find the most effective way to provide an animal with a brain possessing a particular ability and it works, then evolution says keep it and maybe add to it. The *second* principle is about space and energy. Body systems have jobs to do and one of them is keeping the brain well supplied with energy. Our mature brains make up 2 percent of our total body weight but consume about 20 percent of our total energy. Brains are expensive to keep running and costly to maintain. Improvements in brain design must operate within the same energy constraints. It is relevant to our interest in the selective vulnerability of specific neural networks and pathways that the most recently evolved cortical neurons have among the highest energy consumptions of any neuron.

A *first* requirement, therefore, is to build a brain of the greatest size that can be supplied adequately with energy and nutrients. Maximum brain size is limited by the space inside the skull. The maximum size of a human skull has a practical limit—that is, all skulls are limited by the dimensions of the female birth canal. In turn, this is limited by the size of a woman's pelvis, which is further constrained by a woman's need to walk upright (bipedalism).

At first, improvements to evolving primate brains were made by increasing brain size. The brains of modern adult humans are about four times bigger than the brains of chimpanzees, after allowing for differences in body weight. Eventually, bigger brains become too costly to build and run efficiently, so improvements could be made only by reorganizing what the brain already has. This *second* requirement concerns the "self-organizing principle of the cerebral cortex" referred to in Chapter 3. Throughout development and aging, the brain retains an ability to self-organize. This capacity remains useful later in life because as parts of the brain fail, other healthy parts can adapt and compensate for losses attributed to aging.

The *third* requirement is a bit like shopping and pertains to local networks. When shopping must be done and time is insufficient to travel to less expensive stores at some distance, then it makes sense and saves time to shop locally. Neighborhood shops try to respond to customers' needs by offering "personal service." Over time, customers who frequent local shops find these work very well, and being smaller, they are able to adapt quickly to their needs. They provide high levels of service for those who are "time poor but money rich." Like your neighborhood shop, the brain works quickly and efficiently when information is processed locally without too many transfers between distant brain structures. If we can accept the two principles of brain design and meet these three requirements, then we can start to look at how these conditions could be achieved.

It frequently is repeated that the human brain is the most complex known structure in the universe and that we seem to have too few genes to account for all its complexity.[32] So how is the brain's complexity achieved? How can too little genetic information provide an exact blueprint for something so complicated? As a schoolboy, when first interested in biology, a simpler question puzzled me. How does a tadpole turn into a frog? From a clump of cells, something with an obvious front and rear and then four limbs appears. How does that mass of cells know which way is front or back? Which cells will become eyes or ears?

The answer must lie in the DNA contained in frog spawn. Wrapped up inside the nucleus of every cell, a code must determine when cells will divide, where they belong in the body, and what sort of cell they would become. It must be a problem that can be solved only once we understand the genetic control of development.

Until the early twentieth century, most biologists regarded genetics and development as two sides of the same problem. When the work of Gregor Mendel was rediscovered, however, it was quickly realized that genetics could be studied separately from development. In 1955, when Conrad Waddington was professor of genetics at the University of Edinburgh, he took a Greek word "epigenesis" meaning "a theory of development" and combined it with genetics to coin a new scientific term "epigenetics."[33]

This branch of science first was applied to embryology, which is a useful concept when trying to explain some of the fundamental features of development. One of these concepts is the observation that different types of cell are reproduced faithfully each time cells divide and become part of a specialized tissue. Every cell nucleus contains the complete genetic code that makes up an individual. Therefore, each cell must contain some specialized genes that are turned on permanently and some that are turned off permanently. These controls can be highly specific. For example, the bone marrow contains stem cells that will be differentiated into a specific type of cell, such as a white blood cell. When a stem cell divides, it produces another stem cell and a differentiated cell. To do this, genes are required that switch on in the differentiated cell and that switch off in the undifferentiated stem cell.

It was not until 1975 that the chemistry was worked out to explain how genes (i.e., DNA) could be switched on or off. A limited amount of genetic information can create a three-dimensional landscape. Separate chemical gradients formed along three axes and a newly born cell can find a specific position in that landscape that depends on the cell being able to move and then stop moving. The idea that chemical signals determine the shape of biological tissues ("morphogenesis") has a long history but not until the discovery of DNA and the molecular genetic revolution was a good understanding provided of the genes involved, the timing of their expression, and the actions of their chemical products. Today, genetics can explain why brain cells grow and become so different from each other (i.e., they "differentiate") as well as how they find their final position in the developing body (the embryo) and then later how neurons connect ("get wired up").

The brains of animals with backbones (vertebrates) share many characteristics. They develop at the front end of the spinal cord and, early in development, they look surprisingly similar. Figure 4.4 is a well-known diagram that was popularized in the nineteenth century when embryonic development was understood as a process that moves through successive stages and duplicates the evolutionary history of each species. For a schoolboy this was fascinating but for exams, we were warned to be prepared to criticize this sort of "ancestral thinking." We were reminded that the founder of embryology—Karl Ernst von Baer, an Estonian anatomist—who first reported these similarities probably got the idea by chance. The labels had fallen off his jars of specimen embryos—or so it was claimed—and he could no longer distinguish between species. This made him think that because embryonic development is a process of differentiation, the features that developed in the first phases would be those that are shared among species, while those that distinguish one species from another will appear last. This is relevant to the human brain for which development extends beyond embryogenesis and is not completed until almost twenty years of age.

Von Baer's ideas came to be called Baer's Laws of Embryology and went on to lose credibility during the early twentieth century only to be taken up again once the refined tools of molecular genetics could be applied to embryology. Now, many striking similarities can be found in the expression of genes in early embryos from living things as diverse as fruit flies and mammals. Genes expressed first in development ("early onset genes") are shared among almost all multicellular organisms, suggesting a common origin. The term evo-devo was coined to sum up the idea that "development recapitulates evolution."

This idea harks back to the engineer who refuses to reinvent the wheel. If so much effort was used to design expensive proteins and to assemble these into the "protein engines" (the workhorses of intracellular machinery), and when these do their jobs accurately not for a lifetime but for hundreds of millions of years, then these must be conserved. For example, the first genes to be expressed in the fruit fly will determine the domains within which later genes will express themselves. First an anterior-posterior gradient is established in the embryo. Next, sections are set up within that gradient. The same early onset genes are conserved in evolution and probably determine the anterior-posterior axis in all vertebrates. Here, the brain and nervous system are formed from a strip of cells along the surface of the human embryo. Genes expressed during this phase are

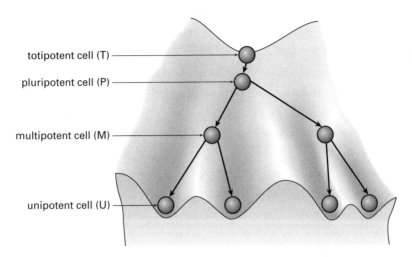

totipotent cell (T)

pluripotent cell (P)

multipotent cell (M)

unipotent cell (U)

FIGURE 4.6 During differentiation, cells pass through stages of development. These are shown as a landscape falling toward the front of this schematic. At first, each cell has great potential to differentiate into one of any different type of cell (i.e., it is totipotent). Next, its eventual form is limited, as it becomes pluripotent. As it develops further, choices are further restricted so it can become multipotent, and then finally it can only be one type of cell that is unipotent. This is the "epigenetic landscape" model of cell differentiation. The figure shows a graded landscape down which the cell passes but cannot return "uphill" to a former state without "reprogramming."

highly conserved between insects and vertebrates showing that Baer's ideas are not so old fashioned after all. Later in development, the genes involved become much more specialized and specific to individual species. This applies particularly to the vertebrate brain.

The next puzzle seems obvious but so far unsolved. How do cells know that they must differentiate at a specific developmental stage? Does some sort of central developmental clock orchestrate all aspects of development? But how would a single clock communicate with more than 100 different cell types each wanting a switching signal at different stages of development? Figure 4.6 shows in diagrammatic form how the differentiation of cells progresses as an imagined descent along a gradient toward final differentiation. As differentiation progresses, so opportunities to change a final form ("destination") are lost.

Testing these new ideas on the regulation of gene expression either singly or as part of "genetic regulatory networks" is not straightforward. One aim of this research is to analyze the complete human epigenome. Eventually, this analysis will detail the exact pattern of gene expression in each cell. We then will know what genes are switched on to make a daughter cell of a progenitor cell differentiate, for example, into a heart muscle cell and not into another a heart blood vessel cell.

As new information about gene expression in the human brain is collected, so it becomes possible to understand the relationship between genes, brain, and behavior. We already know that there are many different types of cell in the brain, so we can be sure that a first step will be to locate the expression of specific genes in specific cell types. The next step will be to link specific genes to specific neural networks. Until recently, this type of analysis could be done only in large brain regions, making the data difficult to interpret at a cellular level. The Allen Brain Atlas[34] is an ambitious and ultimately successful genomics project on a huge scale. It involved the sort of cooperation between teams of scientists outlined in the Introduction.

When gene expression is studied in single neurons, this reveals complex relationships between neuronal activity and gene expression. Very different gene expression profiles are detected even between adjacent neurons. Some genes, however, that control energy production, maintenance, and repair are expressed in nearly all cells.

Once genetic expression patterns for specific brain regions are established, it is possible to link the functions of that brain region to the expression of specific genes. So far, it seems that genetic expression is similar within clusters of neuronal cell bodies, but it is not consistent across larger brain structures. The six layers of the cortex described in Chapter 3, The Well-Connected Brain, show different patterns of gene expression that vary from one cortical region to another. Many of the genes identified in this way have no known function, although some others are involved in cortical synaptic plasticity.

Critical information that would firmly link a specific gene to a specific brain function often is lacking. Often, associations can be stated only in general terms along the lines that such and such a gene affects brain function in general. Without supporting experimental data, no specific link between a single genetic difference and a single aspect of brain function can be accepted. Some additional criticisms of these ideas question the assumption that any genetic differences between humans and primates must be attributed to evolutionary pressure when the case remains unproven.

Genetic regulatory networks were the critical building blocks of developmental programs. These advances displaced the long-held ideas that evolution of anatomy depended on old genes mutating into new genes to provide novel proteins or that multiple copies of the same gene might disperse around the genome.

Sean Carroll elucidated the key genetic principles that govern the evolution of body form and the nature of underlying changes in genetic regulatory networks that are responsible for the resulting anatomical diversity. He emphasized how identical genes could perform different functions in different genetic regulatory networks. Any single gene in a genetic regulatory network, once evolved, he argued, will remain under great pressure to stay the same: Any change would have such widespread effects that it would be mostly negative, and potentially catastrophic.

Evolutionary biologists, therefore, have established that the same family of genes can be shared among species as different as insects and mammals. These are mostly genes that control signaling between cells and the synthesis of proteins. Most remarkably, these genes achieved their present form more than 500 million years ago when the origins of modern groups of species were first established. The evolutionary biology of organs as different as eyes, heart, and lungs was once thought to have followed independent pathways but are now known to be regulated by similar genetic regulatory networks. The current picture is of a very wide array of target genes controlled by a relative small number of regulatory proteins. Seen in these terms, anatomical evolution is determined by the evolution of genetic regulatory networks.

The *Sonic hedgehog* protein shapes structures as diverse in the domestic chicken, for instance, as the number of claws, the structure of the cerebellum, and feather bud formation. The *Sonic hedgehog* gene provides another example of a gene that has undergone accelerated evolution in primates.[35]

All of this came as a surprise, as there was no reason to expect that the same protein would be involved in so many different developmental pathways. The term pleiotropy is used to describe a gene or protein with many different functions. When these functions are dispersed among many different tissue types, the term "mosaic pleiotropy" is used.

Genes that act only in late life are influenced only slightly by evolution. Once the reproductive phase has passed, all selection effects are weakened in offspring born before the first disadvantages of aging can be detected in their parent. Aging also could be caused by mutations in genes that act only late in life and are without effects on reproductive success. So far, with the rare exceptions of premature aging syndromes, no genes are known that actually cause aging.

Nevertheless, aging can be linked confidently to accumulated genetic mutations. Their effects may be countered by the actions of genes that promote fecundity but prove disadvantageous in late life. This is an example of one gene having different and opposing effects at different stages of development and aging (technically "antagonistic pleiotropy"). This could produce quite a complicated outcome at the level of whole populations where the net effects on rates of aging result from both late genetic mutations and a subset of genes that enhance reproductive fitness but in later life predispose to diseases that cause chronic disabilities with some features of premature aging.

One idea is that aging is programmed genetically into each type of tissue. But what is meant by "programmed"? The term originates in machine tool design where "programs" of instructions to be completed in sequence by the machine were punched onto cards that controlled the machine. Contemporary usage of the term "genetic program" implies the actions of factors that control gene expression. These factors can be other genes that code for regulatory proteins or more direct effects of the environment on genetic mechanisms that "switch" genes on or off. Running alongside genetic programs that affect brain-specific aging are numerous biochemical changes that compromise neuronal function without causing neuronal death. The net outcome is cortical circuits damaged by aging, which corrupts synaptic density and impairs learning and memory.

Evolution favors characteristics that improve opportunities for successful reproduction and has little or no effect on postreproductive aging processes. Cognitive development is one such characteristic. Better cognitive function among humans is linked to a choice of partner with similar cognitive abilities and better access to material resources.

Many age-related changes are linked to changes in tissue-specific gene expression, and a common set of genes may change equivalently in different tissues. In the AGEMAP project, the effects of aging were investigated in 8,932 genes found in sixteen different types of tissue from the mouse.[36] Patterns of age-related change could be classified by tissue type. There were different patterns for brain tissue, blood vessels, and tissues that responded to cortisol (a stress hormone). Because the elements of genetic regulatory networks are highly conserved during evolution (as described earlier), it is practical to compare species and then extrapolate from these findings to humans. Studies of this type show consistent under- and overexpressed genes that are involved in inflammation, immune responses, energy metabolism, apoptosis, and cellular senescence. Discovering aging-relationships in this way has helped understand how some diseases, including late-onset dementias, also are age related. There is an approximate three times greater than expected overlap between aging genes and disease genes. Centenarians and their relatives are not spared the effects of disease-associated genes; instead they carry genes with enhanced capacities to counter the effects of disease genes.

VI. BRAIN AGING AND THE EVOLUTION OF NEUROLOGICAL DISEASE

One approach to the evolution of neurological disease was followed by Tomislav Domazet-Los and Diethard Tautz,[37] who ranked genes according to the stage (age) that each gene is activated in each species. They showed how the same genes can be sorted according to a hierarchical classification of animals that carry each gene. This type of specifying which genes are shared among species and which are unique to one class of animals allowed Domazet-Los and Tautz to construct an evolutionary "tree of life" made up of genes. Some genes were identified that are absent from life forms that evolved earlier, and other genes had remained stable from the origins of multicellular organisms up to the complex nervous system of modern primates. Their work draws together strands of research that strengthen the view that susceptibility to certain aging-related diseases of the brain and nervous system may be unique to humans[38] and affect the most recently evolved brain regions.[39]

From this work, it was summarized that genes that had evolved earliest in evolution were expressed earlier in development followed by evolutionarily younger genes that were expressed during maturation to adult forms. Of relevance to our interest in aging is their observation that in late adulthood, genes with the longest evolutionary history began to be expressed. These findings argue against the idea that "old genes" mutate into "new genes" to take over their functions. Rather than displacing old genes, new genes work alongside old genes to develop new forms and functions that increase reproductive success of their offspring.

These studies help explain what is happening in the aging brain and the dementias. In aging, certain proteins are susceptible to aggregation as insoluble clumps of protein inside and outside brain cells. As shown later in Chapter 6,

The Biology of the Dementias, these clumps are toxic to brain cells so their chemistry is a focus of intense research. Are there drugs that will stop these clumps from forming? What are the chances of breaking up clumps once they have formed, thus preventing damage to nerve cells?[40] The study of these proteins has prompted questions about their healthy functions. Why do we make them and what do they do for us? Some of these proteins appear to have evolved very early in evolution, and the genes that code for each of these proteins are first expressed in the developing human brain during the first months in the womb. In old age and in some dementias, the same genes also are expressed and may be involved in the causal pathways that lead to AD and Parkinson's diseases.

Widespread throughout the cortex in AD and affecting selective populations of nerve cells, there is aggregation, accumulation, and deposition of abnormally folded proteins. In AD, these are peptide fragments of the β-APP produced by clipping APP into sections by enzymes called proteases (β- and γ-s). These subunits are highly disposed to form aggregates known collectively as β-amyloid fibrils. These aggregates are believed to cause death of neurons, cause mental impairments, and eventually lead to premature death. In a similar manner, the aggregation of α-synuclein causes Parkinson's disease.

Human longevity is influenced strongly by energy metabolism as suggested by the effects of dietary restriction and the insulin–insulin growth factor signaling pathway (abbreviated to IIS). The importance of the IIS pathway in aging was first shown in the roundworm (*Caenorhabditis elegans*) and later found to be highly conserved in the mouse. Recent reports show that in some centenarians (e.g., Ashkenazi Jews), there is lower activity of the IIS pathway explained by a mutation in the insulin growth factor 1 receptor (IGF-1). Other mutations in the IIS pathway are reported in centenarians studied in Germany, Italy, and ethnic Japanese in Hawaii. What would happen if the mutant APP gene that causes early onset familial AD was implanted into a mouse strain where IIS activity was already lowered using genetic engineering? This experiment showed that this genetic manipulation could delay the onset and the extent of β-amyloid deposition (typical of AD) in mice and supported the idea that reduced IIS activity protected the brain from β-amyloid aggregation.

One line of clinical research supported by studies on evolution of brain diseases in humans is the finding that some aboriginal populations in the Americas[41,42] (but not Australia) may have lower than expected AD rates. Molecular pathways involved in relative protection from AD could prove relevant to the design of drugs to slow AD progression, perhaps by targeting critical components of genetic regulatory networks that are specific to humans[43,44,45,46] or candidate genes involved in processes that lead to the formation of abnormal protein aggregates in AD.[47]

VII. BRINGING IT ALL TOGETHER

The evolution of higher general cognitive ability provides sufficient mental capacity that not only leads to advanced human cultures but also protects against the effects of brain injury and age-related pathologies. Later, in Chapter 10 (Dementia Risk Reduction, 1: Concepts, Reserve, and Early Life Opportunities), the concept of cognitive reserve helps explain how this capacity protects or "buffers" the aging brain from cognitive harm. Higher cognitive abilities are involved in almost everything we find cognitively effortful and contribute to cognitive reserve. There are major between-individual differences in general cognitive ability and these are associated with a wide range of health outcomes. For example, higher general intelligence is linked to later onset of dementia and longer survival before death as well as lower risks of some diseases. There is substantial evidence from twin, adoption, and family studies that general cognitive ability is influenced by many heritable factors. It is estimated that genes account for between 50 and 70 percent of the heritability of general cognitive ability. In light of these sizable estimates it is reasonable to ask: What genes are involved in the inheritance of intelligence and to what extent, if any, are these genes involved in the pathology of dementia?

Investigations of known variations in the structure of a gene (a specific genetic polymorphism) and a disease state or cognitive trait are known as association studies. Finding an association may imply causal significance but in reality can imply no more than that the gene lies close to another gene of major causal importance. Evolutionary studies provide a rationale for selecting genes for tests of association not only on general cognitive ability but also on dementia. Comparisons between humans and other higher primates can help identify genes associated with brain structure and function, with the higher energy metabolism of neurons preferentially lost in AD and genes involved in neuronal differentiation, maintenance, and repair. Genome-wide association studies (GWAS) reflect the need to test as many genes as feasible, as well as reveal the many major disappointments of earlier small-scale studies. When evolutionary perspectives of reported gene-disease associations are taken of the results of association studies, the principle of "biological plausibility" (see Chapter 1, Introduction: Life Histories and the Life Course Approach) should be applied. Genes that are preferentially expressed in brain regions associated with cognitive functions that are specifically human are of great interest in dementia research.

So far, no single genetic polymorphism has been confidently associated with human intelligence.[48] To the outside observer, this seems surprising given the very large numbers of genes tested and the great numbers of individuals studied. The problem seems to be that even when "brute force" is applied to the problems of sample size and numbers of genes to be tested, the bar for "statistical proof" is set so high that even larger and better-designed studies will be required. Genes that influence human longevity and dementia risk appear likely to be involved and some may also influence human intelligence. To be successful in the search for genetic influences on dementia, the forces of evolution should be considered in experimental designs. These will involve careful longitudinal studies of cognitive development and aging and investigation of cognitive trajectories leading to dementia.[49]

5

THE AGING BRAIN

I. INTRODUCTION

Interest in aging increased rapidly during the twentieth century. With hindsight, it seems easy to understand how over a relatively short time aging became such an important topic in biology, sociology, psychology, and medicine. Life expectancy increased, economic prosperity allowed older adults to choose to retire or to work, and more old people in democratic societies seemed prepared to use their political weight. A great many distinguished writers and scientists reflected on the prospects for old age, and new journals were founded, each devoted to specific academic disciplines in the study of aging. Popular books offered advice on how to enjoy old age, dietary tips, the importance of keeping active, how to use "tricks" to improve memory, and much more. Important changes in attitudes toward old people gradually took hold. Perhaps the most influential of these was the principle that old people could achieve happiness, satisfaction, and confidence when positive steps were taken. There was no justification, it was argued, to maintain a negative stance toward aging ("ageism") about which nothing useful could be done. Contemporary thinking emphasized the benefits of adequate assessment of the predicaments of old people that enhance safety and security and that strengthen activities rather than hold onto age-old prejudices about aging.

In the face of wide-ranging improvements in the conditions of old people, doubts remain about individual prospects for a successful old age. If it is inevitable that the brain will age, that key structures will shrink, and that mental performance always will decline, then what is the value of improving the lives of old people when their capacity to hold onto these benefits will never be better than temporary? Chapter 3, The Well-Connected Brain, described the contributions made by enriched experiences in shaping neural networks. Optimum brain development proceeded through critical periods when individuals were sensitive to the presence (or absence) of specific types of learning experience. Aging was recognized as a developmental stage that could be properly understood only when the biological and cognitive underpinnings of brain development and aging were considered in their entirety. The many life experiences that have positive and negative effects on the aging brain are set against these considerations of the determinants of normality in old age. These experiences include physical and mental activity as well as the parts played by stress related to illness or emotional distress or personal loss.

In health, the brain and body must make frequent effective adjustments to meet the challenges of a changing environment. The term "stress response" is used to describe what happens when these changes threaten the integrity of any of the bodily systems, although this usage is too restrictive to include transformation of potential "threats" into positive benefits. The mature brain senses environmental change and coordinates appropriate bodily responses. With aging, the efficiency of brain-mediated sensory, motor, and cognitive processes begins to decline.

WHAT IS AGING?

Does aging exist? On one side of the argument are those who reason that the study of aging distracts from an intense hypothesis-driven exploration of mechanisms of age-related disease.[1] On the other side of the argument is the view that the biology of senescence is fundamental to understanding the pathophysiology of Alzheimer's disease (AD).[2] These are not trivial issues. Some attribute the physical changes that occur universally among old people to an accumulation of harmful effects of a lifetime of injury and disease.[3] In these terms, aging represents a progressive failure to repair damaged tissues, to maintain physiological regulatory systems, and to secure sufficient nutrients to maintain life. Against this "damage-repair" argument, many recent scientific discoveries have shown how the maximum life span of simple organisms (such as the roundworm, *Caenorhabditis elegans*), fruit flies (*Drosophila melanogaster*), and mice can be increased through genetic, nutritional, and pharmacological interventions. The first discovery was that a mutation in a single gene could extend life span in *C. elegans* and suggested that the "problem of aging" would be solved using molecular genetic techniques. This led to a distinction between

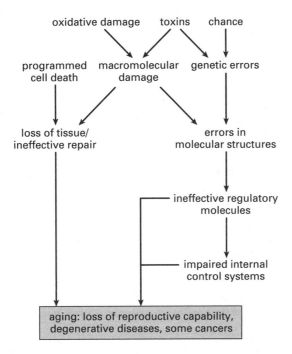

FIGURE 5.1 The principal pathways toward aging. This is a simplified schematic and omits many diverse possible contributions from the environment (e.g., diet) and interactions between factors that modify rates of aging.

"intrinsic" and "extrinsic" causes of aging. In Figure 5.1, intrinsic pathways arise in genetic programs and include cumulative errors in translating and transcribing genetic information. Extrinsic aging summarizes the many harmful influences on body structures and physiological systems encountered over a life course that eventually impair resistance to disease and stressors and make the old person more susceptible to illness.

The pace of brain aging suggested by the rate of mental decline is difficult to judge. What is observed reflects the net effect of opposing processes. This net effect implies that successful aging can be prolonged when repair and compensation of mental function are maximized. Although this could involve promotion of physiological adaptability (as with aerobic fitness), successful aging always involves retention of brain functional efficiency. The balance between opposing forces on mental function is at the heart of understanding brain aging and dementia. On the one hand, aging has an effect on brain tissue and brain blood supply. On the other hand, intrinsic repair and damage-limitation processes maintain brain function in the face of aging. This point is emphasized when the capacity of the brain to reorganize functions to make good or compensate for age-related damage is introduced. The "self-organizing principle" of cortical maturation that was so important during early development returns to the fore in the face of aging. If we accept that aging is a distinct biological process that can be distinguished from the lifelong accumulation of damage that begins to exceed capacity for repair and that the process is similar in all organisms and begins at about the age when optimum reproductive fitness is reached, then many aspects of the biology of the aging brain become tractable problems. Not the least of these aspects are explanations of observations that show that the general physical and cognitive health of the offspring of long-lived parents is better than those born to parents with average life expectancy. Family studies of this type that explore the biological basis of these associations provide many opportunities to understand the effects of aging on brain function and, vice versa, the role of the brain in the regulation of biological systems that are affected by aging processes.

The close relationships among aging effects on a wide range of bodily and cognitive functions has suggested that a single cause or small group of causes is driving rates of aging in observable functions.[4] Effects of this type are placed in broad groups that are either cognitive or noncognitive (e.g., lung function, grip strength, and cardiorespiratory fitness). Links such as these prompt contrary arguments either (1) that noncognitive decline is driving loss of cognitive abilities or (2) that decreases in cognition are responsible for diminished physical functions. Some experts regard these arguments as superfluous, however: They consider relationships between variables grouped in this way as statistical artifacts of their shared associations with age. Most experts agree that mature brain functions include maintenance of relationships between cognitive, sensory, and physical competencies. The aging brain becomes less efficient in the integration of overall control, leading to failed regulation of stress responses.

These issues make research on brain aging quite tricky to summarize. Before we proceed, we make explicit four preliminary cautions. First, in brain aging, it is sometimes unsafe to compare results from different laboratories when the selection of old people for study has relied on different criteria to exclude the presence of cognitive impairment before death. Second, techniques to improve counting brain cells have improved over the past fifty years. Currently and contrary to popular opinion, in the absence of brain disease, there is no substantial loss of brain cells with age. This point will be emphasized and explained. Third, do not assume that brain aging and AD are basically the same phenomenon that differs only in degree. Fourth, structural studies on the postmortem brains of old people are limited by the information available to establish cognitive performance in the final months of life, with often-threadbare accounts of medical and personal histories.

These cautions notwithstanding, we will tackle two questions. One arises when we want to see how an aging brain differs from that of a young adult. The other concerns the underlying physical changes that account for age-related differences found in the brains of old people. To meet the first, we will investigate new contributions to the study of brain aging that provide fresh ways of understanding brain aging. Improvements in brain structural and functional imaging have begun to reveal how the living brain is organized in health and how neural networks from small to large scale are affected by brain aging (this relates to the "connectome" mentioned previously).

Researchers have taken different approaches to understanding relationships between brain aging and age-related changes in cognitive function. The scheme adopted by Tim Salthouse[5] divides relevant studies into (1) brain volume, (2) brain white matter lesion density, (3) brain diffusion tensor imaging, and (4) brain functional activation experiments. Salthouse's scheme is certainly attractive, but it does not allow space for the now-considerable molecular genetic literature or adequate discussion of the possible role played by stress responses in brain aging and cognitive change. To address these issues, therefore, underlying physical changes in the brain are divided here between two questions: (1) What happens to brain cells and brain blood vessels? and (2) What molecular changes occur in the aging brain?

This second question has drawn molecular biologists to the study of the aging brain with a wide array of molecular tools at their disposal. The great wealth of molecular data almost defies fair summation. Evolutionary theory is probably the best way to try to organize their findings. Chapter 4, Aging, Evolution, and the Brain, identified molecular pathways involved in the regulation of gene expression in development and aging, and it showed how these were conserved in animals as diverse as nematodes, fruit flies, and mammals. We raised the possibility that these genetic regulatory networks (GRNs) may prove accessible to intervention, with effects on aging processes relevant to susceptibility to aging-related brain disease. New methods are now available to analyze the control of gene expression (through microarray technology), and we now can envisage how technologies as different as brain imaging and molecular analysis might be combined to study the effects of potentially beneficial interventions. We glimpsed this opportunity when we discussed recent data from the Allen Human Brain Atlas Project (http://www.brain-map.org; a growing collection of online public resources integrating extensive gene expression and neuroanatomical data, complete with a novel suite of search and viewing tools).

THE NAKED-EYE VIEW OF THE AGING BRAIN

There are many detailed and helpful accounts of changes in the human brain across the life course. The first accounts were based on postmortem materials that later were strengthened by imaging of the brain at work in life. The appearance of the aging brain differs from the brain of a young adult. The sulci are wider and the gyri are narrower (Figure 5.2). Brain cells should be affected by aging, and shrinkage

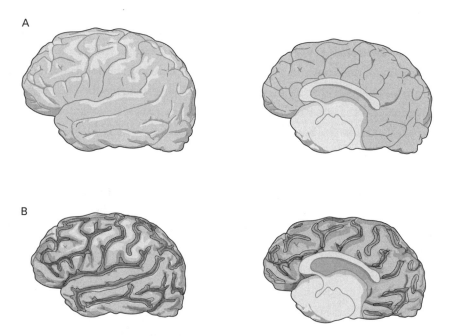

FIGURE 5.2 The appearances of two brains. Figure 5.2a is that of a young adult in which the sulci are tightly folded between gyri. Figure 5.2b is the brain of an adult who is about eighty years old with no history of dementia or brain disease. The gyri appear narrower and the sulci wider than those of a young adult.

of cortical folds seems to imply substantial loss of brain cells. This assumption was difficult to test, and results were contentious. Obvious problems emerge when comparing brains between those who survive into their ninth decade with those of people who died in early adulthood and may have experienced quite different educational opportunities, health care, and diets. Also problematic is age-related brain disease, which if unrecognized may be explained in terms of brain aging. This is especially true of a disorder like AD, which begins at least twenty years before the onset of symptoms and which in its early stages may be difficult to distinguish from aging without dementia. This may be equally relevant when considering the effects on brain aging of common disorders, such as hypertension.

Postmortem Studies

With aging, using just the naked eye, it is easy to notice lower brain weight and volume, increases in size of the cerebral ventricles, and expansion of the sulci between the gyri. Figure 5.3 is a composite of magnetic resonance imaging (MRI) images of two mature adult brains obtained at sixty-eight years of age and again at seventy-three years of age. Brain differences attributable to aging are visible, although only to a slight in degree, and they mostly affect only those structures lying toward the center of the image. These reductions in shape are identified as parts of the hippocampus, where they are shown as white bands on the edge of the darker gray body of the hippocampus. This is of great interest because the hippocampus is critically involved in memory processing.

In a healthy mature man, about twenty years of age, the brain weighs about 1,400 grams (g). Among women at the same age, brain weight is a little less, at around 1,300 g. From around forty years of age, a slight decline in brain weight begins to accelerate, so that by about sixty-five years of age, the weight of a male brain has fallen to about 1,300 g. By around the age of ninety, a healthy male brain weighs around 1,180 g.

Slicing through the brain in postmortem studies, the great white matter tracts appear pallid with age by comparison with the younger brain. Deeper, beneath the cortex, are patches of pigmented fatty material (lipofuscin).

THE MICROSCOPIC APPEARANCE OF THE AGING BRAIN

Using a light microscope, the hippocampus, cortex, and cerebellum seem to have fewer neurons, but this will be shown to be misleading. Myelin (the white fatty

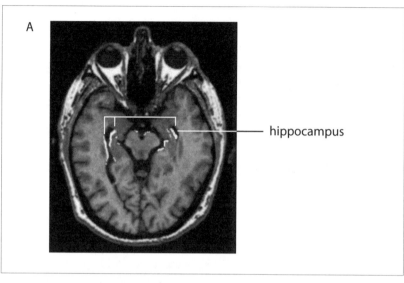

hippocampus

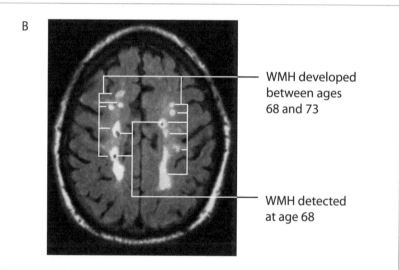

WMH developed between ages 68 and 73

WMH detected at age 68

FIGURE 5.3 These show MRI findings from examinations of two individuals.

(A) combines two images from one individual to indicate how the hippocampus and some adjacent tissue has shrunk from the age of sixty-eight to seventy-three.

(B) combines images from a second individual to show where white matter hyperintensities (WMH) developed between ages of sixty-eight and seventy-three. The white areas show where WMH were detected at age sixty-eight.

The two individuals were participants in the Aberdeen longitudinal study of brain aging and health and were examined on two occasions at ages of sixty-eight and seventy-three. Neither person had suffered from a dementia syndrome.

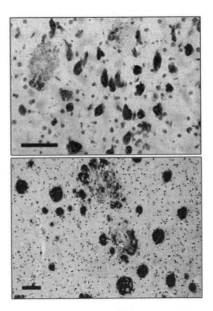

FIGURE 5.4 A photomicrograph of thin slices of brain tissue stained to reveal presence of amyloid (lower panel) or neurofibrillary tangles (upper panel). The images are not to the same scale: the solid bar represents 100 microns on either image.

Provided with the kind permission of Dr. Charlie Harrington, University of Aberdeen.

coating of the white matter tracts) is reduced in the cortico-cortical fiber pathways. On closer inspection, neurons are smaller than expected, and their dendritic connections seem to be less elaborate. Synapses on these dendrites are sparse in number. At greater magnification with an electron microscope, the cumulative effects of mitochondrial damage and defective DNA repair are discernable.

The detection by light microscope of senile plaques (SP) and neurofibrillary tangles (NFT) is regarded widely as the gold standard by which to establish a definite diagnosis of AD. These structures are illustrated in Figure 5.4. Their origins and metabolism in AD are described in Chapter 6, The Biology of the Dementias. At this stage, it is helpful to provide some basic information about their formation, because they are not confined to AD and are found in the brains of cognitively healthy older adults.

SP and NFT are detectable after appropriate staining of brain cortical sections by light microscope. SPs are made up of an aggregation of diverse compounds that include aluminum silicate, zinc, immunoglobulins, apolipo-proteins E and J, cellular debris from dead brain cells, and β-amyloid from which they get their alternative name "amyloid plaque." The term "amyloid" derives from the starch-like staining properties of the amyloid protein (*amylum* is Latin for starch). "Plaque" was used because the aggregates were first described in or close to the walls of cerebral blood vessels in a manner likened to "atheromatous plaques" in vascular disease. NFTs were described by Alois Alzheimer and at first were believed to be found uniquely in AD. They are found, however, within the cell bodies of dead or dying neurons and in extracellular space where they have been likened to the "ghosts" of dead cells. They are made up of abnormal protein deposits that are formed as paired helical filaments, including assemblies of abnormal microtubular assembly protein.

Because SP and NFT are found in normal aging and AD, a distinction between the two has rested on counting the density of SP and NFT in repre-sentative cortical sections. This distinction by quantification, however, is highly problematic for *three* reasons. *First*, AD usually develops over many years before symptoms are noticed, and during this preclinical (presymptomatic or prodro-mal) stage, SP and NFT gradually accumulate in the brain. It is plausible that some individuals who die without a history of age-related cognitive impairment are in the preclinical phase of AD. Much discussion, therefore, centers on how best to distinguish between brain aging without dementia and the progressive brain changes typical of AD. Some experts conclude that the difference is largely a matter of degree of SP and NFT formation and that the areas of brain affected are the same in normal aging and AD.[6]

Second, some experts have reached a different conclusion and argue that it is the location of SP and NFT deposition in the brain that defines the presence of AD.[7] In normal aging, SP and to a lesser extent NFT generally are found to be distributed at low density throughout the cerebral cortex. Among the oldest-old without dementia, NFT are comparatively rare, and SP density (that particularly affects the cerebral blood vessels) can be as great as that found in AD among younger adults. In AD, SP and more so NFT are found to be in greatest numbers in brain structures that support higher cognitive functions and are seen in much lower numbers elsewhere in the cortex in the early stages of AD. From an over-all perspective, some aging changes seem generalized throughout the nervous system, but some appear quite localized, affecting only specific brain regions or structures. These findings reveal postmortem studies at their best: identifying affected brain areas with great precision but sometimes without sufficient clinical history to explain findings fully.

Third, some skeptics purport that AD patients who undergo postmortem examination are truly representative of the wide range of AD brain changes seen in clinics, research programs, and community-based studies. The argument goes like this: Patient samples taken from clinic populations are biased toward inclusion of those people with behavioral problems or more severe or atypical forms of dementia. Postmortem findings, therefore, could be attributed to these symptoms and not be specific for AD. A major research program led by David Bennett at Rush University Chicago (United States) has made important contributions to this question. In a series of related reports,[8] Bennett's team has shown that the brain's appearance at postmortem differs between AD patients identified in community or clinic settings. Importantly, although very few (less than 5 percent) of normal elderly without dementia met postmortem criteria for AD, some features of AD in the brain were detectable in about 40 percent of old people without cognitive impairment.

Additionally, major issues have arisen following the invention of methods to image amyloid in the living brain[9] and the quantification of the amount of amyloid present. These are now critical aspects of the investigation of possible links between amyloid "burden" and cognitive function, and their interpretation and application has become highly complex. Appropriate Use Criteria[10] will help compare new findings and establish the value of amyloid imaging in the prediction and diagnosis of AD.

The pathological diagnosis of AD requires the detection and measurement of abnormal proteins in specific brain regions. The converse is also true: Specific absence of abnormal proteins precludes a diagnosis of AD. Postmortem diagnosis of AD requires the presence of brain amyloid. Therefore, a method that detects amyloid in the living brain would improve the specificity of a clinical AD diagnosis. This is possible during life using compounds that bind to amyloid and by attaching a radioactive label to the compound that can be detected in a positron emission tomography (PET) brain scanner. At least two such amyloid-binding compounds are in development: (1) Avid Radiopharmaceuticals Inc. ^{18}F-AV-45 and (2) the Pittsburgh Compound B (PiB). AV-45 is more readily available than PiB, although reports of PiB are more extensive. PiB is a fluorescent analog of a thioflavin-T derivative of thioflavin, with high affinity for β-amyloid in AD brains. It was developed in the

University of Pittsburgh (United States) by William Klunk and Chester Mathis and was taken forward in partnership with the University of Uppsala (Sweden). PiB was the second compound from Pittsburgh that the Swedish team investigated, so they called it Pittsburgh Compound B (or PiB). By 2004, Klunk and his team had shown that PiB was retained in the brains of patients with AD about twice as much as in healthy people without AD. This novel PET amyloid-imaging technique is a step forward in understanding the detection, progress, and effects of possible treatments on β-amyloid deposits in the AD brain.

VOLUMETRIC STUDIES

The brain is encased inside a rigid skull with many easily identifiable landmarks to locate correctly all parts of the brain image. Two important age-related features are the thickness of the cortex and the volume of brain regions or structures. Interpretation of early MRI brain images was problematic at first but became easier as semiautomated methods were developed and the presence of brain–blood vessel disease was recognized more accurately.[11] With aging in general, in the absence of brain disease, reduced volumes of the prefrontal cortex are progressively less marked in a posterior direction. The temporal cortex is reduced only slightly and the parietal and occipital cortex even less so.

Differences in volume between brains and within the same brain are not simple matters of size. It is unsafe to assume that underlying brain structures are essentially the same and that volumetric differences are only a matter of scale. To maintain brain efficiency, internal brain structure is optimized to save energy and time by making the shortest possible connections between neurons. The geometry of cortical connections varies according to brain size and gives insight into how brain connections are organized during development and are reorganized during aging.

Measures of complexity of cortical folding first used to study brain development are now applied to brain aging. Postmortem studies reported by Susanne van Veluw (Oxford, United Kingdom)[12] show that with aging, complex interrelationships exist among cortical complexity, original intelligence (intelligence quotient), cognitive decline with aging and AD, and the density of senile (β-amyloid) plaques.

BRAIN CELLS

Harold Brody[13] was the first to attribute age-related shrinkage in human brain weight to the loss of neurons throughout the cerebral cortex. These data were later regarded as unreliable because of poor sampling design and technical problems when counting numbers of neurons. By the 1980s new methods were applied to counting numbers of neurons.[14] Essentially, the problem faced by the first researchers was that they tried to count neurons in a two-dimensional flat image when the structure (a tiny piece of cortex) was in fact three dimensional. To deal with the problem of depth of the microscopic image, new methods that relied on the true number of objects being counted and not their relative sizes were introduced. The problems of counting brain cells in tiny samples of brain tissue are summarized in Figure 5.5.

It is no longer commonly assumed that normal aging (without dementia) results in substantial losses of neurons and that the structure of surviving

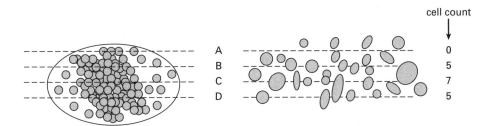

FIGURE 5.5 The problem of counting brain cells in the aging brain. A brain specimen is made up of a mixture of brain cells each of different sizes. The cells form part of structure with boundaries that can be measured. In this example, the density of the structure differs from surrounding tissue, so that as density shifts from outside to inside the structure, the density changes abruptly. This is the boundary. When the structure is set in a hardening compound (e.g., wax) very thin slices can be made across multiple planes (a)–(d).

This figure shows how brain cells of different sizes can be distributed across planes (a)–(d) in the center of the specimen. The (a)–(d) pass through single brain cells. Some adjacent sections misidentify a single large cell as two separate cells in the same structure. If a structure contains more large cells than small cells, cell counting may be biased toward overestimation of cell number in the whole structure. Conversely, as brain cells shrink with age, adjacent sections may miss small cells completely.

———

Suggested by Morrison and Hof, who also provide an elegant solution to the problem of counting cell density in the aging brain. J. Morrison, and P. R. Hof, "Life and Death of Aging Neurons in the Aging Brain," *Science* 278, no. 5337 (1997): 412–419.

neurons somehow is compromised. This misunderstanding probably arose from technical problems (see previous section) in early studies and the inclusion of people with dementia syndromes in study groups. In the absence of dementia, whatever age-related cognitive decline is detected, it should not be attributed to neuronal loss. The same may not be true, however, for the structure of individual neurons. Although there is no consistent pattern across the cortex of neuronal change, there are some localities where neurons appear to have lost significant numbers of dendrites.

Some evidence, however, indicates that certain groups of brain cells are selectively lost with aging. This type of loss is reported in the frontal lobes (particularly the prefrontal motor cortex, PFC) and the subcortical nuclei known as the basal ganglia. Intriguingly, when brain activation studies are performed in older adults with and without evidence of cognitive aging, increased PFC activation is detected. It is speculated that this is a compensatory response to cognitive impairment.

II. THE FRONTAL LOBES AND AGING

The distribution of higher cognitive functions across cortical regions suggests that when deficits arise in executive control, these deficits are linked to deterioration in the frontal lobes. Evidence for this influential idea comes from four sources:

1. Cognitive aging is most marked in deficits on cognitive tasks that require intact executive functioning. This requires healthy frontal lobes, specifically located in the PFC.
2. Brain imaging studies show reductions with aging in frontal lobe areas, most marked in the PFC.
3. The white matter tracts that connect the PFC and other frontal lobe structures with other cortical regions are affected preferentially by age-related deterioration.
4. Brain transmitters in the PFC are reduced selectively with aging. This is most marked for dopamine.

With aging, the dendritic structure changes, but these changes are much more subtle than the considerable decline in branching previously reported.

It was surprising to find that among many old people without dementia, dendrites appeared longer in some regions with corresponding increases in complexity of branches and, by inference, numbers of synapses. These changes in dendritic branches implicate changes in numbers and strength of synapses in the maintenance of cognitive function in older adults. Animal studies show that decline in spatial ability can be linked to reduced synaptic numbers in brain areas that contribute to spatial navigation. Functional studies in these neural circuits suggest that in response to reductions in synaptic number in old animals, there seems to be a compensatory increase in the sensitivity of postsynaptic neurons. Underlying changes in gene expression can be detected by microarray technology.

Microarray methods measure the expression of many genes by extraction of all the RNA from young or old neurons after which complementary DNA is synthesized and compared between young and old animals. Animal studies have ranged from the roundworm (*Caenorhabditis elegans*), fruit flies (*Drosophila melanogaster*), mice, rats, primates, and humans. Microarray techniques reveal age-related alterations in the expression of many hundreds of genes, some of which can be linked to changes in spatial learning and memory. It seems that with aging, genes involved with inflammation, stress responses, and calcium metabolism are up-regulated with aging, while genes involved with energy metabolism and the formation of new synapses are down-regulated. In contrast, old people with AD demonstrate an increased expression of genes involved in energy metabolism.

III. BRAIN BLOOD VESSELS

The blood supply to the brain is established during fetal development and the first few years of life. Small blood vessels (capillaries) maintain the microcirculation of the cerebral cortex and distribute blood containing oxygen and nutrients supplied by the arterioles of the cerebral vasculature. With aging, the density of this capillary bed gradually is reduced and may be sufficient to explain memory loss and other cognitive impairments of old age (vascular cognitive impairment, VCI). More severe forms of VCI often are referred to as "vascular dementia," although the term sometime is used as a catchall for an assortment of dementia syndromes. Major vascular contributions to risk are not simple. Aging of brain blood vessels (i.e., stiffening of brain arterial walls) plays a role in the initiation of vascular disease in the brain and through their reciprocal relationships with oxidative stress and an inflammatory response in the origins of atherosclerosis that contribute to AD.

At rest, the brain consumes about 20 percent of the total energy used by the whole body, although it makes up only 2 percent of the total weight. Energy usage requires a consistent blood supply that meets all of the requirements of brain tissue at work. Two important factors secure a sufficient blood supply to the developing brain. These are (1) the availability of growth factors that stimulate the outgrowth of capillary beds to meet the needs of increased brain size, and (2) the autoregulation of cerebral blood flow (CBF).

The developing brain releases vascular growth factors to initiate the growth of blood vessels into developing brain tissue (*neuroangiogenesis*). These growth factors were first studied in cancerous growths in which rapid uncontrolled cancerous growth can be achieved only if blood supply to the cancer is increased correspondingly. The first growth factor was called tumor angiogenesis factor (TAF) and was extracted from various tumors by Judah Folkman (1933–2008) at Harvard Medical School (United States).[15] Later, nontumor vascular growth factors and cytokines were identified among which vascular endothelial growth factor (VEGF) is best known. Other brain-specific growth factors include transforming growth factor-beta (TGF-β), epidermal growth factor (EGF), and platelet-derived growth factor-beta (PDGF-β).

During brain development, vascular growth factors such as VEGF are expressed abundantly in the brain. In the mature brain, vascular growth factors remain essential to the repair and regeneration of the cerebral vasculature. Older adults without dementia show slight reductions in the density of the capillary bed but significant reduction (about 25 percent) in the thickness of the capillary walls. No data are available on the development (angiogenesis) of blood vessels in the aged human brain. Some suspicion remains that the capacity to form new cerebral capillaries is much reduced in old brains in contrast with younger adults in whom a capacity for angiogenesis is retained.

In the absence of dementia, the walls of the cerebral capillaries experience some thinning, but otherwise these capillaries cannot be distinguished from younger adults. This type of deterioration of blood vessels requires continuous repair and relies on an adequate supply of vascular growth factors. In AD, however, Raj Kalaria and colleagues at Newcastle upon Tyne (United Kingdom) found that more than 90 percent of capillaries were abnormal with many loops and kinks not seen in the absence of AD. These observations prompted intensive research into the vascular origins of AD, leading to the "neurovascular hypothesis of AD."[16] This proposition includes abnormal angiogenesis, defects in the endothelium that separates blood from the extracellular brain space around neurons (the blood-brain barrier), defective clearance of β-amyloid from the brain, and the death of brain cells. Tests of the neurovascular hypothesis of AD specifically have examined the expression in AD brain tissue and age-matched healthy brains of genes (such as mesenchyme homeobox 2) that regulate the proliferation, differentiation, and migration of vascular smooth muscle cells.

Presently, although the hypothesis is not accepted widely, some agreement supports a unifying hypothesis that includes roles for β-amyloid and NFT deposition and impaired repair of the cerebral vasculature. These interacting contributions may prove to be interdependent or partly caused by other molecular changes in the aging brain. Primary age-related increased permeability in the blood-brain barrier also may be relevant. Interest in a possible deficiency of vascular growth factors in the cerebral blood vessels has prompted plans to test how administration of relevant growth factors or stimulation of brain-specific vascular growth factor receptors might be of benefit to AD patients.

The autoregulation of CBF ensures that the brain's blood supply is maintained within set limits (between 50 and 160 mm Hg). With aging and when certain drugs are used, autoregulation of the brain's blood supply is impaired. In health, brain blood pressure is controlled by the diameter of the arterioles as determined by constriction of smooth muscle cells in the arteriolar wall. These cells are controlled by the activity of their nerve supply, the actions of peptides, and some neurotransmitters. When blood pressure is raised continuously (as in untreated hypertension), the arteries and arterioles supplying the brain thicken and stiffen to contain the increased pressure and the overall diameter of the blood vessel is reduced. Adjacent blood vessels simultaneously can compress intervening brain tissue, which, if sustained, leads to the death of those brain cells. The microscopic appearance is often of surviving blood vessels without intervening brain tissue ("lacunar infarcts"; see Figure 5.6).

In addition to damage to brain–blood vessel walls and localized compression of brain tissue, chronic high blood pressure can initiate chronic brain hypoperfusion. Eventually, the cumulative effects of hypoperfusion and ischemic brain

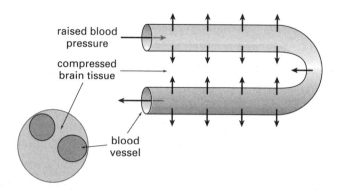

raised blood
pressure

compressed
brain tissue

blood
vessel

FIGURE 5.6 The schematic shows a loop in a single brain blood vessel. When blood pressure is continuously raised, this exerts mechanical pressure on the surrounding brain tissue. The brain area contained within the loop is compressed simultaneously from two sides. Eventually, these brain cells die, leaving a space—a lacunar infarct—within.

tissue impair the functions of neurons and their supporting cellular matrix. This uncouples the autoregulation of CBF, worsening the precarious health of neurons and leading to reductions in CBF and decreased supply of oxygen and nutrients to the brain. Any steps that will avoid this potentially catastrophic sequence of events have the potential to delay or avoid completely the cognitive impairment associated with age-related disturbances in brain blood flow.

Currently, modern drugs designed to lower blood pressure will not impair autoregulation of CBF and do not impair cognition. Control of blood pressure in old age, therefore, has the potential to prevent age-related cognitive impairment. Clinical trials that aim to prevent AD through effective treatment of hypertension thus far have been encouraging.

WHITE MATTER HYPERINTENSITIES

The most obvious change in white matter with age is the appearance of bright spots on MRI (white matter hyperintensities [WMH]). These now are attributed mostly to tiny microvascular lesions in the white matter and are linked with presence of extracranial blood vessel disease associated with high blood pressure, diabetes mellitus, and damage to the blood vessel wall. WMHs in late life also are linked to childhood socioeconomic disadvantage.[17]

The aging brain often contains multiple microvascular lesions in the white matter.[18] Although a small number of such lesions are compatible with retention of cognitive functions, it is now fairly certain that increased density of white matter microvascular lesions increases the risk of progression to dementia and probably toward AD. The nature and type of these lesions vary widely and not all are detectable by brain imaging.[19] Diverse terms used (often interchangeably) to describe different types of lesion are as follows: cortical microinfarcts, subcortical gray matter, microbleeds, deep white matter lacunes (see Figure 5.7), periventricular, diffuse white matter demyelinations, and focal or diffuse gliosis of old age. Although there is much debate about the precise role of each lesion subtype, there is general agreement that the presence of these lesions in large numbers in the aging brain places an additional burden on brain function. In conjunction with other unrelated brain pathologies, this additional burden can be a major determinant of cognitive decline. Current interest focuses on the molecular biology of white matter changes with aging. One promising approach examines links between the molecular genetics of blood vessel disease outside the brain (e.g., coronary heart disease) and the detection of white matter lesions by MRI (Figure 5.8).

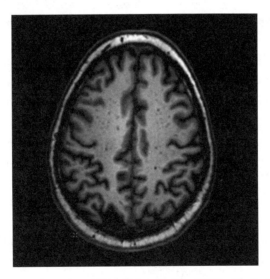

FIGURE 5.7 White matter hyperintensities detected using MRI in the brain of an older adult without dementia.

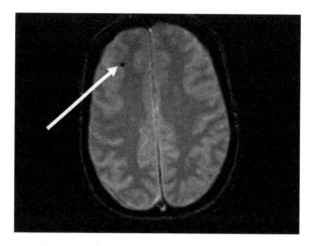

FIGURE 5.8 An area of lacunar infarction detected at MRI examination.

IV. GENES, DIET, AND BEHAVIOR

The overall picture of changes in the aging brain has become just as complex as predicted by the pioneers of aging brain research. One issue, however, has attracted widespread attention—often repeated in popular health magazines—that the processes involved in brain aging are open to modification by genes, diet, and behavior. Central to these propositions is the fact that neurons can be modified by experience and by their nutritional environment. The landscape of the aging brain has three major structural features at cellular, molecular, and neural circuit levels. Much as a landscape changes in winter, with aging, the effects of molecular cascades are detectable as they lead toward the dementia syndromes (so giving "the brain in winter" metaphor).

An important proposition that springs from studies on diet and dementia is that as our diets have changed, the numbers of old people living with dementia have varied. Such differences in rates might not be detected in all cultures. For example, a study of Japanese-American men in the Honolulu Heart Program showed the prevalence of AD was about 2.5 times greater than men and women of comparable age and circumstances living in Japan during the same time period.[20] Investigations of this type introduce quite difficult problems, not the least of which is that dietary preferences for specific foodstuffs change with age, income, and cultural pressures, including advertising by food manufacturers. William Grant with Sunlight, Nutrition, and Health Research, San Francisco (United States) used a combination of approaches to study how diet and lifestyle factors may have influenced dementia trends.[21] After introducing the effects of a possible delay between dietary change and dementia onset, he suggested that temporal trends in AD will continue to rise in Western countries if diets remain "energy dense," with high consumption of saturated fats with contributions from smoking and obesity. The maintenance of healthy neural networks is critical if adaptive responses to stress and disease are to remain effective. These networks include defenses against neural damage and fast effective repair. At one level, these mechanisms are described in terms of molecules that promote neural differentiation, cell survival, and repair. These are the neurotrophins and anti-inflammatory compounds. Nerve cells also have intrinsic antioxidant defenses as well as molecules to "switch off" signals that initiate apoptosis. At the same time, the neuron seeks to preserve the integrity of its DNA using DNA repair enzymes and safeguarding chromosomal telomeres. When neurons are lost with aging and disease, neural stem cells can be mobilized to replace and reconnect with existing neural circuits.

The implantation of neural stem cells into brain regions damaged by stroke or AD is one of the most exciting areas of research in restorative neurology. Understanding how neural stem cells differentiate into specialized mature neurons, migrate to precise brain areas to repair damaged networks, and reconstruct exact patterns of connections in those networks are among the greatest (and most pressing) challenges in clinical neuroscience (referred to at the start of this chapter).

It must not be assumed that these restorative mechanisms are sufficient to make good the effects of aging on the brain. Great variability among individuals determines their success. These are recognized most obviously in genetic mutations in genes that cause familial AD (including amyloid precursor protein and presenilins) and Parkinson's disease (including α-synuclein and Parkin). Certain mutations are sufficient to overwhelm intrinsic defenses against neural damage.

When damage is attributable to oxidation of complex large bioregulatory molecules, two defenses are available. These are the intrinsic enzymatic antioxidant defenses, such as superoxide dismutase, and the extrinsic ("nonenzymatic") defenses provided by dietary antioxidants. From this, it is argued that by improving dietary habits, the aging brain can be protected. Recommendations include caloric restriction and folate and antioxidant supplementation and increased intake of fish oils. In addition, because neural networks have the capacity to change with experience and to be enriched through increased use, additional recommendations favor certain behavioral (intellectual and physical) activities.

Success, it is argued, will be assured if hormones are activated or supplemented. The reasoning behind the use of hormones derives from observations that some sex hormones (e.g., estrogen or testosterone) promote neural differentiation during development, and the neural circuits formed in response to hormone exposure remain reliant on continuing exposure to the same hormone to remain healthy. Similar arguments have been made in favor of strategies that increase production of neurotrophic factors in response to intermittent mild stress (e.g., physical activity). This argument is extended to the advice that greater demands in terms of physical activity will stimulate neurogenesis, which could replace neurons lost to the effects of aging. Nevertheless, other possible explanations for a link between physical exercise and retention of cognitive function are plausible. For example, childhood factors may be important predictors of maintaining habits of regular physical activity in adulthood,[22] and these same childhood factors could be important in the retention of cognitive function in late life.

V. STRESS RESPONSES AND THE AGING BRAIN

The aging brain and body need to adapt to changes in their physical and social environments. When an older adult responds to stress, this is not always considered a negative, potentially hazardous process. Making good use of resources available, mastering the task in hand can be a source of pleasant, reassuring sensations. Stressful responses, however, make demands—sometimes quite considerably so—that have lasting detrimental effects. The brain holds center stage in the management of stress responses, so it is correctly anticipated that the aging brain with or without age-acquired sensory deficits, becomes less effective in providing a coordinated stress response that is appropriate in nature, extent, and duration. The "fight-or-flight reaction" is an intense form of stress response; most stress responses are less extreme. The fight-or-flight component, however, does emphasize that in certain circumstances, it will be appropriate for this phase of the stress response to be activated through brain pathways that arouse the autonomic nervous system.

Age-related impairments of stress control by brain probably are related to reduced efficiency of brain in the coordination of linked brain regulatory systems.[23,24] Most often, with aging, typical stress responses are greater than required for the level of stressor and are unduly prolonged. Perhaps more than any other unpleasant sensation, the physical and psychological sequelae of stress place more self-imposed constraints on the activity of older adults. It probably is not as well recognized that when some components of the stress response "feedback" onto brain structures (e.g., hippocampus) that initiate and regulate stress responses, they can impair the healthy function of these structures, thus causing stress responses to be greater than necessary and, by way of a vicious circle, feedback and further damage the same structures.[25]

Not all stressors arise in the external environment, and aging brings greater demands to respond to failures in regulation of some physiological systems. In the examples of immune surveillance and cardiorespiratory functions, impaired stress responses can increase the likelihood of progressive disorders becoming chronic and can predispose to premature death. In this specific context, the aging brain not only can fail to provide an effective stress response but also what it does provide may be insufficient to contain the beginnings of a disease process with possible fatal consequences.[26]

Contemporary societies make great demands that can extend across the waking day. Many encounters are stressful and, for older adults, can be accompanied by activation of pathways already susceptible to disease because of age. These are (1) immune function, (2) cardiorespiratory function, (3) glucose metabolism, (4) inflammation, and (5) brain structures not involved in stress responses. These systems all are linked intimately along pathways that are only partly explored and understood. Potentially, other systems also controlled by brain structures are implicated in this complex array. An important candidate system includes the control of internal "body clocks" that anticipate regular changes in environmental demands. The best-known example is a rapid increase in the stress hormone cortisol about two hours before waking to better prepare the individual for activity.

To make these types of predictions and to optimize coordination of stress response, the brain has developed mechanisms that identify stressors before these threaten. The capacity of the brain to learn from experience also allows repeated exposure to stressors to reset levels of detection and amplitude of stress response. For younger adults, these attributes are valued as the ability to "bounce back" and remain resilient in the face of a potential hazard. From an evolutionary standpoint, it is likely that stress control systems are as yet insufficiently developed to manage the demands of contemporary twenty-four/seven living.

The term "systems biology" was outlined in the Introduction to unite the rich data sets provided by different disciplines in biology, psychology, and sociology. Aging sciences are beginning to benefit from these integrated approaches. Important contributions from the molecular genetics of longevity show a high degree of conservation from *C. elegans* to fruit flies and mammals of insulin/Insulin Growth Releasing Factor (IGF-1) and sirtuin signaling pathways.[27,28] There are increasing deficits in energy metabolism, maintenance of complex protein structures, and DNA repair. The internal world is faced with critical needs to keep internal temperature constant and to take adequate action to counter transient changes in oxygen saturation with appropriate cardiorespiratory responses. Seen in these terms, age-related inability to protect brain structures from failure of oxidative damage[29] and to sequester potentially damaging misfolded proteins would accelerate age-related changes in brain.[30]

The proposition that stress hormones can damage the brain is not accepted universally. Cortisol is the best-known stress hormone and the one most often implicated in age-related stress-induced brain damage.[31] It is an essential hormone that must be kept within an optimum range but allowed to vary as needs arise. These needs can include perturbations in the internal environment (e.g., salt/water balance, temperature control). Most cortisol enters the brain from the blood. Exposure to stress (either external or internal) will stimulate the adrenal (a gland lying just above each kidney) to release cortisol. This is the first part of the stress response and is regarded as general or "nonspecific." The second part of the stress response is highly specific and determined by the pattern of release within the brain of small molecules called neuropeptides. One view of this two-phase stress response is that the first part anticipates potential harm caused by the second phase. Cortisol thereby protects tissue from damage caused by the overall response to stress.

Cortisol enters the brain where it can dampen its further release. An important consequence of cortisol release early in the life course is that cortisol appears to be able to program the sensitivity of the brain cortisol release system (the hypothamic-pituitary-adrenal axis). This axis changes with age. Many older adults (but certainly not all) release more cortisol, beginning their first daily cortisol pulse some hours earlier than younger adults. Increased release of cortisol is implicated in impairments of memory largely because brain defenses against cortisol are reduced with age. To understand fully the effects of aging on stress and, potentially on cognitive aging, it is necessary to involve not only cortisol but also other factors that defend against cortisol. The hippocampus appears most susceptible to cortisol-induced damage but not all hippocampal areas are affected. Cortisol damage probably is not a direct effect of cortisol but rather is an indirect action that makes some cells in the hippocampus more susceptible to harm (e.g., from lack of oxygen). Cortisol also may inhibit neurogenesis throughout the brain.

CORTISOL SECRETION

Cortisol is a steroid hormone synthesized and secreted by the adrenal cortex. It is released in response to stress and to low blood concentrations of glucose. To maintain internal controls, it raises blood glucose, suppresses immune surveillance, generally stimulates metabolism, and decreases bone formation. In summary:

1. Cortisol is released in distinct pulses nine to fourteen times daily. These pulses are more frequent in older adults.
2. Pulses begin before waking from sleep and are controlled by genetic factors.
3. Early life adversity has enduring effects on cortisol reactivity to stress.
4. Sex hormones affect cortisol release.

The fight-or-flight reaction must contribute to survival by encoding the precise circumstances that triggered this response such that if these recur, an appropriate reaction will immediately ensue. Increased cortisol promotes this type of learning, probably through neural networks in the amygdala. Cortisol, therefore, enhances the learning of emotion-laden memories and may be coordinated with the release of amines like noradrenalin in the brain, as occurs in tissues outside the brain.

Some clinical evidence shows that cortisol can damage brain structures involved in memory and that this type of damage is relevant to age-related cognitive decline. The best known example is provided by posttraumatic stress disorder, which includes memory problems and a smaller than expected hippocampus for total brain size. In a clinical syndrome with excessive cortisol production (Cushing's disease), frequent cognitive problems resolve when cortisol is controlled.

The consensus view is that it is premature to conclude that excess cortisol can damage the brain. Adaptive stress responses depend on the release of cortisol, and this is clearly most often beneficial. It is now widely accepted, however, that high levels of cortisol can make the brain more susceptible to other, more damaging agents. Some experts have proposed that optimum clinical care of excessive stress responses may reduce long-term cognitive problems and may even reduce dementia incidence. This topic is discussed further in the context of management of depressive illness in Chapter 11, Dementia Risk Reduction, 2: Midlife Opportunities to Delay Dementia Onset.

VI. BRINGING IT ALL TOGETHER

The human brain has evolved to improve survival chances of individuals and their offspring. The frontal lobes are recently evolved brain structures with many functions not shared with our closest relatives the chimpanzees. With aging, frontal lobe structures and functions are preferentially lost. These changes have a central role in changes in mental functions in old age but are less important in emotional aging.

The overall effects of aging on the brain are illustrated in Figure 5.9. Two opposing sets of forces are at work: The first promotes aging-related changes that disrupt normal brain function by damaging large molecules that are expensive for brain cells to make and repair and that constrain the capacity of high-energy-consuming neurons to maintain healthy function. The other set of forces utilizes a wide range of neuroprotective systems to prevent and repair brain cell damage and has the potential to support brain compensatory pathways to make good age-related damage.

Many brain structural changes are typical of aging. These affect the brain blood vessels, the richness of connections between neurons, and the accumulation of aggregates of abnormally folded proteins that are toxic to brain cells. Contrary to popular opinion, loss of neurons is much less severe than once thought, and most of the reduction with age in brain volume is attributable to the loss of white matter that insulates large bundles of nerve fibers that connect brain regions and to shrinkage in the size of individual neurons.

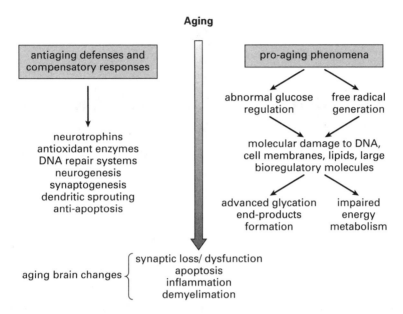

FIGURE 5.9 The effects of aging on the brain. On the right are aging-driven processes that contribute to increased damage to cell membranes, DNA, and other large biogregulatory molecules, including advanced glycation end products (AGEs). Not shown on this diagram is the stimulation of inflammatory responses in microglia through actions of AGEs. On the left hand side of the diagram are the powerful antiaging effects of intrinsic repair and anti-inflammatory systems. Brain responses include compensatory remodeling of neural networks. This schematic diagram summarizes the balance between harmful age-related damage to brain function and the helpful effects of systems that can counter these effects. These systems include pathways that help remodel neural networks damaged by aging or dementia.

For an aging individual to remain efficient and independent, brain cells need to remain healthy and when lost to damage or disease, systems are required that can compensate for that loss or that can regenerate dead and dying brain cells. As in all other tissues, an adequate supply of nutrients and oxygen are vital to brain cell health, but almost uniquely among body tissues, brain cells are differentiated terminally with little capacity for regeneration. Strategies to promote brain cell health will be discussed later in this book. In this chapter, the importance of nutrition was emphasized and a critical role for the potential harm to brain cells that is triggered by stress responses was identified.

6

THE BIOLOGY OF THE DEMENTIAS

I. HISTORICAL TRENDS

Before the molecular genetic revolution opened a treasure trove of new data about Alzheimer's disease (AD), clinical subcategories had accumulated over the twentieth century to eventually include the following: (1) early onset familial AD (rare); (2) a sporadic (nonfamilial) form of early onset AD and without significant brain vascular disease (common); (3) highly penetrant AD in Down syndrome; (4) mixed late-onset sporadic AD, symptomatic after the age of sixty-five and with significant brain vascular disease (common); (5) AD following repeated head injury ("dementia pugilistica" or "punch drunk syndrome"); (6) dementia with Lewy bodies (DLB, common); and (7) Parkinson's disease (PD) with dementia (common).

Even among current clinical scientists, it is not unusual to hear talk of Alzheimer spectrum disorders to convey the clinical impression that AD overlaps with any of the seven subcategories and, possibly, with (8) amyotrophic lateral sclerosis (Lou Gehrig's disease, common), (9) frontotemporal dementia (FTD, common), and (10) cortical basilar degeneration (rare).

What we refer to as AD is the most frequent form of late-onset dementia. Together with PD, these two diseases account for the most common causes of age-related selective loss of brain neurons. Although AD can occur at younger ages (thirty to sixty years old), most often it does not become apparent until late adulthood or old age. Until about forty years ago, what is now called AD usually was referred to as "senile dementia" or "arteriosclerotic dementia" and only later as "senile dementia of the Alzheimer type."

Most medical students who graduated before 1980 were taught that senile dementia was a consequence of brain aging or "atherosclerosis of the brain." U.S. textbooks of psychiatry up to that time placed AD among the rare "presenile dementias" that affected about 1 in 500 people age forty to sixty-five years old. "Senile dementia" was attributed to a "hardening of the arteries in the brain" (arteriosclerosis) and was regarded as both untreatable and irreversible. Palliative therapies were widely prescribed. For example, in the former West Germany, the sale of Hydergine (a mixture of ergot alkaloids believed to reinvigorate brain blood supply) was among the highest selling of all prescribed drugs, although no substantive evidence supported its efficacy.

In part, the idea that AD met criteria that warranted its status as a separate disease arose from observations on the brains of middle-age people with early onset AD (EOAD; before the age of sixty-five, or presenile dementia, occurring in about 1:700 people age forty to sixty-five years old). The case that AD met criteria for a diagnostic category rested on the following: (1) frequent positive family history of EOAD, (2) characteristic brain pathology, (3) similar patterns of early symptoms, and (4) similar progress from onset to death among most EOAD. Then, as now, no biological tests could confirm an AD diagnosis. Studies by the Newcastle group led by Martin Roth (see following paragraph) extended the disease concept of EOAD to the much more common dementia (1:4 age more than eighty years old) occurring in late life. It is much more difficult to show that late-onset AD (LOAD) meets criteria for a distinct disease category, although much neurobiological research presumes that it does.

EOAD brains rarely showed the brain–blood vessel changes of premature aging, and there were no signs that these individuals had suffered a blockage or burst of blood vessels (as happens during a stroke). All that could be seen were the characteristic pathological features described by Alois Alzheimer in 1906, when he presented the case of a 55-year-old woman with profound and progressive deficits in memory, orientation, and language. Discoveries in the molecular genetics of familial early onset dementia (summarized in the following paragraphs) have established that LOAD is not the result of a single genetic condition. About 30 percent of EOAD is caused by *known* mutations in any one of three genes totaling more than 100 possible mutations in early onset familial AD pedigrees. These mutations are found in less than 0.1 percent of people living with LOAD.

Working in a rural mental asylum after World War II, Martin Roth (United Kingdom) noticed how much improved the outlook had become for old people admitted to mental hospitals.[1] Before the war, most old people admitted to the asylum died within about two years. After the war, their outlook improved; a significant proportion now survived and were discharged home within a few months. Roth and his friend Eliot Slater attributed this improvement to the successful introduction of electroconvulsive therapy for the treatment of severe depressive disorders during the war years.

Later, working in an academic unit with Bernard Tomlinson and Gary Blessed in Newcastle upon Tyne (United Kingdom), Roth described in 1968 the typical changes of AD in the brains of old people with dementia dying in mental hospitals. The pathological changes included the loss of cortical tissue (the brain was about 60 percent of its expected weight) and accumulation of abnormal proteins in and among surviving neurons. These were the typical changes of AD previously considered to be relevant mostly in early onset dementia. The team proposed a strong positive relationship between the density of one abnormal protein (β-amyloid plaques) and the severity of dementia before death.[2] Notably, this plaque–cognition relationship was never satisfactorily replicated in independent studies.

Toward the mid-seventies, there was excitement among neuropathologists and neurochemists. New techniques in brain chemistry developed in the fifties detected abnormalities in the activity of enzymes involved in synaptic function. These methods revealed (1) the fine structure of amyloid plaques and neurofibrillary tangles, (2) major deficits in enzymes that were critical in PD and Huntington's disease, and (3) deficits in AD of (cholinergic) neurons that used acetylcholine as a neurotransmitter. The most urgent question at that time asked whether the reported deficit in cholinergic neurons could be reversed by drugs that increased the availability of acetylcholine in the brain. This approach already had proved effective in the treatment of PD. The excitement generated in 1961 Vienna by Oleh Hornykiewicz and Herbert Ehringer's successful treatment of PD by L-Dopa is conveyed by Hornykiewicz's eye-witness account:

> The effect of a single I.V. administration of L-Dopa was, in short, a complete abolition or substantial relief of akinesia. Bed-ridden patients who were unable to sit up; patients who could not stand up when seated; and patients who when standing could not start walking, performed after L-dopa all these activities

with ease. They walked around with normal associated movements and they even could run and jump. The voiceless, aphonic speech, blurred by pallilalia and unclear articulation, became forceful and clear as in a normal person. For short periods of time the patients were able to perform motor activities which could not be prompted to any comparable degree by any other known drug.[3]

The year 1976 proved to be pivotal for Alzheimer research. The United States "discovered" AD when Robert Katzman,[4] writing in the *Archives of Neurology*, in "The Prevalence and Malignancy of Alzheimer Disease: A Major Killer," declared that AD was grossly underestimated as a leading cause of death in the United States and that it probably should be ranked fourth or fifth, and yet U.S. public health records did not include AD among its recognized causes of death. It was also the year that Peter Davies and Tony Maloney reported the cholinergic deficit in AD.[5]

At that time, few cholinergic drugs were available for use in medicine. The Edinburgh clinical group led by Iain Glen immediately sought to test the "proof of principle" that dietary loading with lecithin containing phosphotidylcholine (a precursor of acetylcholine) would relieve AD in a manner akin to the dramatic effects of L-Dopa in PD. This trial failed, and the group turned to a search for more direct interventions to enhance brain cholinergic function.

Experts in chemical warfare knew of several powerful (quasi-irreversible organophosphates) cholinergic agents like Sarin, but because these were almost always fatal, they did not seem to be a good place to start a search for an Alzheimer therapy. What were needed were reversible, long-acting cholinesterase inhibitors with few if any side effects. Intravenous physostigmine (a reversible inhibitor of cholinesterase that breaks down acetylcholine; see Figure 6.1) had been used in physiology since 1935. In experimental use, it improved visual memory in AD for about twenty minutes. As a "proof of principle" these were important trials, but they were unlikely to support the use of physostigmine as a practical treatment of AD. Clinical researchers like Ken Davis and Richard Mohs[6] in New York and Janice Christie[7] in Edinburgh were the first to try to reverse the cognitive deficits of AD using cholinergic drugs.

The anticholinesterase drugs remain the principal therapies to slow dementia progression (Table 6.1). Their development was based on the discovery of cholinergic deficits in AD. Although widely used, they are widely regarded as palliative, largely because their benefits are slight and fewer than 70 percent of

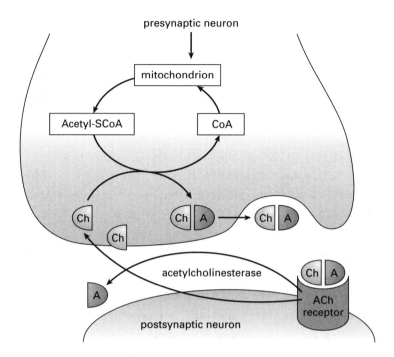

FIGURE 6.1 A cholinergic synapse. Acetylcholine (Ach) is synthesized in the presynaptic neuron from acetyl CoA (A) and choline (Ch). It is then transported in a presynaptic vesicle to be released as ACh into the synaptic space. Here, it attaches to the postsynaptic membrane at a highly specific cholinergic receptor. Acetylchoinesterase is a membrane-bound enzyme that degrades acetylcholine into acetyl (A) and choline (Ch) groups.

people living with dementia are helped. In the face of no other effective interventions and the wishes of families and patients to obtain as much help as possible, however, perhaps delaying progress to high dependency, their use is both welcome and encouraged.

Other longer acting cholinesterase drugs were available and tested. These included tetrahydroaminoacridine (tacrine or THA), which was sold for a short time as a "memory enhancer" in AD, but its marginal benefits were outweighed by its severe gastrointestinal side effects. Within five years, pharmaceutical companies had identified and developed much less toxic cholinesterase inhibitors. These include rivastigmine, gallantamine, and donepezil, which continue to be the principal therapies approved worldwide for the treatment of memory deficits in mild to moderate AD. Memantine belongs to a separate class and is useful

TABLE 6.1 Acetylcholinesterase Inhibitor Drugs for Alzheimer's Disease

	GENERIC NAME (BRAND NAME)
	Donepezil (Aricept) Galantamine (Reminyl) Rivastigmine (Exelon)
Effectiveness	Will produce slight improvements in memory and overall general competencies in about 30% of patients and can be linked to temporary slowing of disease progression in about 30% of others. Within about 18 months these drugs cease to be effective.
Use	Used to treat the symptoms of AD and occasionally used in dementia with Lewy bodies and vascular dementia.
Drug Choice	Choice is based on dosing schedule, ease of use (e.g., Exelon can be given as a skin patch), and frequency of side effects (mostly stomach upsets).

to reduce behavioral symptoms caused by excitotoxic neuronal (N-Methyl-D-aspartate mediated) damage in AD and PD. In skillful hands, the combination of memantine and one of the acetylcholinesterase inhibitors can prove more effective than either drug alone in carefully selected patients. All cholinesterase inhibitors are introduced slowly ("start low, go slow") at small doses and gradually are increased until benefits are established or side effects occur, which often prompts dose reduction. Not all AD patients can benefit from cholinesterase inhibitors, and once these are no longer helpful, most patients deteriorate to the point they would have reached had they not received the drug.

Renewed interest in the genetics of AD arose about this time. Len Heston[8] and his team in Minneapolis (United States) examined associations between early onset AD, hematologic malignancies, and Down syndrome among the pedigrees of early onset AD families. His work led others to reexamine possible links between AD, Down syndrome, and abnormal aging. Cytogenetic studies soon followed, and although these proved uninformative, Heston's genealogical studies helped locate the amyloid precursor protein (APP) gene on chromosome 21 (see the following paragraphs).

As research momentum gathered in the eighties behind the push toward a better understanding of dementia syndromes and the prospects of effective treatment, neuropathologists were encouraged to produce improved descriptions of AD. Arne Brun at the University Hospital in Lund (Sweden) was a pioneer of this approach.[9] He detailed the exact regional patterns of AD changes in the cortex. His careful work showed that contrary to popular belief, AD was not a diffuse disorder spread across the entire cortex, but rather it could be defined by a regional pattern of neuronal loss and disrupted function. Rather than accepting the view promoted by Martin Roth and the Newcastle group—that increased density of cortical plaques and tangles was linked directly to AD severity—Brun showed that an initial phase of increased plaque density was followed by a decrease. Brun also (1) emphasized that automated methods of counting surviving neurons were prone to large errors; (2) stressed that in addition to plaques and tangles, other brain pathologies were important; and (3) suggested that a focus on β-amyloid deposition in AD was unlikely to explain the disorder.[10]

Brun's work anticipated three major strands in AD research that continue to the present day. In addition to plaques and tangles, he anticipated an important role for synaptic loss and the importance of inflammatory processes. Additionally, a fourth major concern in current research—the interplay between brain vascular changes and AD pathology—owes much to his pioneering work. Recently, there is renewed interest in his interpretation of decreased β-amyloid deposition in late-stage AD. If neurons are dying in large numbers in AD, and if β-amyloid is produced largely by neurons, then it may be reasonable to expect reduced amounts of β-amyloid once significant numbers of neurons are lost to AD. With the advent of better methods of brain imaging and the quantification of β-amyloid in the living brain, Brun's finding of regional variation in cortical pathology is seen as evidence that AD is a disorder of selective neuronal loss. If β-amyloid has a major role in AD, then it seems likely that those neurons that produce most of its precursor will be those that are selectively lost in AD. This is a potential confounder of interpretation of amyloid imaging in the aging brain.[11] An alternative view is that selective neuronal loss in AD could be caused by deficiencies in nerve growth factors or selective neuronal toxins. This view was first proposed by Stanley Appel,[12] and selective neuronal loss remains a critical question in AD research. The problem of selective neuronal loss is that this features in both AD and PD. In AD, susceptible cells develop abnormal deposits of altered microtubule-associated tau protein. In PD, intraneuronal inclusion bodies predominantly composed of misfolded α-synuclein protein occur as DLBs. Both AD and PD have a characteristic pattern of brain pathology.

Alzheimer recognized that pathologists already knew some of the pathology in his patient's brain. At that time, a new technique in which silver salts were applied to microscopic sections of brain tissue had revealed the structure of neurons. In brains from old people, these salts had revealed clumps of what looked like the debris of damaged neurons. These aggregates often were found in the walls of brain blood vessels and so were called plaques, much as fatty deposits (atheroma) within hardened blood vessels were called "atheromatous plaques."

II. THE MODERN ERA: FOCUS ON β-AMYLOID

After experimenting with various dyes, pathologists found that dyes that stained starch (Latin: *amylum*) were taken up by these plaques, which then were renamed "amyloid plaques" from the word amyloid or "starch-like." Amyloid can be deposited in many types of tissue in addition to brain. This happens in a group of diseases called "amyloidoses" in which amyloid builds up in one organ or throughout the body. These are more common after fifty years of age; people with amyloidoses are more often men; and some have rare genetic mutations that affect proteins in the kidneys or nervous system. This model of age-related deposition of amyloid provided the basic line of enquiry when the chemistry of β-amyloid was examined by George Glenner and Colin Masters (see the following paragraphs). The discovery of the molecular structure of β-amyloid was a pivotal time in AD research: After this point, the research focus switched so dramatically that within ten years, research papers on amyloid came to dominate the scientific literature on the biology of AD.

With improvements in electron microscopy and brain chemistry, it became possible to analyze the structure of neurofibrillary tangles (NFTs) and plaques. The fine structure of plaques was composed of the remnants of damaged neurons, with a silicate compound at its core, garlanded with β-amyloid characteristic of AD. In 1984, George Glenner and Caine Wong from the University of California, San Diego (United States), first extracted and sequenced this structure from the blood vessels of AD patients and individuals with Down syndrome.[13] Within a year, the same peptide was shown to be a major component of the senile plaques of AD patients' brain tissue.[14] This work became one of the fastest growing areas in modern science and set the stage for the proposal that amyloid-β (Aβ) is toxic to brain cells and a likely explanation of all other AD brain pathology. In addition to amyloid deposits, inflammation and oxidative damage also are present, although in the flurry of Aβ-induced excitement, neither were thought at that stage to be causally important.

Once the structure of Aβ was established, Colin Masters from Australia and Konrad Beyreuther in Germany showed that it was a major component of senile plaques; this work was quickly followed by the identification of the much larger APP,[15] which led to cloning of the APP gene and confirmed its predicted (from the Down syndrome association) location on chromosome 21. Painstaking studies on the physical chemistry of β-amyloid[16] revealed several different naturally occurring forms, ranging in length from thirty-six to forty-three amino acids. The monomer Aβ40 occurs more often than the aggregation-prone Aβ42 type.

By 2005, evidence was sufficient to support neurotoxic effects of Aβ42. Extensive studies had shown how easily Aβ42 could agglomerate into fibrils and β-pleated sheets, forming the fibers of β-amyloid plaques.[17] Local synaptic enzymes, including insulin-degrading enzyme (IDE), regulate levels of Aβ in and around the synapse such that overexpression of IDE will prevent amyloid plaque formation.

Mutations that cause AD and affect the deposition of brain amyloid are rare and thus far have been found in just three genes: amyloid precursor protein gene (APP)[18] and two presenilin genes, PSEN1 and PSEN2.[19-23] A regularly updated list of these mutations is available at http:/www.alzforum.org. Table 6.2 summarizes genetic mutations and variants associated with increased AD risk. Note that apolipoprotein E (APOE) differs from the listed genetic mutations. APOE exists in three naturally occurring isoforms (ε2, ε3, ε4) of which we each possess two copies.[24] The ε4 isoform is linked to increased susceptibility to AD and earlier age at symptom onset.

Aβ is a normal product of APP metabolism that is important in brain development and, possibly, in cell surface recognition. Most mutations in APP are clustered around sites that normally are cleaved by α-, β-, and γ-secretases. These mutations increase the likelihood of Aβ production and promote its self-assembly into β-amyloid fibrils.

Taken together, these observations supported the "amyloid cascade hypothesis of AD."[25] Normal processing of APP is shown in Figure 6.2a. Abnormal

TABLE 6.2 Genetic Mutations and Variants Linked to Alzheimer's Disease

	YEAR OF DISCOVERY	GENETIC FINDING	CLINICAL SYNDROME	FREQUENCY
APP	1991	Variant	Early onset	Rare
APOEε4	1993	Variant	See text	Common
PSEN1	1995	Mutation	Early onset	Rare
PSEN2	1995	Mutation	Early onset	Rare
SORL1	2007	Variant	Late onset	?
CLU	2009	Variant	Late onset	?
BIN1	2010	Variant	Late onset	?
PICALM	2009	Variant	Late onset	?
APP	2012	Mutation	Protects	Rare

Note: Mutations were identified in families multiply affected by AD; variants were found to be overrepresented among nonfamilial (sporadic) AD.

processing, which leads to Aβ formation, is shown in Figure 6.2b. Normal and abnormal *APP* processing is shown in Figure 6.2c.

In AD, more than 170 mutations in *PSEN1* and 13 mutations in *PSEN2* genes are causal. The presenilins are membrane-spanning large proteins that are located on the endoplasmic reticulum. Mutations in either *PSEN 1* or *PSEN2* increase the production of Aβ42. The structure of the presenilins is shown in Figure 6.3.

Amyloid deposits are found in individuals with AD, Down syndrome, and head injury as well as in older healthy individuals. These deposits are complex extracellular structures made up largely of Aβ protein fragments. There are between thirty-six and forty-three amino acids in each type of Aβ (written as Aβ36–43). β-amyloid is derived from the larger APP, which is processed along one of two pathways that cleave APP using three enzymes, termed "secretases" (α-, β-, and γ-secretase). One pathway generates Aβ (amyloidogenic pathway) and the other does not (nonamyloidogenic pathway). Figure 6.2 shows that Aβ is released from APP via a series of metabolic cleavage steps through the combined

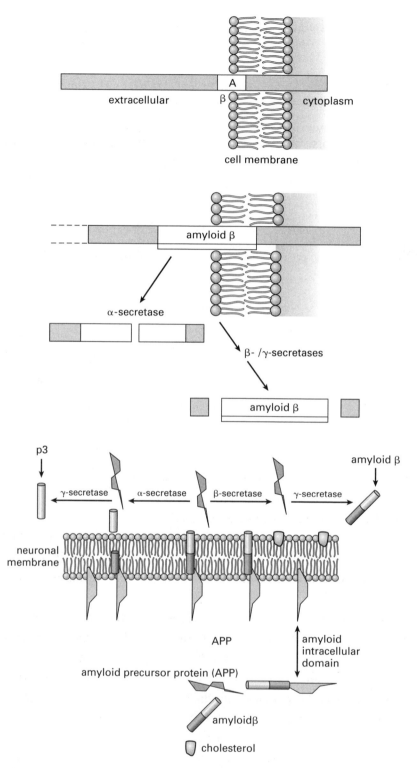

FIGURES 6.2A–6.2C APP exists in various forms but in nervous tissue, it is usually 695 amino acids long. It is a transmembrane protein that is normally cleaved by α-secretase, which does not produce β-amyloid. β-amyloid formation requires the actions of both β- and γ-secretases.

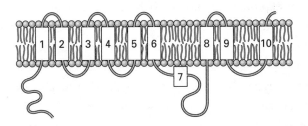

FIGURE 6.3 The presenilins 1 and 2 share very similar structures. They are both membrane-spanning proteins. The diagram represents presenilin 1 (*PSEN1*), showing ten membrane-spanning domains (numbered boxes). Known mutations that cause early onset Alzheimer's disease (AD) lie mostly within the membrane or in a loop extending inside the cell (the cytoplasmic loop).

actions of γ- and β-secretases. β-secretase is also known as BACE (beta-site APP cleavage enzyme). Defective processing of the β-secretase-digested APP fragment by PSEN1-deficient mice, a gene associated with early onset AD, shows that PSEN1 is the γ-secretase.

Although Aβ is released as a soluble protein, it readily aggregates as extracellular deposits in AD. Amyloid plaques contain β-amyloid, and numerous other components that are linked to a chronic inflammatory response and oxidative damage. Brain microglia are activated in the inflammatory response and cluster around the evolving β-amyloid plaque. Inflammatory molecules are abundant inside the plaque, along with apolipoproteins E and J (clusterin) and many metal ions, including zinc, aluminum silicate, antioxidant defense enzymes, and two types of cholinesterase. The presence of most of these compounds inside the β-amyloid plaque is largely unexplained: Some may be attracted as a plaque evolves, and others may initiate plaque formation. *APOE* and aluminum are of possible relevance here. More likely, most plaque constituents are debris left after neuronal death.

By late 2012, many of the major pharmaceutical companies had completed thorough reviews of their AD research programs. After almost 20 years of research with total expenditures in excess of many billions of dollars, the search for an effective drug to treat AD seemed no nearer. At least one major company cut its AD research team by more than 80 percent, and others began serious in-depth analyses of their work so far and why they had so little good news for their shareholders. Science journalists[26] picked up the story and Internet bloggers announced "the amyloid hypothesis is dead."

This section explored the biology behind the amyloid hypothesis and the next considers other, less well-trodden, approaches to the biology of AD. One of many dilemmas facing the drug industry is that, on the one hand, the amyloid hypothesis proves correct, but it was tested too late in the course of AD. On the other hand, the amyloid hypothesis is wrong, and clinicians should be testing interventions in other pathways, one of which might lead to AD.

There are other controversies, too. Many of these controversies were encountered almost as soon as Alois Alzheimer presented his first case. AD is not really a disease at all, some said, and is just a variant of aging in the brain. Furthermore, ever since that first description, some experts never fully accepted AD as a single disease entity but thought of it as made up of several disorders sharing some of their end-stage features.[27] Another distinguishing pathological feature of AD is the formation of NFTs. Some experts regard NFTs as the best distinguishing feature and more relevant than detection of amyloid at postmortem or amyloid estimation in the living brain using PET detection of Pittsburgh Compound B (PiB). These competing views can be related to the loss of synaptic spines in AD, and a third group of experts has regarded this process of synaptic depletion as the defining feature of AD. The simplest way to make sense of the causal importance assigned to these observations is perhaps to consider each as steps on a final common pathway toward AD on which multiple contributory pathways converge.

Studies on young adults certain to develop AD because they carry a pathological PSEN mutation reveal the sequence of these changes. Randall Bateman on behalf of the Dominantly Inherited Alzheimer Network (DIAN) reported the first survey of key clinical features and putative biomarkers in individuals at risk of familial Alzheimer's disease (FAD). Their technique was to group their sample of 128 "at-risk" descendants into two groups: eighty-eight individuals with mutations in *PSEN-1*, *PSEN-2*, and *APP* genes and forty individuals without mutations.[28] In young presymptomatic mutation carriers, cerebrospinal fluid (CSF), plasma, and PET amyloid imaging showed reduced β-amyloid, and increased tau, phosphorylated tau, and other markers of NFTs and neuronal injury or death.[29] Although remaining asymptomatic, these CSF biomarkers of neuronal injury or death later decreased. These findings are interpreted as a slowing of neurodegeneration with symptomatic disease progression. Importantly, this study does not establish which if any of the noted molecular processes is the primary event in AD.

III. MOLECULAR BIOLOGY OF TAU

Tau proteins belong in the family of microtubule-associated proteins (MAP). These proteins are found predominantly in neurons where they bind a lattice-work of microtubules to provide an internal transport system for the neuron and are critical in neuronal development and the outgrowth of axons and dendrites (morphogenesis). In addition to stabilizing microtubules of the neuronal cytoskeleton, they also are involved in signaling within the neuron. When tau proteins are defective, they cannot stabilize microtubules effectively. These actions of defective tau are thought to contribute to AD and may be shown to have an important causal role.

Human tau protein is present in the nervous system in six forms that range in length and are all derived from the protein product of a single gene: microtubule-associated protein tau (MAPT). These were discovered in Marc Kirschner's laboratory at Princeton University (United States). The different forms of MAP have different functions.

The NFTs appear first as intraneuronal structures that are confined to specific neuronal subgroups. These subgroups share some common features among which are their high-energy usage and their capacity to support higher cognitive functions, including language. Each NFT is made up of aberrant forms of tau proteins. These join up together to form polymers that not only are difficult to degrade but also can extend to form lengthy paired helical filaments (PHF-tau).

The PHF-tau burden on the brain in AD is more than fifteen times greater than that found in healthy brains. As PHFs accumulate in AD brain, the concentration of the normal form of tau falls. It is so far unknown how processes that lead to amyloid plaque and NFT formation could be connected. Although an early report noted that amyloid plaque density was directly linked to the severity of dementia before death, this finding was not satisfactorily replicated. Currently, the association between dementia severity and loss of synapses has greater support (see the following paragraphs). Braak and Braak first characterized the spread of abnormal tau during the course of AD. Figure 6.4 shows how the spread begins in the hippocampus and entorhinal cortex and progresses to the temporal, parietal, and frontal cortices. In some studies, the tau pathology precedes β-amyloid accumulation by more than twenty years.

The MAPT gene on chromosome 17 is implicated in chronic, progressive age-related brain diseases other than AD. These are a variety of FTD syndromes

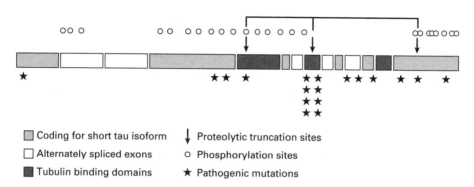

FIGURE 6.4 Schematic of the tau protein gene. There are six isoforms of tau in the nervous system (gray) obtained by splicing exons (dark gray). Mutations that cause frontotemporal dementia and related disorders (see text) are shown as solid stars.

and two relatively uncommon disorders (progressive supranuclear palsy and argyrophilic grain disease). Figure 6.5 shows the location on the MAPT gene of various mutations.

Numerous mutations in the progranulin (PGRN) gene on chromosome 17 have been described since the gene was identified in 2006. These autosomal-dominant mutations typically result in a behavioral FTD phenotype that may be associated with marked impairments of semantic knowledge (FTD, semantic dementia). Radiochemists have developed novel compounds that will detect tau deposition in the brain. Several compounds are in development, but all remain unvalidated. Their availability would help answer key questions about the time course of amyloid compared with tau deposition over the course of AD. At an individual level, differences between rates of tau and amyloid deposition might help identify those individuals who would be more likely to responds to drugs aimed at tau depositions and those who would respond better to anti-amyloid therapies.

The deposition of aggregates containing β-amyloid in and around brain cells is one of two characteristic brain changes in AD. The other is the formation of NFTs made up of hyperphosphorylated tau. The amyloid hypothesis of AD assigns primacy to the deposition of β-amyloid and proposes that the aggregation of β-amyloid triggers that formation of NFTs. Some studies have supported this view by showing that β-amyloid and NFTs interact in selected brain regions. Notwithstanding this support, some critics remain cautious about an interpretation that implicates β-amyloid as the primary abnormality. They point out that

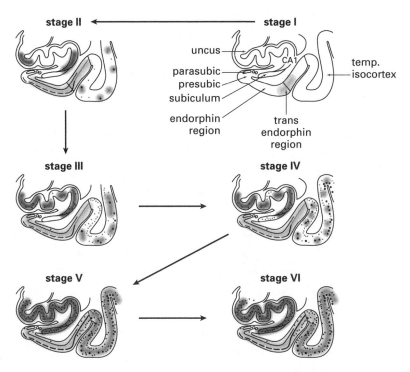

FIGURE 6.5 The spread of neurofibrillary tangles in the cortex follows a consistent pattern. This diagram illustrates the progress of NFT deposition by six stages (I through VI) that typify the development of Alzheimer's disease.

these studies examined the effects of mutated genes coding for both tau and amyloid. When only APP is mutant, tau does not form NFTs. The likelihood is that once small tau inclusions have formed in the brain, they probably become self-propagating and may be independent of any perturbations in the deposition of β-amyloid. This line of reasoning may be directly applicable to a wide range of neurodegenerative diseases.

IV. VASCULAR COGNITIVE IMPAIRMENT AND DEMENTIA

Although cerebrovascular disease contributes to many common forms of late-onset dementia, little explains why brain blood vessels should be affected more than blood vessels elsewhere. The observation that the brain blood vessels are "end arteries" that are the sole providers of oxygenated blood to specific brain areas shows that these areas are certainly vulnerable to arterial blockage.

In general terms, risk factors for blood vessel disease outside the brain are also risk factors for the brain blood vessels. Major risk factors are hypertension, diabetes, and smoking. Lesser risk factors include obesity, poor diet, excessive alcohol, and a sedentary lifestyle.

Interest is great in some unusual genetic causes of vascular dementia largely because molecular genetic studies provide novel insights into the underlying processes and also because, in some rare instances, molecular genetic tests can contribute to diagnosis. The main features of one rare type of vascular dementia called cerebral autosomal dominant arteriopathy with subcortical infarcts and leuko-encephalopathy (CADASIL) are as follows: migraine between the ages of twenty and thirty, followed by recurrent ischemic stroke between the ages of thirty and forty, and dementia beginning after the age of forty. The disorder is caused by a mutation in the NOTCH gene on chromosome 19. The brain is affected by widespread cortical infarcts and finely distributed blood vessel lesions in white matter ("leucoaraiosis"). The brain blood vessels also can be affected by β-amyloid deposits caused by a mutation in the APP gene ("cerebral amyloid angiopathies") and by raised blood concentrations of homocysteine caused by a defective MTHFR gene (Methylene-Tetra-Hydro-Folate-Reductase) or by an ineffective cystatione β-synthetase enzyme. Genetic tests for these genes are widely available.

Vascular disease contributes through unknown pathways to the risk of AD. There are many early inflammatory changes in brain blood vessel walls in AD. These are thought to initiate AD changes in some vulnerable patients through pathways that promote oxidative damage and activation of brain inflammatory cells (microglia). Small lesions in white matter (detected as white matter hyperintensities on MRI) are probably of microvascular origin. Together with AD pathology, these vascular lesions contribute to the overall picture of cognitive impairment found in dementia syndromes. In addition to these "mixed" pictures of vascular-plus-AD pathology, there are relatively rare cases of pure vascular dementia, but these cases never make up more than 10 percent of all dementia cases. To the discerning eye, it seems that the more closely the brain vasculature is examined, the more vascular pathology is detected in older adults.

There are good reasons for investigating the function of the brain blood vessels in dementia. Not least is the possibility that the barrier between brain and circulating blood vessels (the blood-brain barrier) may be defective and allow abnormal proteins to pass either from brain to blood or blood to brain. Not least among the consequences of a "leaky blood brain barrier" is the possibility that β-amyloid released from neurons can diffuse out of the brain to damage blood vessels. So far, there are no treatments of the vascular components of dementia other than those already in place for the prevention of stroke and lowering homocysteine concentrations using folic acid supplements. Although some clinical trials of the effects of homocysteine lowering on age-related cognitive impairment and dementia risk were encouraging, the current consensus is that grounds are insufficient to introduce homocysteine measurement in the routine diagnostic workup of dementia.

V. THE MODERN ERA: FOCUS ON THE BIOLOGY OF AGING

There are two main theories of aging. One proposes that aging is determined by the accumulation of errors during cell replication ("replicative senescence"), and the other proposes that internal or external programs increase the risk of mortality ("programmed cell death"). A major goal in current dementia research is to understand how the biology of aging can inform susceptibility to age-related diseases (ARD), including the dementias. All involved in aging research agree that human aging is a complex, multifactorial process influenced by many unknown genetic and environmental factors. Current research programs are complex and diverse, and many accept that some form of combination of the two main theories of aging is required. Figure 6.6 shows these theories in diagrammatic form.

Nevertheless, several themes appear more promising. These ask what is the importance of protein synthesis deregulation and protein aggregation for aging and dementia? Do age-related alterations in protein synthesis interfere with drugs intended to slow or prevent dementia? How do lifestyle choices or comorbid chronic diseases and their treatment affect rates of aging?

Critical components of contemporary programs address the nature of individual differences in rates of aging by measurement of underlying pathologies as, for example, in variability of protein synthesis as a marker of aging. In turn, once a marker is determined as valid and useful, it then could become a proxy for overall rate of aging in specific tissues (as in brain structures) and as a measure of treatment outcome.

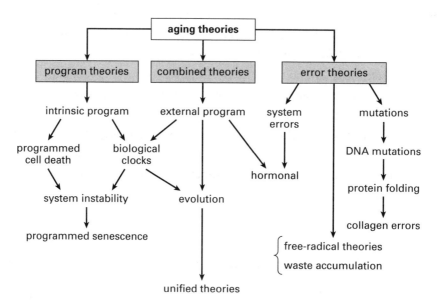

FIGURE 6.6 Examples of theories of aging placed in broad classes. For clarity, only selected main theories are shown.

Research teams aim to identify genes, epigenetic factors, protein, and metabolic networks that control protein synthesis—including its rate, accuracy, modification, folding, and aggregation. Teams then apply these findings to the evaluation of variation in rate of aging and to ARDs, including the dementias. Many human diseases are already known to be associated with protein misfolding that leads to acceleration of protein breakdown, to "improper trafficking," and in AD, to aggregation with the formation of amyloid fibers. To achieve their goals, large-scale collaborative studies are needed that start with clinical examination of samples drawn from the general population. The scale of these projects is such that their success depends not only on the scientific resources (infrastructure) available to each team but also on the efficiency of collaborative links between groups. Molecular genetic epidemiology pioneered this approach with considerable success in the biology of the dementias. It is significant, however, that discovering a specific genetic influence on an ARD (such as one of the dementias) is only a first step. Current research anticipates success and plans to place each new discovery within the context of one or more pathways toward dementia. In plain terms, it is never enough to find that a gene contributes to a specific dementia; therapeutic success depends on finding out what that gene does and how this modifies the risk of dementia. The example of Huntington's disease (HD) is often cited in this context. There are important consequences in HD of autosomal dominant inheritance of a genetic abnormality on mitochondrial function and oxidative damage. So far, attempts to influence the expression of this genetic abnormality or protect against its harmful effects on neuronal metabolism have disappointed. In the face of almost thirty years of intense research since the discovery of the genetic abnormality, there are no known neuroprotective treatments to slow or prevent HD.[30]

Three main signaling pathways influence rates of aging and differences in life span: dietary restriction, insulin/IGF1-like signaling, and the mitochondrial electron transport chain. Genetic regulation of energy metabolism requires close coordination between genes in the cell nucleus and those in the mitochondria, and this can be disrupted in aging. Variations in diet can have profound effects on this type of linkage and suggest ways that dietary habit could influence the risk of ARDs. Abnormalities of protein aggregation are frequent in human populations, giving rise to a wide range of clinical syndromes, including many that are associated with aging. Genetic mutations in components that function downstream of protein synthesis (including those operating on protein folding pathways) are associated with clinical syndromes as diverse as some cancers, hypercholesterolemia, cystic fibrosis, and certain types of neurodegenerative disease. These last diseases include LOAD, which has established links with high cholesterol, insulin resistance, cardiovascular diseases, and chronic inflammation. Some experts anticipate that "metabolic reprogramming" induced by caloric dietary restriction, modification of insulin/IGF1-like signaling, or the mitochondrial electron transport chain could influence the risk of LOAD.

Progress in the development of scientific models to test ideas in aging and the dementias concerning the possible roles of genomics, proteomics, and metabolomics represent a major challenge. Bioinformatic skills and computing platforms with sophisticated statistical data analysis are all required to investigate this level of biological complexity. Eventually, clinical trials will follow but not until high throughput screens to identify potential drugs (small molecules with antiaging properties) are completed and possible drugs are identified.

Like all research ambitions that hold such great promise to benefit mankind, it will be important for societies to devise appropriate legal and ethical frameworks in which to communicate, implement, and monitor the impact of these developments. Not the least of these will be concerns about disclosure of personal risk of ARDs and the impact of this on personal adjustment, job security, and insurance coverage.

CLINICAL GENETICS OF AD

Alois Alzheimer identified NFTs and amyloid plaques in AD in 1906. Twenty years later, Belgian pathologists reported premature aging in Down syndrome (caused by an extra copy of part of chromosome 21) with high densities of β-amyloid plaques in the brain. This association between AD and Down syndrome, although occasionally replicated, remained largely ignored for almost fifty years. It was discounted and attributed to "degeneracy" in Down syndrome. It was not until 1980 that it was recognized as a major clue to AD.

Reports of families multiply affected by EOAD gradually accumulated during the twentieth century. Early twin studies on EOAD showed greater concordance in EOAD between identical twins than nonidentical (fraternal) twins. A consensus was established that genetic factors were important in EOAD, but until the discovery of the molecular structure of amyloid and through a process known as "reverse genetics" leading to the structure of APP, the mechanisms of genetic transmission were unknown. An early report suggested that LOAD was associated with increased expression of the *APP* gene, but this was soon discounted. Currently, in LOAD, no single gene of major causal effect is securely linked to this condition. The strongest data available to confirm considerable influence of genetic factors in LOAD derive from the work of Nancy Pedersen's group at the Karolinska Institute in Stockholm with Margaret Gatz and Chandra Reynolds at the University of Southern California (United States). In broad terms, their collaboration has established that genetic factors contribute between and 60 and 80 percent of the heritable causes of LOAD. Genome-wide association studies (GWAS) suggest that variations in the structure of many genes each of small effect are important. Jean-Claude Lambert recently reviewed all informative studies and, in a meta-analysis, concluded that seventeen genes, including *APOE* (discussed in the following section), were important. The functions of these genes include roles in the immune response, antioxidant defenses, synaptic remodeling, and intracellular calcium trafficking.

MATERNAL TRANSMISSION OF AD

The risk of late-onset dementia may be similar to other conditions that have their origins during embryogenesis or fetal development. These related conditions may be influenced by the inherited genotype and the intrauterine environment, which in turn is strongly determined by the maternal environment and the maternal genotype. If these similarities between late-onset dementia and risk of other conditions (which include birth defects) were supported, then this would be relevant to understanding the causes of late-onset dementia and, potentially, to the provision of genetic counseling.

Recent progress in understanding the heterogeneous nature of the molecular genetics of late-onset dementia has led to a better understanding of the dementia syndromes. Identification of specific genes that contribute to subtypes of dementia has not been as successful as once confidently forecast. This may be because the genetic background may involve not only the inherited genotype of an affected offspring but also that of the mother. Methods to investigate how one genotype may affect another and to understand gene–environment interaction are needed. This information becomes especially relevant when the maternal genotype may influence susceptibility to late-onset dementia through its modification of the intrauterine environment.

Several reports are suggestive of maternal transmission of late-onset dementia.[31] These reports are supported by observations of phenotypic differences between the unaffected daughters of affected mothers that would be consistent with the unaffected daughters having features of a preclinical dementia syndrome. These observations include smaller hippocampal volume in MRI-derived brain regional volumetric studies, lower than expected utilization of glucose in positron emission tomography, and decreased activity of maternally derived mitochondrial enzymes in peripheral blood tissue.[32]

HEREDITARY OR FAMILIAL DEMENTIAS

The inherited dementias follow an autosomal-dominant pattern of inheritance. Although they have many clinical features in common with late-onset dementias, their age at onset is almost always before age seventy years. Depending on the dementia subtype, between 2 and 50 percent of dementia cases are thought to be inherited. Whenever a familial early onset case is investigated, it is essential to ensure that appropriate genetic counseling resources are available to families. This is particularly important if molecular genetic diagnostics are planned. Genetic testing is confined at present to families with a clear autosomal pattern of inheritance or rare cases with age at symptom onset before age thirty-five years. Even when families are selected in this way, it is usual to find that only about 10 percent of families carry a known mutation in *APP, PSEN1,* or *PSEN2.* Although *APOEε4* is strongly linked as a susceptibility factor in dementia, there are no benefits from genotyping and a consensus exists that *APOE* testing should not, with one rare exception, be offered. The exception is that when a causal mutation in *APP, PSEN1,* or *PSEN2* is found, the presence of *APOEε4* is linked to earlier symptom onset by about five years.

FTD can be one of three clinical types. These are behavioral variant (bvFTD or FTD) and two others termed semantic dementia (SD) and one nonfluent dysphasic syndrome. Unusual syndromes can accompany any of these three variants; these are grouped as Parkinson's, corticobasilar, and motor neuron disease. Relationships are complex between genotype and clinical features of the types of FTD. When Parkinsonian signs are present, the genetic defect is in the *MAPT* or less often in the *PGRN* gene. Screening of these genes is undertaken when a family history is consistent with autosomal-dominant inheritance.

The prion diseases include the spongioform encephalopathy, Creutzfeld–Jakob disease (CJD)—known in the popular press as "mad cow disease." Before contaminated beef entered the human food chain, CJD was an unusual cause of early dementia. There was a family history of CJD in about 20 percent of cases. Facilities for genetic counseling and molecular diagnostic tests have been developed in several European centers, but thus far an epidemic of CJD that can be linked to "mad cow disease" has not arisen. The disorder remains important, however, and surveillance of possible new cases is essential. The possibility that effective treatments of the abnormal protein folding that underlies CJD will lead the development of anti-AD drugs also remains a focus in several research laboratories.

Familial AD (FAD) is known to have several distinct genetic causes. A small number of kindreds have one of four relevant mutations in the *APP* gene on chromosome 21. The majority of FAD kindreds are linked in about 500 families worldwide to mutations in *PSEN1.* Mutations in *PSEN2* gene are infrequent. Among the best-characterized FAD kindreds are the five Columbian early onset pedigrees discovered by Francisco Lopera from 1980 to the present. Although all affected family members in each kindred share the same genetic mutation, age at onset (from thirty-four to sixty-two years) and the frequency of specific cognitive deficits varies significantly. To date, around 1,235 members of this single kindred have been genotyped, of whom 480 were carriers of the *PSEN1* mutation. People who carry mutations linked to FAD have contributed enormously to the investigation of biomarkers of AD risk and disease progression. These studies have included measurements of tau and β-amyloid in the CSF that bathes the brain.

These families are recognized widely as providing opportunities unobtainable by other means to evaluate biomarkers and potential AD-modifying treatments. When the age at AD onset can be determined reliably, it is feasible to examine the possible contribution of other genetic and environmental factors that cause variation in age at AD onset.[33,34]

APOLIPOPROTEIN E

The discovery of a major role for APOE among risk factors for AD is one of the most important findings in AD research. The first observations reported in 1991 showed that APOE is present in senile plaques. At first, APOE was thought to be part of an inflammatory response triggered by dead or dying neurons. In 1982, the gene coding for APOE was located on chromosome 19 and present in three forms (*APOEε2*, *APOEε3*, and *APOEε4*). When the *APP* gene was located on chromosome 21, it was anticipated that most types of FAD would be linked to this gene. Association studies did not confirm this linkage but suggested a linkage to chromosome 19. A research group led by Allen Roses based in Duke University (United States) showed that a common variant (ε4) of the *APOE* gene on chromosome 19 is linked to AD. In a series of landmark studies, the team established that the *APOEε4* genotype is associated strongly with AD.[35,36,37]

APOE is unique to humans. It is a lipoprotein involved in the metabolism and transport of lipids. Slight differences in protein structure change the chemical properties of each APOE variant. APOE now has several established roles in the nervous system, which have led to ground-breaking ideas about the role of APOE in processes leading to AD and in the development of treatment strategies. Comparisons between APOEε2, APOEε3, and APOEε4 show that ε4 is least efficient in clearing cholesterol from neurons and is least protective against synaptic damage and against β-amyloid toxicity.

APOEε4 is the best-established susceptibility genetic variant for AD arising without a family history of dementia after age sixty years. In FAD, *APOEε4* is linked to earlier age at onset. Risk of AD is linked strongly with *APOEε4*, not at all with *APOEε3*, and *APOEε2* may protect against AD. *APOEε3* is present in 50–90 percent of people in all populations, whereas *APOEε4* is present in 5–35 percent and *APOEε2* in 1–5 percent. *APOEε4* also is associated with increased risk of dementia following head injury and stroke and in Down syndrome. These associations sometimes are interpreted as "nonspecific actions" of APOE, meaning that common pathways affect neuronal repair after injury (caused by AD, trauma, or stroke), and these are less efficient when *APOEε4* is present. As the exact role of *APOEε4* in neuronal health and disease is uncertain, however, it is possible that the contributions of *APOEε4* to the outcome differ substantially among these three conditions.

About 50 percent of patients with LOAD carry *APOEε4* compared with 20–25 percent of old people without dementia. Carrying one copy of the *APOEε4* increases risk of LOAD by about three times and two copies by about twelve times. In LOAD, one or two copies of *APOEε4* leads to an earlier age of onset by about ten to twenty years compared with noncarriers and to faster rates of progression.[38] Compared with noncarriers, *APOEε4* resulted in earlier onset of dementia in individuals with a *PSEN1* mutation in one large Colombian kindred described by Francisco Lopera. Individuals with *APOEε2* have reduced risk of developing LOAD. Epidemiological studies from various populations have confirmed the increased frequency of *APOEε4* in patients with late-onset disease compared with noncarriers, although the frequency varies among different subtypes of dementia.

The increase of LOAD risk associated with *APOEε4* is consistent in populations examined worldwide. In broad terms, *APOEε4 is* a risk factor for EOAD and LOAD and the pattern of inheritance is comparable to the effects of genes contributing to familial forms of breast cancer.[39]

VI. BREAKDOWN OF NEURAL NETWORKS IN DEMENTIA

The preceding sections presented the molecular pathologies of AD and related tauopathies and vascular cognitive impairment. An obvious question arises from these descriptions: How do clinical dementias develop and spread from the first seeds of molecular change? It is clear from these brief descriptions that the two most important features of the dementias are the formation and deposition of abnormal protein structures[40] and characteristic patterns of brain cell loss across the cortex and some subcortical brain structures.[41] Although it is intriguing to find that the spread of neuronal damage tracks the evolution of uniquely human cortical structures, by itself this does not explain the pathology. Current understanding of brain aging and the dementias lacks a framework that can predict observed patterns of damage. So far, it is not possible to trace the pathway from molecular pathology to the clinical symptoms of dementia.

In Chapter 3, The Well-Connected Brain, the brain was shown to be made up of neural networks and methods were described that can accurately represent their structure and performance. So far, conventional approaches to the distribution of neuronal loss in the dementias have failed to explain their distinctive patterns. Analysis of neural networks is a logical next step toward a better understanding of the dementias. Novel techniques suggest how abnormal proteins can spread through neural networks along functional pathways. Although it is attractive to speculate how this spread determines the clinical progression of each type of dementia, these methods do not yet explain why abnormal protein deposition should start in one particular place and spread from there along functional nerve tracts. Some insight was obtained through careful analysis of neuronal damage in FTD. These studies identify how specific cellular elements are affected by known genetic mutations in FTD. As new mutations are identified in this and other tauopathies, the molecular principles that determine neuronal spread in each dementia subtype will become clearer. In AD, the first brain pathology typically is found in the entorhinal cortex where APP is produced as a repair response to initial synaptic damage. This may be sufficient to trigger the "amyloid cascade hypothesis of AD" and cause neurotoxic APP fragments to diffuse along nerve tracts leading to the characteristic pattern of neuronal loss.[42]

Dendritic spines are the postsynaptic contact locations of the majority of excitatory synapses in the nervous system. They are extremely small (around 1 μm) and near impossible to see using a light microscope. Dendritic spines are rapidly remodeled by learning and memory through the actions of a contractile protein (actin). Extension of spines increases the number of synapses and enhances synaptic transmission.

A link between cognition and the richness of dendritic spines was demonstrated from the 1960s onward when the brains of children with profound

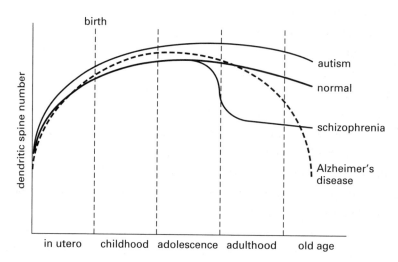

FIGURE 6.7 In the absence of disease, dendritic spine numbers change over the life course. In autism, increased numbers of spines are maintained throughout life. Individuals with schizophrenia have a greater than normal pruning of dendritic spines. Individuals with AD have greater loss of spines in late adulthood and old age.

learning disabilities were systematically examined.[43,44,45] This work provided clues to the importance of the brain's cellular architecture in aging and in a wide range of neuropsychiatric disorders, including AD,[46] schizophrenia, and autism (summarized in Figure 6.7). Although these links do not presume that dendritic spine pathology is the only significant brain change in AD, they do suggest locations where molecular perturbations might be causal and suggest new targets for drug development.[47]

AD progression initially was linked to the density of β-amyloid plaque deposition in the brain but is now more firmly linked to loss of synapses.[48] Impairments of synaptic functions also can be induced experimentally in mice by administration of β-amyloid mimicking physiological changes seen in aging and to a greater extent in AD. *APOE* isoforms (see the following paragraphs) influence the shape and motility of dendritic spines. These effects differ between isoforms and may be relevant to the development of AD.

The dementias of late adulthood are linked to neural networks by the clinical and anatomical progression observed postmortem in patients with dementia.[49] Novel neuroimaging techniques demonstrate that the spatial patterning of each subtype of dementia can be mapped closely onto distinct functional networks found in healthy brain during the "resting-state" fMRI. So far, it is not known if these findings predict regional neuronal loss in disease. There is a strengthening proposition, however, that some age-related pathologies originate within vulnerable "hubs" that previously were shown to play a central role within the architecture of each target network. Although attractive, this proposition does not explain why each subtype of dementia follows a specific network-based spatial pattern. The best guesses are as follows: (1) Those brain networks that carry high workloads of information transfer are more likely to suffer damage that exceeds with time their capacity for self-repair. (2) At the start of a molecular cascade that leads to neuronal death, neurons release one or more toxins that diffuse along network connection. (3) Failure of those brain cells that support neuronal function through the release of neurotrophins is the primary cause of neuronal death. (4) Neurons that make up the network share a specific gene or protein that makes neurons in that network exquisitely susceptible to the effects of an unknown toxin. Tests of these propositions suggest that distinct subtypes of dementia are linked to patterns of neuronal loss that mirror neural networks already identified in healthy younger adults.

VII. BRINGING IT ALL TOGETHER

Over the past twenty years, many attempts have been made to pull together the effects on brain biology of genetic factors implicated in AD,[50] the role of aging on brain cell maintenance and repair, and the possible contributions of inflammation, oxidative damage, and immune responses. Among the many reviews on this topic, the most implicated scenario offers highly complex interactions occurring at molecular levels and seeks to integrate these with claimed harmful effects of the environment and protection provided by a nutritious diet, education, and an active and socially integrated lifestyle.

When the molecular data summarized in this chapter are set out alongside what is known about a number of age-related chronic progressive diseases of the brain and spinal cord, some general points seem to hold up well on examination. The first is that these diseases often are linked to abnormal aggregation of proteins. The first diseases to be recognized in this way were AD and the systemic amyloidoses. Later, other protein aggregation diseases that affect the nervous system were detected. In all of these diseases (the proteinopathies), specific proteins were found to aggregate inside cells and form toxins that caused cell death. These diseases include Huntington's disease and certain spinocerebellar ataxias in which abnormal proteins accumulate in the neuronal cell nucleus; in the cytoplasm as in α-synuclein Parkinson's disease; or in the extracellular space as in prion diseases; or the proteins form both intracellular and extracellular deposits, as do β-amyloid and tau in AD.

These observations lead to two important applications. The first observation concerns the search for factors that initiate abnormal protein folding processes. The second is the opportunity to develop new anti-AD drugs based on the prevention of abnormal folding. Translation of progress in understanding these proteinopathies into therapies targeted at AD will depend on substantial advances in the early accurate detection of AD and the development of improved animal and cellular models of AD. This chapter has raised questions about how to detect the presymptomatic signs of AD and how these relate to the degrading of neural networks that support higher cognitive functions lost in the course of dementia.[51] These are complex problems that can only benefit from recent advances in computational neurobiology and systems neuroscience fully informed by large-scale, population-based neuropathogical studies.[52] In the meantime, much current interest focuses on genetic forms of AD: familial early and late onset AD, AD associated with two copies of *APOEε4*, and the dementia of Down syndrome. All have great potential to improve clinical methods of early detection and may prove critical to the detection of preclinical biomarkers for AD.

7

THE DISCONNECTED MIND

I. INTRODUCTION

The challenge of this chapter is to understand how aging affects mental life in general and what, if any, are the changes that appear specific to aging. This is not an easy task. Some terms—even *mind*—evade simple definitions, whereas others, such as *intelligence*, are so difficult to define that it seems no two psychologists can agree on what they mean. Ideas about *mind* are linked to fundamental ideas about the nature of *self*. The mind distinguishes between what is *self* and what is *not self*.[1]

Underpinning the various levels of competence in these facilities are the capacities to remember and to communicate. Together, remembering and communicating form an array of abilities that function cooperatively in health in a connected, efficient manner. In aging and mental diseases, some of this efficiency is lost. For most people who survive into their eighties, normal degrees of loss of efficiency seem to have little impact on their functional capacity for independence. When these abilities are evaluated, the tests used may be designed to find the best level of functioning. Extreme levels of functioning rarely are required in day-to-day life, during which time it seems that most people carry on with familiar routines and seldom are expected to perform at their best or with maximum effort.

The metaphor of the "disconnected mind" conveys the sense of loss of connectedness between mental processes. This is not a disruption of mental performance in ways specific to aging or dementia. Other types of mental disorder are associated with apparent disconnections between mental functions. For example, Paul Eugen Bleuler coined the term "schizophrenia" to convey the disconnection in the mind of a patient with schizophrenia—a so-called schizophrenic split between an individual's known (cognitive) world and their soul or spirit.[2]

Thus far, it is not possible to decide on four key issues: (1) Is cognitive aging a gradual decline from early adulthood or does it reflect a sudden drop in abilities around a critical age (say, seventy years of age, give or take three years)? (2) Does everyone who experiences "nonpathological" cognitive aging decline in the same way, with decreases in the same range of cognitive abilities at similar rates? (3) Is aging kinder to those young adults with superior mental abilities? (4) What are the main sources of differences in rates of cognitive aging that commonly occur between individuals? If we dwell on these questions, we are reminded of the public criticism surrounding psychological tests. For example, to determine an appropriate test to measure the threshold of a "first responder" to act quickly and appropriately in an emergency, set a test that is as close to a real emergency as possible and observe the outcome. Psychological tests are different, however. It is fair and practical to score an individual on a test of vigilance when the core demand of an occupation is continuous watchfulness. Do psychological tests of cognitive abilities typically reflect capabilities in the natural world? If better tests could be devised, could they be used in large populations to determine what to expect as individuals from diverse backgrounds age? These are not frivolous points and are made here before consideration of the results of cognitive tests from old people. To gain wide acceptance, tests of cognitive aging need to be based soundly on large samples that are truly representative of the populations from which they are drawn.

II. COGNITIVE AGING

What we know relies entirely on our cognitive abilities. These include powers of reasoning and understanding, the use of language, the ability to attach meaning to our perceived sensory information, the ability to pay attention—sometimes to several matters simultaneously—to form and recall memories, and some extraordinary abilities to place ourselves in another's position. The first question about cognitive aging is whether it takes place at all. Timothy Salthouse at the University of Virginia probably has provided the most comprehensive and well-constructed account of the scope and limitations of research methods on this question.[3] If cognitive aging is defined as normal (i.e., nonpathological) aging in the absence of dementia, then we should find evidence that over the final decades of life cognitive abilities diminish in a characteristic fashion. Four types of tests evaluate this proposal: The first type is composed of personal accounts about what it is like to grow old. Most of these accounts are provided by individuals of high average mental ability who describe a loss of "mental energy" or interests. On close inspection, these individuals rarely distinguish usefully between (subjective) failing mental powers and their feelings about becoming old, the impact of loneliness, and acquired physical disability. A second type of test compares old and young people on mental tests. Interpretation of these (so-called cross-sectional studies) is not straightforward and is confounded by differences, for example, in life experiences, types of education, and general health. A third type of test is the "longitudinal" study in which a selected group of individuals is followed up with repeated mental tests over many years. This test seems like the "gold standard," but firm conclusions can prove tricky.

Repeating a mental test over many years can give the impression of improved performance over time. Test anxiety falls with familiarity, people reconsider tests on completion, and test takers may use a better strategy even after intervals between tests that can last years. Observed performance will account for the net contributions of positive effects of practice and decreased test anxiety as well as the negative effects of brain pathology and aging. The fourth type of test examines the performance of nonhuman species on tests of memory and learning. These tests show that, with age, animals kept in uniform conditions demonstrate decreases in learning new material in behavioral tests. It is notable that those who age more successfully tend to remain in longitudinal studies and thus reinforce the impression that cognitive aging is of slight consequence in the absence of dementia.

Taken together, these four approaches appear to show that in the absence of dementia, cognitive abilities fall with advanced age. To date, there is no established age at which cognition begins to fail. The picture is of slowly evolving changes that develop over many years. In the absence of reliable measures of the original level of mental ability, many studies cannot reach firm conclusions about the age of onset of decline.

The next question about cognitive aging is that if it does occur frequently, is it the same for everyone? Are there typical patterns of decline such that specific abilities always begin to fail first, or does "it all go together when it goes"?[4]

This point is highly relevant to the detection of the earliest stages of dementia. Underlying the idea that cognitive abilities decrease with age is the problem that the root cause is shared among all cognitive abilities. For example, it is sometimes argued that different mental abilities tend to be similar within an individual such that, for example, someone who is good at math tends also to be good at other things like vocabulary. If the effect of aging on cognition was mostly on a single component shared between abilities, then aging may be acting more on that shared component and not separately on each ability. This leads to a very important issue: Do aging effects on sensory and motor abilities determine the rate of cognitive aging? Before we can examine this question, we need to look again at how cognitive abilities are organized (see Figure 7.1).

Imagine you are lost, all alone in a strange town, trying to find your way back on foot to your hotel and you realize how desperate you have become. Nothing seems familiar, not a landmark in sight. Finding your way around the science of

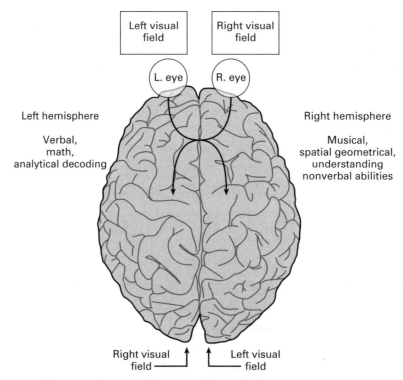

FIGURE 7.1 The principal cognitive functions of cortical regions.

mental functions is at first quite like being lost. None of the words seem familiar, and when the words are familiar, they seem to be used in a special way that only experts can understand. The job of making sense of this information and linking what you know to brain biology is difficult. Chapter 3, The Well-Connected Brain, summarizes major brain structures involved in cognition and emphasizes the importance of the major nerve fiber tracts that connect local neural networks. The next step involves naming the main cognitive functions provided by these connected networks. Many excellent comprehensive accounts of the organization of cognitive functions are available. The well-illustrated textbook by Michael Gazzinaga and his colleagues is especially helpful and very accessible.[5]

In this context, we rely on words in common usage like "perception," "memory," and "comprehension." In Chapter 3, The Well-Connected Brain, Figure 3.6 diagrams the names of the main regions of the cortex. As a general principle, the diagram shows that the sensory systems are linked to regions toward the back of the cortex. These regions are where sensory signals from the eyes and ears first reach the cortex and thus are called "primary cortical areas." Output functions like speech production, voluntary motor control, and decision making (executive function) are in the front half of the brain.

The stranger lost in town needs a map to get around. The map in Figure 7.2 shows how the main cognitive functions are organized. Like many similar diagrams in behavioral neuroscience used to explain how information moves around the brain, the boxes and arrows suggest the directions of information flow. These directions, however, should not be understood as having the accuracy of the wiring diagrams that show exactly how electrical circuitry is made up or is even comparable to diagrams of brain connections.

Another way to make sense of complicated situations is to devise a metaphor that represents only the essential parts, removing anything irrelevant. Aristotle thought that memory was like the wax tablet he used for writing. When new, the wax tablet was soft and easy to use; when old it became hard and difficult to impress. Perhaps a young memory is just so impressionable. Similarities between the capacity of memory and that of a fluid container are also made. "Old people complain about failing memory because they have more to forget" is a well-known cliché. Both observations reflect the idea that human memory has a finite capacity and, once reached, it begins to fail.

Figure 7.3 makes the functional map in Figure 7.2 appear a little simpler. The figure is a model of how aging affects the recall of verbal information. The figure

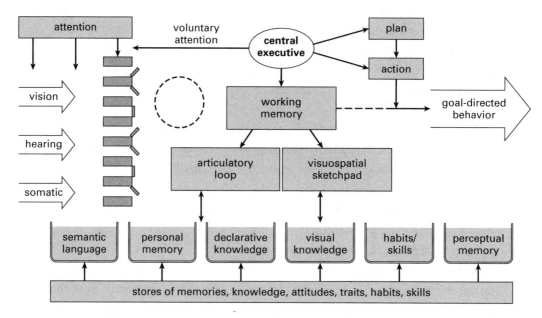

FIGURE 7.2 The organization of cognitive functions. Sensory information is presented on the left side of the diagram. Information from the sensory organs (eye, ear, etc.) enters awareness (the broken circle) through gates that can be opened and closed or hold sensory information briefly in temporary stores (rectangles). Selective attention is shown as downward arrows above the sensory input labels. Working memory is a temporary store of memories needed to organize responses required to meet immediate goals. Although shown as a box below the central executive, working memory involves the central executive, the articulatory loop, and the visuospatial sketchpad. The central executive exerts voluntary control over learning and retrieval of memories, planning, and voluntary attention. In everyday language, the central executive has top-level control over behavior. It can organize all the mind's resources to plan and achieve goal-directed behaviors while remaining flexible and adaptable. The broken circle represents a mental area of consciousness.

The template for this diagram was radically adapted from B. J. Baars & N. M. Gage, *Fundamentals of Cognitive Neuroscience* (Academic Press, 2013).

illustrates how, when an old person is tested on verbal recall, two methods commonly are used.

In one method, a list of unconnected words is read out loud to the subject (see bottom left of Figure 7.3). A test example would be twelve words read at a natural speed. For example, a list could be "repeat after me: elephant, cabin, pipe, silk, giant, pillow, teacher, drum, bell, curtain, chair, . . . parent." The test score is based on the number of words correctly recalled either immediately or

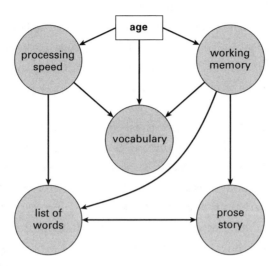

FIGURE 7.3 The word list and the Audrey Brown story explained. A model of the effects of age on relationships between two types of verbal memory test: recall of a list and recall of a prose story. The important difference is that speed probably does not predict recall of a prose story. This recall is much more affected by working memory having access to automatic processes that aid understanding of the story. Age has major negative effects on speed and working memory and, until very old age, it has many positive effects on vocabulary. At the foot of the diagram, the double-headed arrow shows that performances on lists and prose recall are interconnected so that people who do well on one tend to do well on the other.

after some minutes' distraction. In another verbal test, a short piece of prose (see bottom right of Figure 7.3) is read carefully again at a natural speed. For example:

> Audrey Brown is eighty years old. She is a widow and lives alone. One day when shopping in town she realized that she had lost her purse. She returned to the two shops she had visited but could not find her purse. She started to worry. Her purse contained her name and address, her house keys and $55. She asked the shop assistant what she should do. She was directed to the local police station. The duty officer listened to her story but said there was nothing he could do; no purse had been found. Mrs. Brown sat down and said she was too frightened to go home. People passed by and some asked the duty officer why she was waiting. After an hour or so, a police officer came up to her. He said that he was

about to go off duty and that he lived near her home. She accepted his offer to take her home. When he got there, her house was locked and she said he could break in. After looking around, he said he would come back with a new lock for her door. The next day, she wrote to the police commissioner to thank the police officer and to recommend him for promotion.

After a brief interval, the testee is asked to repeat as much of the story as possible. Test scores are based on the number of elements of the story correctly remembered. Tests that ask subjects to recall lists of words are quite different from tests of prose recall. It is very likely that these two types of recall task use mostly the same verbal processing pathways but with aging, prose recall holds up better than remembering a list of words. This difference is probably caused by list recall being more affected by age-impaired loss of processing capacity. Each word on the list must be recognized, a meaning found and stored. The prose story requires the same abilities but deficits can be made good by remembering the "gist of the story," assuming links between elements and recollection of the logical links between elements given in a sequence.

An individual tested in this way uses two distinct pathways to complete the task (Figure 7.3). Recall of a list of words depends heavily on processing speed and is affected by age. Prose recall relies more on working memory. When the story maintains a logical narrative that reaches a conclusion (the police officer was recommended for promotion), it is easier to piece together what the story was about. When the story has a "moral value" as in an animal fable, questions about the reasons for the story ending as it does can be posed. This involves concept formation and abstract reasoning and is referred to again in this chapter's summary of Piaget's stage of "formal operations."

Working memory (WM) is used to maintain sensory information during the performance of cognitive tasks that require information. It includes attentional and central executive functions as well as the storage of sensory data. Timothy Salthouse has argued that WM holds the key to explaining how memory begins to fail in old age.[6] When older adults lose their ability to maintain a useful WM, they also impair their overall cognitive functioning.

WM is supported by a diffuse network of cortical regions that are shared across the entire cortex. Attentional and central executive functions are largely in the frontoparietal regions. Just how these cortical networks store, sustain, and retrieve WM is largely unknown. Current understanding of the neurobiological basis of WM suggests that the assumption of memory having a finite capacity may be unsafe and that other aspects of memory such as freedom from errors are more important.[7] If this is understanding is true, then the efficiency of memory might be better represented by the number of errors made and not just the number of items correctly recalled. The concept of individuals' possessing only a limited memory store that is depleted by brain aging at one time appeared useful in discussions of cognitive aging. It does not, however, help understand how memory is supported by multiple resources that may be divisible with the facility of separate resources to access additional independent brain networks. Many investigations of visual WM suggest that synchronization of neuronal activity provides the basis for WM. When neurons coordinate their activity during WM tasks, oscillations in amplitude of activity depend on the memory load of the task under investigation. The importance of this type of synchrony is that it is detectable across a diffuse network (i.e., there is long-range synchrony) supporting the idea that WM is distributed widely across several cortical regions and is supported by network "hubs" located in the cingulate and insula.

The basic picture of the effects of brain aging on mental performance suggests similarities in patterns of brain activity in young and old people. It shows that brain structures are maintained from youth into very old age. Although there are many similarities, however, some older adults have less activity in brain areas involved in memory processing and in the control of mental processing, including attention. In these circumstances, other brain areas are recruited (activated), possibly in response to the altered brain activity. These observations suggest that additional recruitment is often compensatory in old people and allows mental performance to be maintained. Other examples of additional recruitment, however, point to a diffused impact on critical brain structures of age-related brain disease.

III. SENSORY SYSTEMS

A distinction is made between the range of cognitive abilities described thus far in this chapter and the functions of the sensory systems, including vision, hearing, touch, smell, taste, and balance (shown as boxes on the left of Figure 7.2).

Their interrelationships with cognitive aging are an integral part of healthy and successful aging. It is worthwhile considering how sensory systems age.

VISION

Vision changes are one of the most noticeable changes in aging. Structural changes in the eye begin at about forty years of age and changes in the retina (the light sensitive membrane at the back of the eye) begin at fifty years of age. The first thing older adults notice is that they need more light when reading. This need for light is caused by changes inside the eye that impair the passage of light through the lens and onto the retina. As a consequence, for example, many old people try to avoid driving at night, which is more difficult because they become more sensitive to glare with age. Older drivers feel the effects of bright headlights from an oncoming car for some minutes after the encounter, and this alone may convince some to cease driving at all after dark.

These experiences are accounted for by changes in the lens of the eye, making it more difficult to accurately discriminate between colors in the green-blue-violet end of the visual spectrum. The lens also stiffens, making it more difficult to change focus quickly from near to far objects or even to see close objects clearly at any time. The lens normally is clear but can become cloudy with age. These areas inside the lens are called cataracts and make reading and car driving quite difficult. The retina lies at the back of the eye, covering about two-thirds of its surface. With age, the retina becomes less sensitive to light, and this is most marked in the macula. About 20 percent of people older than seventy-five years have some degree of macular degeneration. Other causes of retinal disease are not as common. About 2 percent of people suffer from diabetes, and this can cause macular degeneration as well as changes in the retinal blood vessels.

Visual problems are an important contributor to impaired driving safety for older adults. One visual problem that is difficult to remove when driving a car is the reduced "useful field of view" that affects many old people. This term covers the extent of the visual field that is available to an individual at a brief glance. Fortunately, problems with visual attention caused by a limited useful field of view can be improved through training.

Assessment of the risk of highway accidents has focused on driving strategies learned through experience by older drivers and includes measures of selective attention, divided attention, and reaction time. Attention is one of the characteristics of consciousness and is placed among our cognitive abilities. When older drivers are assessed in driving simulators for driving safety, the tasks that demand attention are those most likely to cause concern. It is misleading, however, to rely on the results of these tests and extrapolate directly to a decision about driving safety. Older drivers are just as conscientiousness as they were when younger, possibly more so. These drivers always will make efforts to change their driving style in keeping with their limitations. It is much more informative in the clinical assessment of driving risk to identify obvious causes that might compromise driving safety. These causes include the use of prescription medications, detectable impairments of vision, and the presence of cognitive impairment. The assessment of driving safety is a critical component of clinical assessment when a diagnosis of dementia is suspected.

HEARING

Hearing loss is a well-known effect of aging. The eventual effects include age-related failure to detect high-pitched tones. This hearing loss has various causes, including the loss of healthy cells in the receptor area of the inner ear, stiffening of the mechanical systems that detect and translate sound into nervous impulses, and loss of myelin in the auditory pathway from the ear to the primary auditory cortex.

Hearing loss can disrupt social behavior, making communication difficult, and, when accompanied by increased sensitivity to rapid changes in volume, can cause irritability as well. From the listener's point of view, deficits in hearing can lead to misinterpretation of what was said. In some sensitive people, especially those who are easily slighted, hearing loss can trigger paranoid or persecutory ideas. Depressive reactions are probably the more frequent emotional responses to hearing loss and, in turn, this can disrupt close relationships with family and friends.

TOUCH, WALKING, AND BALANCE

Sensations of touch and temperature are well preserved in late adulthood, although some old people complain more often of feeling cold. Pain is felt just as easily among the old as the young, but when someone is living with dementia, they may not be able to describe painful sensations as clearly as they would like. When communication is impaired, all the caregiver may know about painful suffering is that the care recipient has become anxious and perturbed.

Fear of falling often is accompanied by feelings of giddiness on standing. Although age-related changes in the inner ear and in the brain stem (the cerebellum is relatively spared from aging) seem to be likely culprits, this is not the whole story. To maintain balance, an individual needs at least two of three types of information: vision, inner-ear data, and bodily feelings (somesthetic) that indicate positions of body parts in space. When any one of these is lost, even slight impairment of another system can have catastrophic consequences, leading to falls, broken bones, and fatalities. Balance and walking can be improved by regular light exercises, such as provided in group sessions to music that promote ease of movement and retention of muscular strength among old people.

Elizabeth Maylor at the University of Warwick (United Kingdom) investigated postural stability in volunteers without dementia.[8] She used a force platform to measure personal adjustments to balance while her subjects performed five cognitive tasks. She found that even after correcting for the large effects of age and perceptual speed, the mental effort of performing tasks of varying difficulty affected balance. She understood this effect in terms of competition between the need to maintain balance and that needed for effortful cognition. She reasoned that both tasks required the use of the visuospatial sketchpad of working memory (Figure 7.2).

TASTE AND SMELL

The ability to detect different types of taste declines slightly in old age though the exact extent is not well understood. Pleasure from food and alcohol certainly declines with age. This may be due to the fact that alcoholic drinks increase the fear of falling or even urinary urgency or, as seems more likely, the opportunities to enjoy good company with food and drink are restricted for many older adults. The sense of smell declines from around sixty years of age. Some research has suggested that decreased ability to detect differences between odors may be linked to early preclinical Alzheimer's disease (AD). Tissue samples taken from the olfactory nerves (accessible through the nose) have detected some of the abnormal protein clumps (neurofibrillary tangles) that are typical of AD. Thus far, however, this has not provided a reliable aid to diagnose AD. Of greater concern is the possibility that impaired odor detection may jeopardize the safety of some old people exposed to potential hazards of fumes and gas leaks.

IV. WHAT INFLUENCES RATES OF COGNITIVE AGING?

It is well established that general health is very important. Other than this fairly obvious statement, there is too little to be sure about. This situation arises because it is so difficult to design and conduct the ideal study. The requirements are daunting: Participants must be human, they must be randomly assigned to different levels of exposure to the proposed moderating influence on cognitive aging, the study should start before aging begins and continue until aging has reached its greatest effect, and repeated measures of multiple cognitive functions should take account of practice effects, incident disease, and the impact of major life events over the course of the study.

It is safe to state that the major moderators of cognitive aging await discovery. There are associations found in cross-sectional studies between cognitive aging and general health, duration of formal education, sensory abilities, and the amounts of physical or mental exercise. These cannot be accepted as the primary causes of differences in rates of cognitive aging because the research designs do not allow for the distinction between cause and effect. For example, general health is linked closely to socioeconomic status, which in turn, is linked through multiple pathways to occupational complexity, original childhood level of general mental ability, and health literacy—any of which alone or in combination could influence rates of cognitive aging. Likewise, poor visual acuity might be more frequent with age either because of the effects of age or because relative poverty precludes regular eye exams and the purchase of prescription glasses. Current longitudinal studies of cognitive aging may resolve some of these issues, but so far, what we do know must be treated cautiously. Reliable statistical methods to examine longitudinal cognitive data are not yet fully developed and accepted. Until these are applied satisfactorily, no firm conclusions about the sources of individual differences in rates of cognitive aging can be reached.

David Bennett and colleagues at the Rush Alzheimer's Disease Center Memory and Aging Project and Religious Orders Study (Chicago) took an informative approach to this question.[9] They examined 856 deceased participants on whom they had collected an average of 7.5 cognitive exams over a period of up to seventeen years before death. Around 95 percent remained in the study, and 80 percent had agreed to autopsy. Using the pathological criteria for dementia summarized in Chapter 6, they calculated that, together, these could account for 41 percent of the variation in cognitive decline before death. In the context of moderating influences on rates of cognitive decline, 41 percent is a high figure, given that estimates around 1 to 2 percent are more usual. Nevertheless, the conclusion is inescapable: A large proportion—around 60 percent—of the variation in cognitive aging is unexplained. When we consider that the search for interventions (often drug based) to prevent progressive age-related cognitive decline and subsequent dementia is focused on the pathological hallmarks of AD, dementia with Lewy bodies (DLB), and cerebrovascular disease (CVD), the urgent question becomes "What is the neurobiological basis of the large proportion of cognitive aging not explained by AD, DLB, and CVD?"

Moderators of differences in rates of cognitive aging and progress to dementia are of huge interest to dementia prevention. The work of Bennett and his colleagues has the potential to identify hitherto unknown sources of differences between individuals in rates of cognitive aging. Understanding these sources may suggest novel approaches to slowing or preventing age-related cognitive decline. "Risk factor" studies are no better than association studies that inform the choice of interventions for the prevention or delay of cognitive aging or progress to dementia. Currently, none are conclusive and many seem plausible, but all lack convincing evidence that these interventions might slow cognitive aging and possess the potential to delay or even prevent dementia onset.

The following sections describe how specific cognitive functions change with age. These changes are largely detrimental, although in the absence of pathology, they do not significantly impair an individual's capacity for independent living.

V. ATTENTION

Day dreaming again? You really need to keep a handle on what you are doing. You need to pay attention either to the page in front of you or to what you stored in your short-term memory a few pages back. Or maybe you need that information from an article you read in a magazine some weeks ago? Either way, you need to be able to call up memories from long-term or short-term stores or from the present in the form of what you are reading now or while listening to an "inner voice." All this contributes to your conscious experience of reading this page. Maybe, some emotions also are flitting around your consciousness. In step with all of this going on at the same time, you seem to be able to work out what the book is about. You can work out the reasons why the brain works the way it does and then conclude that you understand it. You even may feel confident enough to explain to someone else what this book is about. How do you do this?

Attention decides which messages make it through this mixed-up world. Some rules apply, or a message will not get through. First, the message needs to last long enough to be picked up against the background noise. Second, attention needs to be directed. This can occur through two routes. One occurs under voluntary control (via the thalamo-cortical neurons) and is called "top-down" attention. The second occurs when the signals coming in are so strong or emotionally laden that they cannot be ignored. This is called "bottom-up" attention. Importantly, attention is not the same as consciousness. Rather, attention directs information into consciousness either by calling it up ("top-down") or by allowing it in because it is so strong ("bottom-up"). Attention is easiest thought of as a "gatekeeper" into consciousness. This section focuses on the ways aging affects three types of attention: selective attention, divided attention, and sustained attention.

There are many metaphors for attention and consciousness, some of which lead to engaging illustrations for students. A simplified "theater metaphor" is shown in Figure 7.4.[10] Imagine a darkened lecture theater with a central stairway dividing a class of several hundred students. The lecturer becomes the narrator who maintains a continuous logical commentary on events to follow. A spotlight picks out a single player speaking out loud. In the shadows, a dimly lit chorus of voices murmurs incomprehensibly.

The narrator explains that the "bright spot" represents the power of attention and that they, the audience, are the conscious brain. "What's missing," says the narrator, "is a theater director . . . let's put one in the audience so you decide how the spotlight will move." The single player recites from memory:

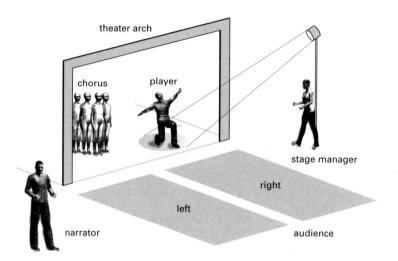

FIGURE 7.4 This figure shows how attention is directed by the central executive (the "stage manager" shown holding the spotlight) on a player selected by the stage manager to be shown in consciousness to the audience, including the brain regions concerned with self-awareness and discrimination among different types of experience arising from internal (body and mind) and external worlds (the environment). Logical continuity between successive experiences is maintained by a narrator (or "inner voice") who explains experiences to the self. A chorus is shown off–center stage, not at this moment, in the spotlight. The chorus represents the emotions associated with immediate or recent experiences or as recalled from emotional memory of experiences relevant to the moment.

Mind in its purest play is like some bat
That beats about in caverns all alone.
Contriving by a kind of senseless wit
Not to conclude against a wall of stone.
It has no need to falter or explore;
Darkly it knows what obstacles are there,
And so may weave and flitter, dip and soar
In perfect courses through the blackest air.
And has this simile a like perfection?
The mind is like a bat. Precisely, save
That in the very happiest intellection
A graceful error may correct the cave.[11]

At points during the recital, the stage director switches the spotlight from player to narrator and to the chorus. The narrator is speaking, but the audience see only moving lips. The chorus is louder and speaks of the sense the player is making of the poem. The audience can understand what is happening only with difficulty. On ending the recital, the narrator becomes audible once more. He explains what the poem is about but is contradicted by the chorus who complain that is not how they feel. The narrator explains that the chorus represent the emotions felt by the player and advises the chorus cannot be trusted.

In the student discussion that follows, some pick up on the problems that arise from the metaphor and some reflect on the many layers of understanding required to appreciate the metaphor, but all identify the central role attention plays in consciousness.

SELECTIVE ATTENTION

Selective attention is the ability to organize relevant inputs from sensory systems and from memory stores and, at the same time, to ignore what is irrelevant to the task in hand. The two types of selective attention are "top-down" or voluntary attention and a "bottom-up" reflexive attention in which one or more sensory systems demands and captures attention. Selective auditory attention is sometimes identified as the "cocktail party effect," when in a noisy setting, it is possible to focus on a single conversation. Notably, this is something that many old people find difficult and sometimes can be so embarrassing that they shun such company.

One of the problems that emerges when explaining the cocktail party effect is the ability to decide when the filtering out of unwanted noise has occurred. Is it early or late? In the early model, the information is not fully analyzed before being ignored. Investigating the early model, David Broadbent from Cambridge (United Kingdom) proposed[12] that some sort of gating mechanism operated early in the selection of incoming stimuli. Models of late selection propose that all incoming information is processed to the same extent that meanings (semantics) are attributed before selection. The likely correct position is that much information that is ignored by consciousness remains available for analysis if needed but only in some sort of limited temporary store (possibly equivalent to the rectangles adjacent to the sensory gates in Figure 7.2).

These ideas about selective attention can be folded into an "information processing theory" of attention. According to this theory, with aging, impairments of selective attention occur when the amount of information to be sorted is large and complex. Older adults find it difficult to do this quickly and accurately because, for some unknown reason, their information processing capacity is too small to handle the large amount of data. With aging, the older adult has difficulty deciding what is relevant or not to the task in hand. This problem may be attributable to the failure to inhibit irrelevant input presumably because of impairment in central executive control of attention (Figure 7.2).

DIVIDED ATTENTION

Divided attention is the ability to divide attention between two or more sources of information. Divided attention concerns the extent to which one set of information competes for entry into consciousness with another set of information. Undertakings that demand divided attention involve the ability to perform multiple tasks simultaneously. With aging, studies in attentional capacity measure the amount of information that can be attended to at the same time. Older adults typically report that tests of divided attention are more difficult than reported by younger adults. In part, this difficulty arises because older adults expect to perform less well and say so when self-evaluating. Experimental research is unconvincing, and it is possible to be confident only that older adults do less well than younger adults on some but not all tests of divided attention. Differences really are obvious only when tasks become very difficult, with multiple sensory inputs demanding correct responses from a complex array of possibilities.

Interestingly, even the most difficult tasks can be mastered by older adults if they are allowed sufficient time to practice, sometimes with supervision or training. This is probably relevant to the pleasure that older people derive from ambiguity in literature, art, and music. When cognitive development continues into old age, positive benefits in abstract reasoning sometimes involve the ability to tolerate uncertainty and to allow contradictory ideas to be developed in novel and more interesting ways. These positive aspects of cognitive aging are discussed later in this chapter as aspects of wisdom.

SUSTAINED ATTENTION

Known in common speech as "vigilance," sustained attention is the ability to maintain attention over extended periods of time. This type of ability is used in occupations in which the detection of rare and infrequent stimuli (e.g., radar blips by an air traffic controller) can be critical. Vigilance has two elements: accuracy of detection over time and changing error rate over time. Factors that contribute to age-related decreases in accuracy and increased error rate include age-related concerns of effects of medication, general fitness, and bladder control. Quite simply, physical and medical issues affect the concentration of older adults to a greater extent than younger individuals because the issues are more frequent with age. When vigilance tasks require responses to infrequent stimuli, older adults perform as well as younger adults. This response is particularly well marked when the observer knows where the stimulus will be detected and is not required to search for it.

VI. INFORMATION PROCESSING

LANGUAGE PROCESSING

Is it a word? Is there a meaning? The first step in language processing is for the auditory system to pick up the signal. Some sounds are more difficult to hear as we get older, especially after eighty years of age. This is most marked when the background is noisy. Much depends on context. When speech is rapid and the context is appropriate, most old people can follow even up to speeds of around 300 words per minute.

Knowledge about the subject being discussed eases speech processing and comprehension among old people. It appears that when the spoken word refers to things that are familiar to the older person, they have an advantage over younger adults. When the words or the topic is unfamiliar, the old person is at a disadvantage that even experience cannot overcome.

VII. GENERAL MENTAL ABILITY

General mental ability (intelligence) can be separated into two broad but overlapping categories proposed by Raymond Cattell from the University of Illinois (United States) and John Horn from the University of Denver (United States), which they termed fluid (Gf) and crystallized (Gc) intelligence. Gf refers to the capacity to solve problems of a type an individual has never before encountered and for which previous experience is of little use. Gc reflects a body of knowledge that an individual has retained and organized over time. Cattell and Horn thought that Gf contributed importantly to Gc, so that people with the same Gf but different sociocultural experiences could differ markedly in Gc. This simple model that distinguishes between Gf and Gc, however, is insufficient to handle other major components of overall mental ability. Information processing speed (Gs), language perception (Gl), and visual perception (Gv) are needed to complete a comprehensive model.

The term "intelligence" is not universally accepted but remains in wide use largely because it is such a useful measure of a group of attributes that continue to have high predictive value about the future behavior of an individual.

The efficiency of connections between cortical regions is degraded by the effects of age on white matter tracts (illustrated in Figure 7.5). Disruption of pathways across a neural network, as shown in Figure 7.5b in cartoon form, causes information to find an alternative but slower route. Inputs to the network can be inhibitory or excitatory, and the network functions as a "decision tree" that is open to modification. If an inhibitory pathway is followed across the network, it is possible to identify connections that, if broken, would prompt the network to reestablish an alternative inhibitory or excitatory route, much like plotting a path through a maze. The network is performing at a much more complex level than this, with each neuron making an average of 10,000 connections with other neurons. When disrupted by degenerative disease, the network undergoes remodeling to allow compensatory pathways to be established.

AGING AND INTELLIGENCE:
THE SEATTLE LONGITUDINAL STUDIES

K. Warner Schaie[13] and his wife Sherrey Willis have spent their working lives aiming to understand the sources of individual differences in intellectual abilities that arise over the life course. In 1956, Schaie, assisted by his wife, began his doctoral dissertation at the University of Washington. This became the Seattle Longitudinal Study (SLS), which now provides the basis for much that is now known about aging and intelligence. Schaie realized from the outset that studies that simply compared people of different ages—young vs. old—did not give the same result as those that compared the same individuals at different ages. His

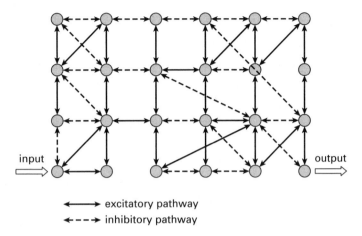

excitatory pathway

- - - → inhibitory pathway

FIGURE 7.5A A diagram showing a neural network made of neurons (circles). This is highly oversimplified and does not show the high density of connections made between neurons. Some connections are inhibitory and some are excitatory.

Adapted from B. J. Baars & N. M. Gage, *Fundamentals of Cognitive Neuroscience* (Academic Press, 2013), 80.

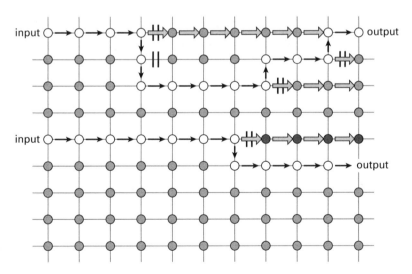

FIGURE 7.5B In one pathway, successful transmission of signals across the network is shown by a succession of open circles. When a link is broken, the signal takes a detour and thus takes longer to complete the route that increases from eleven neurons to twelve. On another pathway there are four breaks so that detours increase the route from eleven neurons to sixteen and so substantially increases the time to cross the network.

work continues—currently for more than fifty years—eventually including six cycles of data collection from the same cohort to which new recruits were added as each new cycle began at seven-year intervals. More than 5,000 people participated and provided data about a great range of life course variables. Like many other follow-up studies, SLS was faced with the fact that people who remain in the study tend on average to perform better on mental tests than those who drop out. His team used six core measures of mental ability: inductive reasoning, spatial orientation, perceptual speed, numeric ability, verbal ability, and verbal memory. In addition, measures of socioeconomic status and health were added with other more sensitive cognitive tests as the study progressed to maturity.

SLS showed that no simple pattern of age-related increases followed by a decline exists for any single mental ability. Confounding observations are the effects of gender, demonstrating that women tended to do better on tests of verbal meaning and inductive reasoning but that men do better on spatial ability and numerical skills. Complications were added when Schaie examined the effects of birth cohort within the sample. The social, educational, and work experiences of the postwar baby boomers differed from earlier generations and possibly accounted for differences in overall ability and patterns of change with age.

SLS data have been discussed widely, and since their first publication, they have informed U.S. public concerns about age discrimination, mandatory retirement, and population projections of future numbers of dependent elderly. With the single exception of word fluency, which begins to decline from about fifty-three years of age, decrements in other abilities do not reliably occur until about sixty-seven years of age. Even at this age, decrements remain modest until about age eighty, by which time most abilities have begun to decline. Verbal ability is a notable exception, so much so that comparisons among individuals between the age of twenty-five and eighty-eight reveal no significant differences.

One of the most interesting findings of SLS was the discovery of major differences among birth cohorts in patterns of age-related change in specific abilities. These differences are so large as to conclude that earlier studies that compared young versus old had overestimated cognitive decline identified before sixty years of age. This finding could be attributed to increased formal education in later recruited participants and differences in exposure to more complex lifestyles.

SLS has provided important data on factors that influence patterns of age-related cognitive change. These factors are as follows:

1. Lifestyles associated with good health, especially a reduced risk of heart disease and stroke.
2. Relatively better access to material resources through higher socioeconomic status and longer formal education.
3. Better mental flexibility and adaptability by midlife.
4. An active and socially engaged lifestyle with mentally effortful leisure pursuits, such as reading, travel, and interest in cultural activities (e.g., drama, music).
5. Being married to a spouse with high cognitive status. This is important when a person of lower functioning acquires a spouse with greater mental ability.
6. Maintenance of high response speeds in perceptual-motor tasks; this tends to generalize to other cognitive abilities.
7. Satisfaction with lifetime accomplishments by middle age.

When these predictors are considered together, poor maintenance of cognitive functions typifies a male with less education who is dissatisfied with his lifetime achievements. The SLS has provided life course data from youth to old age and has offered explanations of the sources of differences in patterns of change. Schaie's studies have raised important questions about patterns of age-related decline. When an aging individual begins to fail on one type of test, does that mean that failure on other tests is certain to follow? In other words, "does it all go together when it goes?" What actually seems to happen is that in the absence of dementia, very few people have declined on all tests even by the age of ninety.

This finding has led some researchers to focus on the relationships between types of ability and how these change with age. Schaie showed that the structure of interrelationships remained largely unchanged over the life course, but this could be explained by all of the abilities measured in SLS relying on the same internal structures. These findings can be related to the approach to cognitive aging developed by Horn and Cattell, who proposed that within a range of primary mental abilities, the two that are most important in mental development and aging are crystallized (Gc) and fluid intelligence (Gf).

Gf and Gc are measured in different ways. A test of Gf would involve reasoning, so previous knowledge would not improve performance. For example, what comes next in the series 25, 24, 22, 19, 15, . . . ? The answer is 10 because the series is progressing by steps of "– n," where n is the number of each step, and the missing entry is the fifth step, so the decrease is –5 from the last number. A test of Gc depends on previous knowledge. For example, what comes next in the series R, O, Y, G, B, I, . . . ? The answer is V for violet. The series is the initial of the name of each color of the rainbow in the order that the colors appear. It cannot be solved by reasoning and relies on recognizing the order of the colors from their initials. Obviously, if you were not interested in knowing about the visual world, you may not have noted this pattern. An "interest in the world" is associated with Gf, so measures of Gc partly reflect Gf.

When the results of cross-sectional studies of cognitive functions at various ages are systematically combined (meta-analysis), as much as 79 percent of variability in subtypes of cognitive functions can be explained by processing speed. It remains uncertain, however, if processing speed is a stronger influence than working memory. Some of this uncertainty may be caused by the use of cross-sectional rather than longitudinal data. Longer observational periods in some longitudinal studies allow more complex models of cognitive aging to be devised and tested. For example, Martin Sliwinski and Herman Buschke[14,15] from the Albert Einstein College of Medicine, New York, using longitudinal data found that processing speed could account for only between 6 and 29 percent of cognitive aging. Other approaches have shown that cognitive aging can be separated into two trajectories: one is a gradual slow decline from late midlife and the other is an accelerating rate of decline from about age seventy onward.

When two traits—gradual and accelerating decline—share common causes, it is possible to examine environmental and genetic influences on cognitive aging. Nancy Pedersen from the Karolinska Institute (Sweden) and her colleagues at the University of Southern California (United States) have found evidence for the role of genetics in determining the rate of acceleration in cognitive aging through its effects on processing speed.[16]

Cattell and Horn showed that Gf and Gc follow quite different courses in development and aging (Figure 7.6). Note that Gf declines throughout adult life, although the reasons for this are not fully understood. It may be that the effects of aging, injury, cumulative exposure to toxins, and lack of practice are all relevant. Although we all use our language and knowledge every day, we rarely are asked to complete serial completion tasks once formal education is behind us. Improvements in Gc probably are related to the accumulation of knowledge across the life course, almost on a daily basis. The differences between Gf and Gc at different ages show up clearly when individuals of different ages

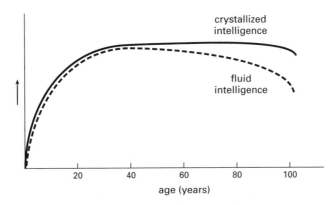

FIGURE 7.6 The associations between age and crystallized and fluid intelligence. Note that crystallized intelligence (Gc) is maintained after reaching its maximum in middle adulthood and, in some people, might continue to improve until the ninth decade of life. Fluid intelligence begins to decline around the age of forty-three and is significantly lower by about the age of sixty-seven.

are compared at a perceptuo-motor learning task. For example, a thirteen-year-old girl can increase her competence quite quickly in a foreign language. Her seventy-three-year-old grandmother will have more difficulty, even though her verbal ability is significantly better than her granddaughter's ability. Higher Gf at thirteen will make the task of learning that much easier.

Information processing speed is an important explanation of differences between Gf and Gc. Using the schematic neural network (Figures 7.5a and 7.5b), it is possible to devise a scenario in which the time taken to complete a processing task is slowed by breaks in connections along the optimum route. We should think about the role of the central executive that controls and keeps information available in working memory while the task is processed. With increasing age, it also becomes more difficult to exclude competing thoughts that might distract from the task in hand.

Figure 7.7 demonstrates individual trajectories of cognitive performance found in the Aberdeen 1921 and 1936 birth cohorts data sets. Compare these plots of real data with the schematized aging curves of crystallized and fluid intelligence given in Figure 7.6. At this point, it is helpful to consider why individuals should vary as much as they do and why it is so difficult to spot among the data the individuals who might comprise subgroups. It is reasonable in an unselected group of volunteers to consider the possibility that a sizable subgroup is represented therein who are in an early "preclinical phase" of AD. Others will

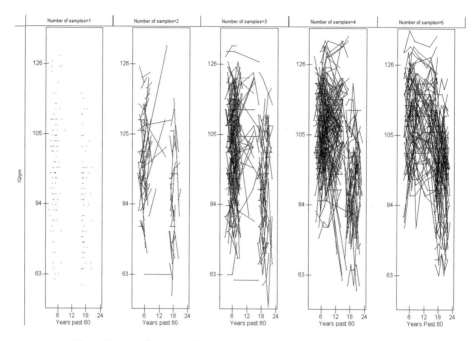

FIGURE 7.7 Plots of longitudinal data from 791 participants in the Aberdeen 1921 and 1936 Birth Cohorts Study. Data are cognitive scores on between one and five occasions from the ages of sixty-three to eighty-two. The lines connect data from one person over each occasion. These scores show the slight negative effects of brain aging and the stronger positive effects of repeated practice.

provide cognitive test data that are on some occasions reduced by intercurrent illness or the effects of prescribed medication.

COGNITIVE AGING AND OCCUPATIONAL COMPLEXITY

Social engagement is known to promote retention of general mental ability with aging. At a basic level, different kinds of work require different skills and competencies. Level of mental effort and engagement also varies substantially among jobs. With age, many older adults make an effort to retain mental ability. For many, retirement from work closes a chapter centered on work, time with colleagues, and opportunities for active pastimes. Not surprisingly, the same people after retirement share many misgivings that a life of leisure with few intellectual demands could accelerate cognitive decline. It is quite tricky to analyze what is going on: Mental and physical work changes, health can fail, and serious unpleasant life events tend to be more frequent with age.

The first studies of cognitive aging took place long before Schaie began his studies in Seattle. In Paris in the early twentieth century, school teachers were compared with the fisher folk of Brittany.[17] Older retired teachers were significantly better on mental speed tests than fishermen of the same age, but on repeat testing, both showed the same rate of decline. Teachers not only were better than fishermen on memory tests but also declined with age much more slowly. By contrast, the fishermen (few of whom were fully retired) performed much better on tests of muscular strength, speed, and endurance and declined more slowly on these as well.

The U.S. Dictionary of Occupational Titles (http://www.occupationalinfo .org) lists the complexity of each job category in three areas of work: (1) with people, (2) with data, and (3) with things. Obviously, educational attainments will regulate job entry and in-service training when employed in more complex jobs, and selection for complex work certainly implies higher levels of mental ability were present before aging would be detectable on test performance. When intelligence and education in youth are taken into account, better retention of cognitive functions with age is associated with complexity of work with people and data but not with practical tasks. At one level, this would seem to reflect the contributions of acquired confidence in established levels of cognitive ability.

PIAGET, MENTAL DEVELOPMENT, AND AGING

Discussion of cognitive aging most often focuses on what can be measured. The results of psychological tests of mental performance tend to be regarded as the gold standard of cognitive changes, but there are other approaches. Jean Piaget (1896–1980) produced what remains the most influential account of how children see the world, and this account is recognized as relevant to cognitive aging. Students of psychology who prefer approaches that attempt to synthesize social-emotional-cognitive theories of cognitive development often think of Piaget's work as being rather abstract and distant. One student remarked that he preferred a "touchy feely approach" to what seemed a "cold and aloof approach" by Piaget.

Jean Piaget transformed how we think about the intellectual development of children. Because he published entirely in French and used methods that varied with those that prevailed in the United States and United Kingdom at the time, his work at first remained little known and hardly discussed. His early interest was in the growth of intelligence and initially he worked with Binet on the first intelligence tests. He soon realized that Binet's aim was simply to count correct or incorrect answers, but Piaget's interest was in how children arrived at an answer. After many years of careful observation, Piaget put together his theory of childhood intellectual development.

Piaget reasoned that development represented the interaction between each child and its environment. He presumed that every child was born with innate capacities that were organized in such a way as to be preset to respond to whatever was encountered at specific times during development and in quite specific ways. By concentrating on how cognitive structures were assimilated and organized at each stage of the life course, Piaget placed organization on center stage alongside adaptive capacity in intellectual development. He saw development progressing through four major stages as an essential element of his theory (Table 7.1). Differences between stages rested on the fact that the child understood the world quite differently on moving from one stage to the next. These transitions occur three times in childhood (i.e., between each of four stages at approximately ages two, seven, and eleven). The exact sequence cannot vary, and a child cannot progress from one stage to the next without having mastered the preceding stage. Not all children complete the final stage. Figure 7.8 summarizes Piaget's stages of cognitive development.

Because we are interested in cognitive aging as a feature of adult development, these early stages are not discussed in detail. Our interest is in the final stage of formal operations from eleven years old onward. As the capacity for logical reasoning develops, as the youngster becomes an adult, problem solving becomes easier. Each problem becomes a series of steps that if solved in a systematic way will provide an overall solution. This approach is effective even with problems never previously encountered. Underlying this approach is a generalizable logical structure, whereby thinking is "one step at a time"; problem solving is "goal directed" heading for a single solution; and, with experience, all possible solutions are limited to what may be practical or realistic. Formal operations can be applied to imaginary situations as might be faced, for example, in writing fiction. Examples of "one step at time approaches" abound in daily life. If your television breaks, you can follow a checklist: first check the power supply, then the television signal, the control panel, the default settings, and so on. Each step you take allows you to check and perhaps modify the television, always doing one thing at a time while holding all else constant.

Comparisons between young and old adults on tasks requiring formal operations have yielded mixed results. Some studies show that older adults

TABLE 7.1 Piaget's Four Stages of Cognitive Development During Childhood

STAGE	MAIN FEATURE
Sensorimotor (0–2 years)	Babies learn about their world through sensory and motor behavior. Cognitive structures are action-based, gradually becoming more complex and coordinated. Late in this stage, actions are internalized as the first representational symbols.
Preoperational (2–7 years)	Children now use words to understand their world. Play includes use of imagination, but the child is able to distinguish fantasy from reality. Most behavior is egocentric until the late phase of this stage when a "theory of mind" begins to form.
Concrete Operations (7–11 years)	Children become skilled in areas that include classification, reversibility, and conservation, showing the ability to mentally manipulate representational symbols. First evidence of logical thought but usually applied only to concrete not abstract thoughts.
Formal Operations (11 years on)	Abstract and logical reasoning appears, and when applied to problems, the individual can evaluate different solutions to problems. Mental life is almost entirely about ideas and not objects.

Note: When Piaget's theory is applied to cognitive aging, an important obstacle is met head on. Many adults never reach the stage of formal operations, and even for those who do, this may be quite patchy, applying only to areas in which they are well trained. For example, some highly skilled engineers can show great proficiency in solving machine tool–based problems but fail to perform as well at tasks of abstract reasoning. Asked about the difference, the same engineers may reply that they have "no interest in problems of no practical significance."

do not do as well; some show that formal operations performance is predicted accurately by Gf, which declines in late life; and some show no differences at all. Unfortunately, the majority of studies are cross-sectional in design. Studies in which the same individuals are retested at regular intervals (as in the SLS) do not explore Piaget's theory in older adults. Cognitive operations possibly decline with cognitive aging in reverse order of that followed when they were

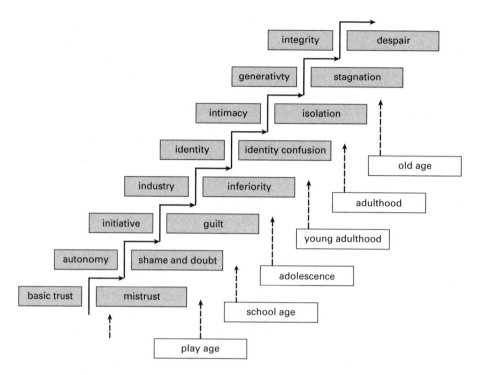

FIGURE 7.8 Schematic showing how Piaget visualized stages of cognitive development. Note his primary focus is on the early stages of cognitive development and how late adulthood is relatively neglected.

acquired. This is almost impossible to test satisfactorily using cross-sectional data, but some evidence suggests this may be the case.[18]

ROLE OF HEALTH STATUS IN COGNITIVE AGING

Growing older is linked to failing health. Warner Schaie and François Bourliére established that poor health—especially poor cardiovascular health—is associated with greater age-related cognitive decline. Comparable data have shown that chronic lung disease increases rates of cognitive aging, as does long-term exposure to some medications. Low socioeconomic status and low average childhood intelligence are firmly established risk factors for premature death, particularly those caused by such "avoidable" causes as failure to quit smoking, poor compliance with health advice, and excessive alcohol use.

These negative factors all contribute to cognitive decline, but are there other outcomes? Are there older adults who continue to make positive progress beyond Piaget's fourth stage of formal operations? Is there such a thing as wisdom? Over the past forty years, serious attempts were made to establish the concept of wisdom on solid foundations. The first obstacle was that conventional methods of psychological investigation failed to capture how adults considered to be "wise" differed from others who were not classified as wise.

WISDOM

Piaget provided a conceptual framework that helps explore positive aspects of intellectual growth in late adulthood. His work was concerned exclusively with children and young adults, but it would be unreasonable to anticipate that intellectual development did not progress further. Many commentators have pointed out that the stage of formal operations is insufficient to account for all adult intellectual activity. At first, it was necessary to show that many adults undergo cognitive growth that extends beyond Piaget's stage of formal operations. Evidence of this type of "post–formal thinking" is good not by just some but by all adults. Piaget suggested that mature adults came to recognize that their own views could be more appropriate (and possibly have greater validity) than those of authority figures. This seems like cynicism, but it is more than that. Adults can adopt a position and hold onto it with confidence, yet at the same time, they can acknowledge that others can hold dissimilar views that have comparable validity to their own.

Reasoning about social and lifestyle issues becomes more reflective as mature judgment is established. Skillful navigation through complex issues requires not just a capacity to absorb the views of others but also the encapsulation of personal areas of expertise. This is not the final stage of growth in cognitive abilities. Eventually, beyond reflection and reasoning, there are stages at which emotional life is fully integrated with logical judgment. As set out in Chapter 8, accurate assessments of social situations cannot rely only on logical reasoning. Everyday problem solving often has an emotional component. This contrasts starkly with the types of problem set in the psychology laboratory. When problems with a substantial emotional element are set, older adults tend to outperform younger adults and adolescents. Young people conform to Piaget's description of formal operations, whereas older adults tended to use "life experience" to solve this type of problem.

Ideas about wisdom are discussed in many cultures. Almost all associate wisdom with growing old and accept that not all old people are wise. In fact, if asked, most adults are as likely to identify as "wise" adults of any age. Paul Baltes and his colleagues at the Max Planck Institute in Berlin have argued that wisdom was composed of elements of an expert system concerned with the meaning and conduct of life.[19] Broken down into useful components, wisdom was found in understanding the human condition, making correct decisions about how to optimize health and well-being, and balancing demands and recognizing achievable goals from a pragmatic standpoint. Baltes proposed five criteria on which to judge wisdom. These are as follows:

1. Knowledge about facts that is extensive in scope and depth and is relevant to the human condition.
2. Knowledge about strategies of good decision making in a range of situations and accepts that there can be more than one "correct answer."
3. Ability to take account of the richness of situations encountered across the life course and to relate these to each other.
4. Capacity to evaluate individuals as living in circumstances that can be unique with personal goals that may be idiosyncratic.
5. Balance judgment between doubt and certainty that offers not one but several solutions that resolve difficulties in achieving lifetime goals.

The concept of wisdom is not universally accepted. One criticism is that the criteria are acknowledged only by individuals who might wish to be considered wise. These include some lifestyle therapists, policy decision makers, expert consultants, religious leaders, and many more. Others feel that wisdom as defined relies too much on language skills and that wisdom of a different sort can be found in the later work of creative artists. A final criticism is that, as defined by some, wisdom relies too much on the ability to develop high levels of expertise in an area related to the management of people. Instead, the pragmatists argue that the retention of older managers in large industrial concerns is based on hard economic benefits of those people who retain intimate knowledge of cycles of trading conditions and those who know how to anticipate and implement solutions unfamiliar to younger managers.

VIII. MEMORY

Decline in memory with age lies at the heart of dark fears about aging. The general public shares a common belief that memory inevitably will decline in late life. Once called "senility," this is more often now called dementia. Many dread a mental decline that accelerates with aging. The truth is that not all aspects of memory decline at the same rate—some may not decline at all—and it will not be until the tenth decade of life that memory will be impaired in almost everyone. At the outset, it is accepted that certain aspects of "fear of senility" reflect a predisposition to expect the worst of old age. Simply because an individual is old, it is not safe to assume that they will be impaired. For some, old age brings benefits that include not needing to be responsible for others. In some cultures, this type of reassignment of social position is a regarded as normal.

VIGNETTE 7.1 THE PROBLEM OF NORMAL COGNITIVE AGING AND ITS DISTINCTION FROM EARLY DEMENTIA

Mary is seventy years old and has lived alone since her husband died eight years ago. Recently, she moved out of her family home into an apartment in the same block as her eldest daughter, Mary-Lou. Mary seems to cope quite well but needs help with shopping. Her son-in-law manages her finances much as her husband had done before. Recently, Mary has felt that she needs more help, but she doesn't want to ask so she worries and does less. She got lost a couple of weeks ago when shopping by herself in a local mall that should have been familiar to her. A friend brought her home. Since then, she stopped driving and her grandson now helps her to get about. Mary-Lou keeps an eye on her medication and thinks her mother forgets her blood pressure tablets and should really go back to her physician for review. She hopes that he will spot her mother's slight mental deterioration and recommend to the family what steps they could take to prevent any worsening. Mary knows her memory for routine things is not as good as it was but stresses that she can remember stories from her childhood "like it was yesterday." At family parties, she seems her usual self. She's always asked to recite a poem from memory and she never fails to be word perfect on a poem she learned when she was just ten years old. Mary-Lou wants to know if her mother is normal for her age or should the family be worried?

To understand what is happening to Mary, we should look again at Figure 7.2. This showed the organization of her mental life into broad areas. Efficient and accurate goal-seeking behavior depends on the main components working together. There are elements that direct attention to selected aspects of the external world, how language is understood, what memories are evoked by what is perceived, and how some thoughts and feelings are retained longer than others. In Mary's case, we can see that some memories for distant events are retained, whereas other more recent events are lost. This suggests a major distinction between short-term and long-term memory. Long-term memory is made of two components: implicit and explicit memory. Those two terms show that some memories are explicit or conscious—as in knowledge of facts—and others are implicit—that is, these are implied, but we are mostly unaware of them. An example of implicit memory would be a motor memory like how to brush one's teeth. Among the facts we know, there are two broad types: (1) episodic memory, the knowledge of events to do with people, places, or times; and (2) semantic memory, the knowledge of facts about the world and language.

In certain types of brain injury or disease, symptoms of memory loss are typical and can be related to damage or disease that is confined to specific brain areas. These clinical observations show that specific brain structures are critical

It is not uncommon for it to be more difficult when aging to remember what has happened recently. Short-term memory is made up of processes that are critical in holding new incoming information in mind for use in decision making, problem solving, and memory retrieval needed to complete a current task. Short-term memory has a limited capacity so that only about seven items to be remembered can be kept in mind at any one moment. An item can be anything: If it is a word, this counts as one item, not the number of letters in the word.

Short-term memory capacity declines with age. If this indeed is happening, then it may be sufficient to explain many of the deceases in separate domains of mental performance seen with aging. This view hinges on the key tasks performed in short-term memory in understanding language, where immediate meanings are attributed to experiences and where transfers begin to long-term memory storage. Making matters more complex is the possibility that different subtypes of memory are affected differently by aging. For example, evidence suggests that spatial memory declines with age to a greater extent than does verbal memory. If this is true, then a more complex model of short-term and long-term memory will be needed to understand the effects of aging on mental ability.

Such a model would have multiple interlocking components in which changes in one could affect another.

VIGNETTE 7.2 WHEN ANXIETIES ABOUT INCIPIENT DEMENTIA BECOME TROUBLESOME

Frank had just turned seventy when his best friend died after living with dementia for almost five years. So Frank thought he should volunteer for our research study on dementia: "Harry would have wanted me to do something," he said, and soon he immersed himself in a study of memory training. Frank talked a lot about Harry and then later about his own mother and her mother, too. What became clear was that Frank had suspected he had a family history of dementia and that his grandmother had died in care seeming helpless but that nobody had spoken to him about it. He wondered if his own forgetfulness was "normal." He'd always thought seventy was the age at which his mental faculties would start to fail. Now, he kept a diary not only to list appointments but also to keep a note of his "mistakes." Looking through his lists, nothing seemed important, and everything he had noted seemed trivial. Frank was befriended by George, another volunteer who was eighty-five years of age. He was a much-decorated war veteran who had taken up electronics as a hobby on his sixtieth birthday. After the death of his wife twelve years ago, he filled their home with computers, short-wave radios, and a well-equipped work-shop. "You're daft," he'd told Frank, "Just get stuck in with something and forget about memory problems . . . there'll be no cure in our lifetime."

Frank remained preoccupied with his memory. He pestered the research team with questions about how memory worked. He read more and was convinced in the face of reassurances that his mind was indeed failing and that dementia was imminent. George would have none of it: "Come and see what you could do if you were prepared to make an effort," he exhorted, and he cajoled Frank into a visit to his home. Later, asked how they had got on, Frank said the visit had taken a most unexpected turn. "It was all true, like he said, he's got every gizmo going but there's more. He's got girlfriends, four he said." And George confirmed that he did have four lady friends, one for each weekday leaving Fridays free. "I asked him how he coped," said Frank, and he said that the first was a favor toward a friend of the family, who said she couldn't get out. "Then I suppose I just got into the habit of driving them around, I liked their company. . . . It helps that they are called Margaret . . . that way I never forget a name or get confused about which one I was with."

Frank was having problems with everyday life. On average, he recorded about a dozen "memory mistakes" each day but, of course, never noted all the things he had remembered correctly. For example, in the research clinic, Frank performed slightly better than average for a man his age on a test of spatial memory. He seized on his few mistakes as confirmation of his imminent dementia diagnosis. When a research student placed him in a local supermarket as part of a map-reading test, he was surprised by his success. He proved error-free completing a grocery item search and retrieval task, having been asked to memorize a shopping list and a location diagram of the supermarket.

Individuals like Frank often are encountered in memory clinics. Sometimes referred to as "the worried well," they emphasize that an opinion a person forms about their own memory reflects not only how they view their memory and their self-regard or esteem but also, in a more complex way, their understanding of how memory works and how it is affected by aging and dementia. Aging affects the way individuals monitor their memory. Older adults not only believe they have made more mistakes but also anticipate making more memory errors than younger people. George was used to success in life; he felt comfortable taking on new challenges and appeared to be mischievous at times (nobody believed about the girlfriends, but there was a grain of truth in his story). He took much pride in showing how he learned new skills well into his eighties. He had devised numerous ways to deal with occasional memory lapses and almost defiantly showed a vigorous and energetic face to the world. He was very good company.

Frank was a different man. Financially, he had been so successful in his oil business that he could retire and live on his savings at age fifty-five. Never married, he sold his business, giving his secretary a handsome bonus: "I thought if she was independent she wouldn't be after my money . . . she bought a flashy sports car and went to live abroad . . . I never heard from her again . . . I've been unlucky in love." For Frank, and many others like him, their self-rating of memory is an important influence on their scores on memory tests in the clinic. Of some relevance is the phenomenon in which successful people who are confident of their powers of memory will seek out mentally demanding occupations and lifestyles without expecting failure. Those who anticipate that their memory will "let them down" find fewer challenges and lead less mentally effortful lives. These differences are related to aspects of personality that are enduring and linked to individual styles of interacting with other people (these will be discussed later Chapter 8, Emotional Aging).

Other sources of differences merit consideration. These are the negative effects of poor health and drugs used to treat illness in old age. Together, assessment of the contributions to decline in mental performance of these factors forms a critical step in the evaluation of older adults who experience memory problems. The second source concerns the role of diseases on the nervous system. These cause mental symptoms and mental impairments that can be transient and amenable to improvement through treatment, as is the case with delirium, or irreversible and progressive, as with a family of disorders called the dementias. Among this family, common variants include AD and disorders caused by diseases of the blood vessels that supply brain tissues. Overall, the process of deciding which if any of these disorders is the cause of age-related mental impairment is a process of excluding potentially treatable causes before agreeing on the specific type of dementia.

IX. VULNERABLE PEOPLE

Not everyone enters old age with the same opportunities or resources to achieve success. Some lifetime experiences can hinder mature development of coping strategies, decrease the capacity to manage adversity, and predispose one to accelerated rates of cognitive aging. In broad terms, people with the greatest vulnerability to the effects of brain aging are divided into four groups. Most important are those who develop chronic disabling illnesses as children or young adults and, although adequately supported in early life, find with aging that the services they once relied on are no longer sufficient for their needs. As many individuals disabled in this way are now cared for largely in the community, their needs as older disabled adults must be accommodated within a larger framework of social care for old people. Second, older adults, who as children had special needs linked to learning disabilities, are probably the largest single group. Third are the large groups of individuals who have experienced severe recurrent enduring mental illnesses who would have remained in institutional care in the last century but currently are supported to varying degrees in community facilities. The fourth group includes those who have the more common neurological conditions, such as multiple sclerosis and epilepsy, certain types of personality disorders and social maladjustment, and to an unknown extent the victims of severe psychological trauma.

MILD COGNITIVE IMPAIRMENT

Notwithstanding these specific clinical subgroups, one major group includes those who are at greatest risk of progressive cognitive decline and for whom the annual risk of conversion to dementia is about one in eight. These individuals are known by numerous similar terms among which mild cognitive impairment (MCI) is perhaps the most widely known. It is notable, however, that not all clinicians accept MCI as a valid concept. They argue that this category includes so many different conditions (including some listed in the following paragraphs) that it is unhelpful and sometimes misleading to assume that those individuals meeting the criteria for MCI will share similar risks of dementia. Their underlying causes are so different and a sizable minority not only fail to progress but also remit fully to an unimpaired condition.[20] Their objections may be accommodated by the terms "preclinical Alzheimer's disease" or "MCI due to AD." In the absence of objective biomarkers for AD, these terms do not represent much of an improvement on using MCI alone.

Central to the concept of MCI is the self-awareness that mental function has declined sufficiently to be noticeable but not enough to impair daily life. Importantly, this decline is insufficient to make a diagnosis of dementia, but it should be severe enough to be recognized by someone who knows the individual sufficiently well to confirm that this decline represents a fall from a previous level of functioning. That person might be a physician experienced in the diagnosis of AD. Scores on cognitive tests should on at least one test fall well below what is expected in a normal population sample of similar age, sex, and level of education. Impaired memory test results are not a critical part of most definitions of MCI, but when it is present along with one other impaired cognitive test result, this gives strong support for a diagnosis of MCI.[21] Change can occur, therefore, in a range of cognitive functions, including memory, executive function, attention, language, and visuospatial skills. Impairments in the ability to learn and retain new information are seen most often in MCI that progresses to AD.

INTELLECTUAL DISABILITIES

There is a well-established association between Down syndrome (DS; the most frequently occurring genetic form of intellectual disability) and AD.[22] Almost all individuals with DS will develop AD if they live beyond the age of forty.[23] The mothers of many DS patients often were age thirty-five years or older when their affected child was born, so their ability forty years later to plan the care of their children cannot be assumed when those children prematurely enter the period of high AD risk. Planned care is vital for these children, and some communities acknowledge the inevitability of AD by arranging respite care in specialized facilities in anticipation of entry into palliative care within a few years. Otherwise, facilities are extraordinarily patchy and older DS individuals may be taken into residential care designed for older adults. This is especially unfortunate as the issue of AD in DS is almost entirely foreseeable.

It is difficult to generalize about cognitive aging in those born with intellectual disabilities. Large-scale autopsy surveys do not show an increased incidence of AD in this group; nevertheless, many gradually and prematurely lose many of their hard-earned skills in self-care and safety and need specialized help. This is sometimes difficult to provide when their learning disability is compounded by physical handicaps or poorly controlled epilepsy.

EPILEPSY

Older patients with epilepsy present an important medical problem. Once epilepsy is established, many patients make satisfactory lifelong adjustments to the risk of recurrences and are aware of the cognitive side effects of antiepileptic drugs (AEDs). This opportunity is not available to the substantial number of older adults who suffer late-onset epilepsy, which often is attributable to stroke or brain tumor, although frequently (up to 50 percent) results from unknown cause. Most patients with epilepsy do not report greater cognitive problems with aging than other older adults. Those who do report problems can describe difficulties with attention (sometimes called "stickiness") when they feel slowed when shifting attention from one topic to another. Memory problems are common in this minority group as well as some slowing in overall mental speed. The major concern about these emerging problems is the possibility that they are caused by unrecognized partial seizure activity or adverse effects of the AEDs. The possibility that chronic poorly controlled epilepsy causes accelerated cognitive aging is supported by many case reports (that seem largely anecdotal) but not by systematic follow-up surveys.

AEDs vary in their capacity to cause adverse cognitive effects. Benzodiazapines, phenobarbital, and phenytoin are more often implicated than other AEDs.

SCHIZOPHRENIA

About 1 percent of the adult population will receive a lifetime diagnosis of schizophrenia. These individuals experience a wide variety of symptoms that overlap with the autistic spectrum in childhood, so it is not unusual to hear about a proposed schizophrenia–autistic spectrum. Within this group of disorders, a significant proportion (about one in three) never fully recovers from schizophrenia and develops a chronic defect state with substantial cognitive impairment. As older adults living with schizophrenia, their defects become slowly progressive, and they suffer a lifelong disabling illness with multiple care needs. These poor outcomes were linked to schizophrenia in antiquity, but it was not until the nineteenth century that they were reliably distinguished from late-onset dementia (senility). Currently, many schizophrenic patients with features of a defect state are not in long-term residential care. Those who live in the community experience not only the stigma of severe mental illness, enduring effects or earlier long-term institutionalization, and persistent side-effects of antischizophrenic medication but also brain changes linked to biological processes that underlie schizophrenia. Their cognitive deficits include problems in memory and orientation often confounded by the coexistence of chronic hallucinations, delusions, and impaired reality testing.[24]

MULTIPLE SCLEROSIS

Multiple sclerosis (MS) is the most common neurological disease diagnosed between ages twenty and fifty years in the United States. The disorder is rarely fatal and, especially after the discovery of new treatments, many patients have a normal life expectancy. The number of older adults with MS is expected to grow in step with overall changes in the age structure of populations.

Importantly, there are distinct subtypes of MS. In early adulthood, there are multiple relapses and remissions. In later life, these are much fewer and most features appear attributable to permanent damage to neural networks caused by MS pathology. Despite this picture of steadier progression with long-term apparent stability of the clinical features, old people with MS present a difficult challenge for neurologists. These difficulties arise because patients and their families tend to accept that whatever happens to the MS sufferer are explained by the disease process. Strokes may be missed and the contributions of bladder infections or subtle changes in mobility may be investigated and treated insufficiently. As drug therapy in MS is developed further, higher expectations of improved quality of life for old people living with MS will become commonplace. It will be important to monitor the cognitive status of MS sufferers not only to detect possible adverse effects of novel drugs but also to identify therapies that have the potential to prevent cognitive deficits in MS. As a group, these individuals remain difficult to categorize. The underlying distribution of brain changes in MS tends to be a diffuse mixture of widely distributed abnormalities accompanied by signs of nervous tissue repair and restoration.

SEVERE PSYCHOLOGICAL TRAUMA

Most experience in the health care of victims of severe psychological trauma has been obtained in the care of the casualties of warfare or younger adults after brain or spinal cord injury. Thus far, with the exception of war veterans, the effects of aging on cognitive functions have not been reported widely. The major exceptions include several small-scale follow-up studies of prisoners of war (PoWs) interned in European and Pacific theatres during World War II (1939–1945). At first just anecdotal, these surveys show greater degrees of cognitive impairment in those who survived imprisonment by the Japanese than the relatively larger numbers held by the Germans. The acute effects of malnutrition and tropical infection in returning PoWs were well recognized on cessation of hostilities and were fully documented at the time. PoWs released from German camps were less undernourished with fewer acute medical problems. PoWs released from Japanese camps were much weaker (their mortality was many times greater than those held by the Germans), with higher rates of dementia that cannot be attributed to more frequent head injuries.

TRAUMATIC BRAIN INJURY

Traumatic brain injury (TBI) is associated with an increased risk of acute and enduring severe cognitive impairment. Less well recognized is an increased risk, often after an interval of many years without noticeable cognitive loss, of late-life cognitive decline. Susan Corkin drew attention to the link between TBI and late-onset cognitive decline in World War II veterans.[25] By comparing those veterans with and without TBI, she showed that accelerated decline in age-related cognitive functions was linked to TBI. Comparable follow-up studies of men who had sustained brain injury more than thirty years earlier in physical contact sport found similar patterns of cognitive loss associated with aging.[26] The acute effects of TBI on white matter tracts involve widespread and diffuse disruption on cortico-cortical pathways. This disruption probably induces compensatory mechanisms that make good any loss of connectivity but at the expense of reduced overall capacity as aging impairs brain function.

These studies are interesting but do not establish a connection between TBI and later AD. One study combined data from seven reports and reached no firm conclusion.[27] This result was unaffected by stratification of TBI victims by *APOEε4* carrier status. The cortical networks shown in diagrammatic form in Figure 7.6 would be damaged by aging and, quite separately, by acute TBI. The overall effect is to reduce the capability of the network to withstand progressive age-related deterioration. This simple model is refined by the suggestion that the brain has a limited capacity to regenerate after injury, and it is this "neuroplastic" response that is diminished after TBI and unavailable to respond optimally to the presence of AD. In these terms, TBI does not induce AD but rather precipitates an earlier symptomatic stage of AD.

In clinical practice, it is not unusual to hear experts suggest that the psychological effects of TBI depend as much on the "nature of the injured head" as on the precise location and extent of brain injury. Here, the expert is weighing the resources available to an individual after TBI to adapt to cognitive deficits. These resources include the pre-TBI acquisition of coping strategies to deal with impairments, the presence of personality traits associated with proneness to anxiety, and critical aspects of long-term social adjustment. These latter traits include impulse control, the ability to tolerate uncertainty or ambiguity, and predilection toward substance abuse. Among some disadvantaged groups of young men, clustering of TBI with these predisposing factors frequently is encountered, and when they present to health services in late adulthood with greater than expected age-related cognitive decline, it often is extraordinarily difficult to reach firm conclusions about likely causes.

X. PERSONALITY AND COGNITIVE AGING

Personality traits are closely related to the life course acquisition of coping strategies and to general mental ability. At present, it is reasonable to assume that personality traits remain relatively stable across the adult life course. Data from the SLS and smaller studies elsewhere, however, suggest that some traits are less stable than others and some may be linked to differences in rates of cognitive aging.[28]

An important confounder of studies of this type is that personality and general mental ability are major influences on health behaviors, including behaviors that increase the risk of illness and those that modify help-seeking behavior. For example, high trait anxiety is increased by intercurrent illness and is linked to greater and more persistent stress responses. Taken together, these increase the likelihood of use of prescription medications that impair cognitive functions.[29] Careful clinical assessment of participants included in studies on links between personality traits and cognitive aging are required but rarely undertaken to resolve these issues. Conversely, the personality trait of "openness" is associated with higher levels of general mental ability. Many studies now show that greater intelligence is linked to better overall health and lower long-term mortality.[30]

XI. BRINGING IT ALL TOGETHER

Cognitive functions are divisible among a wide range of mental abilities. These abilities concern the work done by the brain with aspects of knowing what is happening in the present, how this relates to a sense of what has happened in the past, and how the future can be anticipated reliably. Aging brings major concerns about impairments in memory and reasoning as well as some gains in certain types of judgment. Sensory abilities frequently are impaired but more often in a limited way that does not affect cognition. We have good reasons for understanding cognitive aging in terms of decline in the efficiency of information processing and accepting that this could be an impairment that underlies most if not all other aspects of cognitive aging. The idea that all aspects of cognitive ability decline uniformly in step with one another, however, does not hold up in observational studies. Because of the many individual trajectories of cognitive change with age, the idea of a single unifying process, while attractive, does not hold up. Evidence thought to support an information processing hypothesis of cognitive aging possibly will find consistencies between changes in aging cognition as a result of statistical artifact. Expressed more cogently: "It doesn't all go together when it goes!"

Comparisons between cognitive development in childhood and cognitive aging are too few to be informative. We have solid ground, however, on which to explore how the acquisition of cognitive abilities, as understood within the framework proposed by Piaget, might clarify how mental life changes in late adulthood. Particular opportunities can be used to apply these techniques to the study of older adults whose cognitive development was thwarted by the onset of severe mental or neurological illness. In late life, these individuals may be particularly vulnerable to the ill effects of cognitive aging and can require special measures of support to adjust helpfully to old age.

8

EMOTIONAL AGING

Very few effective interventions can protect against dementia. Among the stronger associations is the finding that a socially engaged or socially integrated lifestyle predicts lower rates of age-related cognitive decline and fewer clinical cases of dementia. The first report emanated from the Kungsholm Island project in Stockholm, Sweden, where Laura Fratiglioni and Bengdt Winblad followed up on old people without dementia and showed that new dementia cases were predicted by fewer social ties and less social activity.[1] Their results were consistent with four later studies,[2-4] including a follow-up study of 16,638 U.S. citizens who took part in the Health and Retirement Study. Karen Ertel and Lisa Berkman from Harvard Medical School and Maria Glymour from Columbia University confirmed[5] that high levels of social integration predicted slower memory decline in a representative sample of individuals who were fifty years old or older. These benefits were most marked in those who had the least education and were without heart disease. This may yet prove to be an example of "reverse causality," in which deficits caused by incipient dementia impair the maintenance of social relationships.

Attempts to treat or prevent age-related memory impairments are usually unsuccessful. The finding that social integration has a protective effect is one of few promising lines of research. If the beneficial components of social integration could be identified, future interventions could be designed based on those components. This chapter sets out to discover what those components might be. It starts with the nature of emotional aging that examines the ways our social relationships are connected intimately to our emotional life. Next, we explore social networks and the concept of "social capital" among older adults. These considerations take us closer to the resources we already have to cope with stress in old age.

I. EMOTIONAL LIFE

The effects of age on our emotions are linked closely to the fact that as we age, most of us get better at recognizing, understanding, and controlling our feelings. We encounter many sentiments that rise and fall somewhere between love, pity, and anger with much in between and everything else to varying degrees. By late adulthood, we hold a library of emotional memories that enable us to analyze not only our own emotions but also those of others. Over our life course, we pick up on emotional cues from other people and gradually increase our ability to understand our reactions to their feelings and intentions.

Many accounts of emotional life focus only on the negative aspects of our emotions. Suffering from emotional symptoms predominates in textbooks of psychiatry. This chapter is not about these clinical emotional disorders. Instead, it emphasizes the positive roles played by our emotions in later life. Our goal is to understand the possible protective effects against dementia and cognitive decline afforded by a socially integrated lifestyle.

Emotions are part of normal healthy development. They figure conspicuously in the strength of our social attachments. Our emotions often determine positive or negative outcomes of critical life choices and experiences. As in youth, our emotions remain remarkably stable. We still feel the pain of social exclusion, and we experience the same changes in our bodies when we become emotional. Consider a grandmother who still says, "my heart skips," when she sees her grandson, just as it had when, as a younger woman, she saw her future husband. Although changes in personality do occur, these are slight. The overall picture is of stable emotions linked firmly to the same preferences to be part of a social group, most often a family. At a superficial level, emotions seem easy to define, but we have many different ways to think about the nature of emotions. No matter how much we might improve with age, understanding the nature of emotions continues to challenge psychologists, psychiatrists, and a good many philosophers.

As we mature, our self-esteem and social regard are influenced strongly by the success or failure of our style of emotional management. To achieve success, we learned how to be aware of our emotions and the emotions of others and to behave appropriately. Eventually, we will use these competences to make positive gains in our emotional life in old age.

One aim of this chapter is to challenge a widely held view that emotional life deteriorates in late life. In Chapter 5, The Aging Brain, you read how aging impairs mental abilities. Deficits are attributable to aging in memory, attention, and mental speed. From this, you might presume that impaired mental functions should include emotional problems, and you would not be alone in this thinking. Until the seminal work of Laura Carstensen and her colleagues at Stanford University,[6] this view remained unchallenged. Carstensen pioneered an approach to the study of emotions in late life from a perspective that included social, cognitive, and motivational factors.

WHAT ARE EMOTIONS?

Emotions have played center stage throughout human history. For most people, emotions are an obligatory part of our human nature, as important as our capacity for rational judgment. Although some people regard emotions as dangerous as they are essential, they often justify human behavior.

Emotions are subjective experiences that are accessible only through introspection. We recognize that they correspond to feelings that differ from sensations like pain or nausea. We know, too, that our emotions can predispose us to understand ourselves and our circumstances in positive or negative ways. With maturity, we understand that this link may prejudice our judgments, and we learn to make appropriate adjustments.

It is customary to define emotions as "subjective feeling states," but this dodges the question. Just what is a "feeling"? William James first posed this question in 1884, and a satisfactory answer remains elusive. Most social scientists have focused on the observable parts of emotions. In other words, when individuals report happiness or sadness, what can be seen that distinguishes one state from the other? What expressions do people use to describe how they feel? Comparisons between individuals, between infants and young children, between different levels of cognitive ability, and especially between cultures all show that most emotional states can be classified using self-reported feelings, use of language, and behavioral changes. This sort of categorization holds up well and supports the opinion that emotions can be defined using self-reports and that these definitions represent the corresponding mental phenomena.

Laura Carstensen's findings are substantial and reproducible and support her "sociomotivational theory" of emotional aging. Her basic premise is that unexpected emotional gains are linked to growing old. Among these gains, we have the ability to reflect on happy times and the facility to avoid unpleasant memories. In keeping with the life course approach taken in this book, this chapter picks up the life course from infancy to late life and charts our emotional growth into a rewarding and accomplished old age. The so-called honorable elder[7,8] is not confined to Eastern societies. Many examples abound elsewhere throughout the world, including North America and Europe.

This chapter explores how older adults reach their emotional goals and emphasizes how this works in the management of social relationships. When these relationships improve a sense of personal well-being, they are reinforced and encouraged. So we continue to invest in social attachments throughout our lives. Not all old people think of their social behavior as part of their emotional lives. If asked, some break down their emotions in terms of "something I want to do," which implies that they manage their emotions in terms of what they want to achieve or control, particularly with respect to others. This type of emotional control sanctions the exclusion of the unpleasant, the disgusting, and all else of that kind. Others carefully manage their mental efforts, directing their energies only toward matters they believe important. These individuals may seem detached to a point at which they appear to be cognitively impaired, but they are energized by the chance to act with a likely positive outcome. Motivation is relevant not only to what we do and feel but also to the case for doing something in the first place.

In later life, we experience a substantial shift toward making sense of our emotions and our goals. These are significant tasks bound up with the awareness that time is running out. The time perspective affects personal choices, motivates our behavior, and identifies attainable goals. We shall consider issues of time management later in this chapter.

Successful aging requires expertise in emotional management. Just like younger adults, old people need people. When social supports are strong, older adults do better than the young controlling or avoiding stressful life events. If older adults can maintain strong social ties, these will not only be a source of happiness but also improves chances of remaining well. Our emotions, therefore, play an important role in healthy psychological functioning.[9]

Sadly, the converse is also true. In later life, when social relationships are few or of poor quality, emotions suffer. Stress management is more difficult without social supports. As a young, self-confident adult, it may have been easy to speak of relationships from a position of inner strength. This becomes more difficult with age. Sparse social networks make older adults more vulnerable, exposing older people to increased risks of emotional illness. Their weaknesses are laid bare by the death of "one true friend," for example, or an unexpected sensory loss like deafness or failing eyesight.

The five strong studies mentioned at the start of this chapter showed how an integrated, socially active lifestyle might protect against dementia. One consequence of the absence of social ties would be a greater vulnerability to stress. Chapter 6, The Biology of the Dementias, considered how stress might predispose to dementia. Similarly, the link between a lack of social ties and greater risk of dementia is attributable to the effects of stress on the brain experienced by those who are more vulnerable to this stress because of social disengagement in late life.

II. OLDER ADULTS AS EMOTIONAL EXPERTS

By the time we reach old age, we know what emotions are like. They featured in our transitions from love to hate that colored some important relationships. We know, too, that the way we handle our feelings affects how we interact with other people. In old age, we aim to feel confident in our self-knowledge. For many people, this knowledge translates into a need to control our feelings. Otherwise, we will waste a great deal of the little time we have left. In our search for the "beneficial components" of a socially integrated lifestyle outlined at the beginning of this chapter, this sort of emotional expertise is an essential component.

This section coves three major competencies of emotional life in late adulthood. First, we explore how aware we should be of our emotional state. When mental powers decline with age, should we expect our self-knowledge to be diminished? Can our capabilities in understanding and regulating our emotions make positive gains extending into late adulthood and beyond?

Most people who believe they are in control of their lives tend to report better health, fewer symptoms of mental illness, and higher self-esteem. Being in control can be equated to feeling that control is located within oneself (i.e., is internal). Feeling as though one is largely controlled from the outside (e.g., by others, by circumstances) places control in the external world. Whether or not control is "internal" or "external" is associated with variables established as risk factors for dementia. For example, women, individuals with low occupational status, and people with low educational attainments often may consider their controls to be mostly external. In contrast, people with strong religious beliefs are more likely to feel in control of their lives. Many variables linked to external–internal location of controls (technically, the "locus of control") are tied closely to personality variables and how these are organized.

Second, we consider whether we should always display our feelings. Some people think that emotional control is about controlling the outward expression of our inner feelings (often defined as "keeping a stiff upper lip" in the face of unpleasant events). Social rules dictate what should and should not be expressed; the most obvious example is aggression. We learn to express our inner feelings in acceptable ways. In some societies, social rules can be demanding and prohibit the excess expression of almost all feelings. As a mature adult, we hope to appear adept at handling our emotions, wanting to show self-confidence and trustworthiness.

Third, as youngsters, we learned how to recognize the emotional states of others. Sometimes this was simple. We learned to interpret facial expressions that communicate sadness, disgust, hatred, love, and happiness and understood how that person was probably feeling. We learned how to use this information to form an appropriate response. As children we learned when a parent could be approached for help or comfort. If mom was cheerful, singing along with the radio in the kitchen, it probably meant she was happy and pleased to see you. If dad was late home from work, silent, avoiding eye contact, and frowning, it perhaps was best to steer clear until mom had sorted things out. As mature adults, we hope to develop the skill of correctly reading the thoughts and feelings of others.

These three aspects of emotional life are extensively investigated. Relevant studies in late life have approached emotional aging from social and biological viewpoints and often have introduced research methods and concepts directly from child development. Some of the social learning acquired during child development is disrupted by aging processes. So it is reasonable to ask whether brain aging in late life is to blame for the breakdown of emotional competences and social judgments that sometimes occurs in old age.

III. EMOTIONAL STATES

Emotions are so familiar to all of us that definitions may appear to be unnecessary. We start by thinking about negative feelings. Most often, when someone is described as "emotional," it is never flattering. Emotions are subjective: Only you can experience your own feelings. Others can only observe your physical responses to these feelings. Emotions are regarded as transient, never enduring, and they show a high degree of specificity. This means that the same event or stimulus when repeated will produce the same emotion, although varying in intensity with each repetition.

Feeling an emotion typically is accompanied by sensations of physical change. In concert with feeling anxiety, fear, or dread, we sense physical constriction as though our body is oppressed by some unseen outside force. Breathing can be difficult, and the heart races with a sense of tightness that may extend around the chest to include the throat. Feelings of alarm involve a sense of something harmful about to happen that can cause shakiness, trembling, and cramps; rapid breathing and a sensation from a fast strong heartbeat; and tingling sensations around the mouth and fingertips. As we become acutely aware of these physical changes, we try to understand what is happening. As we mature, we involve cognitive processes to deploy strategies to escape from or remove the cause of these unpleasant emotions.

With aging, attempts to explain unpleasant bodily feelings are more difficult. At one level, we know that sensory impairments acquired with age are important modifiers of emotional experiences. Hearing loss can seriously hinder social life, and worries, for example, about bladder control can limit social activities. The implications of the physical changes that accompany emotions include the possibilities of undiagnosed physical disease. Fear of a heart attack and concern about having a stroke are just two of many common fearful interpretations of acute emotional states. These feelings accompany biological changes that underlie the physical components of emotional responses. Starting in the first few weeks of life, facial expressions of emotions not only are consistent between families but also among ethnic groups. Newborn babies, even those born blind and deaf, display fear, disgust, and joy in much the same way everywhere, and rather surprisingly, by three months of age, babies can recognize these feelings in the happy or angry facial expressions of their parents. Our emotional expressions, therefore, almost certainly are innate and probably specific to our species.

IV. THE AGING BRAIN AND EMOTIONAL AGING

Mature healthy adults show brain changes that accompany each type of emotion, making it possible to speak with confidence about a "neuroanatomy of emotion." Studies have used functional magnetic resonance imaging (fMRI) and positron emission tomography (PET) to investigate patterns of brain activation induced by types of emotional states. Luan Phan and his colleagues[10] at the University of Illinois (United States) combined relevant studies using these techniques through 2000. They found fifty-five studies suitable for inclusion in their analysis and concluded that one brain region (the medial prefrontal cortex [PFC]) was activated whenever an emotional state was induced, no matter whether it was negative or positive. In addition, fear specifically activated the amygdala, whereas sadness activated the subcallosal cingulate.

Decreases in volume of the PFC are among the most dramatic developmental changes of any cortical region in the absence of disease. There are age-related PFC increases until about twenty-five years of age; thereafter, the PFC reduces slightly in volume until at about fifty-five years of age, when the rate of decrease begins to accelerate rapidly.[11] The PFC is involved with speed of learning and thought and is concerned with social and emotional behaviors and with goal-directed planning. Correlative studies show that age-related reductions in PFC volumes are linked over time to decline in cognitive performance.[12]

Identifying a relationship between the brain structures that are involved in emotional processes and the age-related reductions in the volume of these same structures corresponding with a decline in cognitive performance seemed to be a good test of the idea that the course of changes in emotional well-being run parallel to a decline in cognitive function. Susan Charles at the University of California–Los Angeles and Laura Carstensen at Stanford University (United States) provided a much more complex (and intriguing) relationship between emotions and cognition in later life.[13] Among young adults, emotional aspects of information often are deemed irrelevant or are ignored (i.e., the neural networks involved are inhibited). When age-related brain changes impair the capacity of these networks to inhibit, emotionally laden information becomes more important and cannot be ignored by older adults. Charles and Carstensen have argued that emotional well-being is preserved in older people partly because it is difficult to ignore and also because:

slower processing speed may provide seemingly paradoxical benefits. The function of emotions is often placed in evolutionary terms, which stress rapid responses where "fight or flight" patterns determine survival. In the modern social world, rapid responses may not be the best response. Snapping at someone with a fast retort may not be as wise as pausing before responding to an interpersonal slight. When responding to negative interpersonal conflicts, faster responses may not translate to an adaptive response.[14]

V. THE ANATOMY OF EMOTION

The insula is a region of the cortex folded deep in the lateral sulcus between the temporal and frontal lobes. The insula is involved in self-awareness, pain perception, and consciousness. It is activated automatically when observing the expression of disgust (e.g., smelling an unpleasant odor associated with excrement) and when experiencing the smell directly. The insula also responds to disgusting tastes, and when stimulated during brain surgery, it will induce the urge to vomit. These findings are consistent with the idea that the insula provides a common means to identify unpleasant stimuli and recognize that someone else is having a comparable unpleasant experience.

A complex interconnecting network serves the emotional perception of pain. This network includes the insula and extends to involve the anterior cingulate, the cerebellum, and the thalamus. Parts of this system can share a painful emotional experience between someone suffering pain and someone observing the sufferer. In addition, the sufferer activates part of their sensory cortex, so that they actually feel the pain. The best examples of this phenomena is evident when we witness a traffic accident and feel the victim's discomfort but not his or her pain. In the emergency room, the paramedic experiences the emotional impact of the pain felt by the accident victim but does not physically experience the pain. If paramedics felt the same pain, this probably would incapacitate their effectiveness in an emergency.

Decreased efficiency of PFC functions eventually impairs judgment. Emotional control ultimately becomes inefficient and progressively more haphazard. The failure of emotional regulatory systems with aging sets off a cascade of physiological changes that damage the nervous system by activating the cortisol stress response and overstimulating aged, stiffened blood vessels.

VI. EMOTIONAL INTELLIGENCE

We can ask whether individuals have an adequate capacity to understand and regulate their own emotions, decide on the emotional states of others, express sympathy, and show trust. We are unsure how people should best cope with stress and how to judge the ability of others to manage their emotions. When we start thinking about how emotions affect our relationships, we think about the best way to express emotions in those relationships and to accept our emotional limitations. We arrive at the limits of emotional competence.

These issues are relevant to later life. If older people make errors of judgment, do they do so because of age-related weaknesses in the analysis of complex, interacting causes? Or do particular events or situations trigger long-held strong emotionally laden beliefs that will not accept contrary views? Although some types of judgment can be affected by brain aging, our emotional goals also become more important as we progress toward old age. Our emotions change as our life ahead shortens and becomes more uncertain.

In 1995, Daniel Goleman published his book *Emotional Intelligence*.[15] Many psychiatrists were intrigued, but would the public at large accept the idea of emotional literacy or his ideas about emotional malfunctioning? At the time, his vocabulary was not part of the language used by psychiatrists in their clinics. Over the past twenty years, his ideas gained acceptance and, more recently, have been founded on good studies.

We now accept that there is such a phenomenon as emotional intelligence; that it has its roots in early childhood; and that major differences exist between people, between sexes, and between the old and the young. For many of these individual differences, we are unsure whether they result from important sociocultural effects, brain maturation, or aging—or perhaps from a combination of these effects. For example, parents raised in a household governed by strict practices to enforce "moral development" may grow up without knowing key elements of an emotional vocabulary commonly available in their culture but excluded from their home. As parents, they may be unable to provide the role model for their own children who, in turn, grow up with emotional problems that affect their relationships.

Children who are unable to signal their emotional states to others are less popular and have difficulty throughout their lives in managing secure, intimate relationships. When the child is better able to read the emotional states of others, peer groups regard that child more positively. When these attributes are linked to a preference to make positive comments about others and to cope with anger without expressing aggression, then they often are widely liked and frequently chosen as leaders. Popular children tend to be popular adults, have more friends, and gain more opportunities to develop mutually beneficial supportive relationships.

Is it possible that cognitive aging can impair emotional competence? Does the reverse happen so that awareness of failures of emotional intelligence leads to cognitive failings? In this regard, we can begin to think about the routes we take when we exercise emotional judgments. Do these judgments use the same type of information processing pathways (resources) that support cognitive performance? The following vignette (8.1) sets the scene for a series of social judgments made by a man in his sixties. On the basis of scant information, he immediately saw Tom as a stereotype. His negative first impression on hearing about Tom did not change on meeting him. It is not surprising that no effort was made by Elizabeth, Chrissie, or Tom to change Mike's opinion of Chrissie or Tom. Elizabeth knew too well that "Mike just can't change . . . he makes up his mind very quickly and can't be shifted."

VIGNETTE 8.1. SEX BEFORE MARRIAGE

Mike married Elizabeth in her early fifties. By then, Mike was sixty-seven, had always been single, worked in prison resettlement, and held strong opinions about foreigners, alcohol, and government finances. Elizabeth had married once before and was widowed soon after. Her only daughter Chrissie was about to go to college supported by Mike. After the first semester, Chrissie came home and told her mother about dating an older boy, Tom, at college. Chrissie wanted to start an oral contraceptive before the relationship "got serious." Tom was due to visit and return with Chrissie to college.

Mike's opinion about Tom formed immediately. He was typed as "no good" and called "a loser." Mike told Chrissie that Tom "wouldn't look at you if you weren't so young . . . he just wants to get you into bed."

When Tom arrived, Mike was unpleasant and aggressive. "I know your type," he said, "I've seen lots of your type behind bars and that's where they belong." Tom told Elizabeth he hoped to work as a teacher when he graduated, maybe keep up

his hobby of playing bass guitar in a rock band. Perhaps he would help school kids
do something similar after school. His comments angered Mike further: "So you
like school kids then? Is that boys or girls?"

It is evident that Mike not only stereotyped Tom, but also Elizabeth stereotyped Mike as being fixed in his ways, quick to judge, and impossible or unable
to change. These negative stereotypes concern Mike's views about the sexual
intentions of younger men who date his stepdaughter but extend beyond that to
Elizabeth. She has a negative stereotype of Mike as an old man, set in his ways.
But what in fact do we know? Has Mike always held these views?

Stereotypes are derived from a special type of social knowledge that we apply
when forming new impressions about people we are meeting for the first time.
We do not know about Mike's social knowledge and whether this changed as he
grew older. With Elizabeth, too, she has gotten to know Mike a lot better after
they married, and now she finds that he's not the man she fell for ("He's no John
Wayne type; he's no golden ager") but rather is more reclusive, hides his feelings,
and attributes his failings to his "age, bad memory, and loneliness."

Emotional competence and age stereotypes are closely linked. The ability
of individuals who are aware of their age stereotypes to perform cognitive tasks
effectively and efficiently frequently is viewed with a negative bias by the same
individual and by an observer who makes a stereotypical judgment. The evidence for this bias comes from an intriguing series of investigations led by Joan
Erber at Florida International University (Miami, United States).[16] The team
explored how a memory problem that was the same in nature and extent in
young and old study participants was viewed as more serious in the old than
the young when a young person was asked to compare the two age-groups.
Old people took the same view of both young and old with the same degree of
memory impairment.

We form and apply stereotypes based on our self-perceptions and our views
of other people. We see each other in different social contexts, and the more
we rely on stereotypes to reach judgments, the less we are aware of using them.
This becomes important when we judge the cognitive competencies of older
adults. It is not the stereotyping that impairs performance on cognitive tests but
rather the fear of being stereotyped and regarded as another "old fool" who will
test poorly.

VII. OLDER ADULTS AS ACTORS

The autobiographies of older adults are peppered with self-characterizations. These adults speak as an author would narrate his or her life: "I have always or never been . . . ," speaking as though these were fixed qualities. We tell ourselves "That's the type of person I am . . . it's in my nature" or use a similar phrase that means much the same. We may hear older people say, "I am just like my father" or something similar to suggest that they feel the way they are is a mixture of heredity and upbringing. Some older people seem more likely to add a profession of faith, "I'm a Baptist, just like my daddy," as though this declaration provides a special insight into their true character.

Dan McAdams at Northwestern University, Illinois, has followed a life course developmental perspective on personality.[17] He has teased apart features from the broad range of traits that distinguish an individual from other people in terms of individuality and a sense of identity. These features include the individual traits (like temperament) detectable in infancy, ways of adapting to situations that characterize an individual, and the "life narratives" we construct to give meaning to our lives. McAdams suggested that the first of these features account for the large internal dimensions of personality that make us behave in consistent ways and enable us to maintain particular attitudes and styles of thinking and feeling. In this sense, he has suggested that older individuals continue to display distinct emotional styles and behavioral repertoires.

VIGNETTE 8.2. PERSONALITY DISORDERS IN LATE LIFE

"You know I've learned a lot from life . . . it's got so I'm really good at it . . . I can say one thing and mean another at the drop of a hat . . . someone else will think I'm listening but you know I'm not . . . I just look calm . . . when inside I feel like I wish they would just shut up. I'd say it's one thing I really got better at as I got older!"

Linda was not a patient. She was seventy-one years old, the wife of a man age seventy-three, and talking about forty-four years of married life. Her husband had recently retired and had left her to take a vacation from which he seemed unwilling to return. She said she needed his medical records but didn't explain why.

As she talked about herself, she emphasized how important she had been in his business. She spoke without prompting about her dreams of a successful legal career, although she held no legal qualifications, adding that "I could have been a doctor but he needed me in the shop . . . Of course you would have needed to see me then . . . Men used to turn and stare when I walked into a room . . . women were so jealous . . . people say even now I would pass easily for fifty!"

Linda said that at first when he retired, they'd seen a lot more of the grandchildren, but she didn't get on with either of her daughters-in-law. Then he spent more time playing golf with a friend. A sudden death on the golf course stopped him playing. She continued: "One day he just took off . . . he said I had no feelings for anybody just myself . . . after all I've done for them! He went to our home up at the Lake and then he phoned to say he would take the trailer over to a dealer he knew who would give him a good price . . . he still phones . . . not as often now . . . I think he's somewhere in the southwest . . . he's bought a dog . . . he's just so selfish . . . why doesn't he come home and look after me?"

A week later her husband phoned to say that if his wife ever asked for information about him, nothing should be said: "She doesn't care how many people she hurts or uses . . . Her temper just got too much."

According to her son, Linda was well known to be "over-sensitive" to criticism with exaggerated opinions about her appearance and abilities. She seemed unaware of her husband's unhappiness, although her sons at one time feared he might be suicidal.

Her husband filed for divorce the following year, and she met a retired lawyer on an Internet dating site and no further contact was expected. Five years later, two years after her death, the family lawyer was instructed by her eldest son to contest her last will. She had bequeathed her share of the family business to a medical charity and excluded all other claims on her estate. Her psychiatric records were available to the court. These showed she had entered psychotherapy at about forty-two years of age and remained in intermittent contact with her therapist for eleven years. He had recorded a diagnosis of "narcissistic personality disorder" and identified depressive mood swings alternating with periods of unreasonable behavior, which possibly was made worse through excess use of alcohol and prescription medications.

Vignette 8.2 typifies the sketchy nature of the information available in a clinical setting to reach judgments about mental disorders or personality types. Linda certainly suffered from her mood swings, and in the opinion of her therapist, her personality traits made her vulnerable to emotional distress. It was certain that many other close family members suffered because of her behavior and that much of this was linked to her personality traits. Her substantial assets would help her grandchildren through college, but she disregarded their claims. Instead, in the view of her sons, she aimed to hurt them even after death. Her son said, "All she ever wanted to do was to control our feelings . . . I think she was always angry inside . . . she really hated us."

The question of stability of personal characteristics is relevant to understanding the emotional life of older adults. To sort out the issues raised by this type of question, it is helpful to think about the ways personality differences are studied. First, we need a definition of personality. We assume that what is meant by personality is relatively stable and is made up of traits that are longer lasting than our emotions, which we think of as temporary. We view personality traits as different from emotional states. These matters are never clear-cut, however, and some emotions (like anxiety) are more likely to be experienced by individuals with long-lasting anxious traits.

Some traits tend to be linked, such as the shy person who also may be introspective and dislike change or uncertainty. We begin to think that some traits go together more frequently in the same individuals and that these traits will emerge in whatever context an individual is placed. This viewpoint has encouraged the idea that some traits are clustered along a single continuum or dimension. When traits are analyzed in this way, groups of traits are found to cluster in a consistent fashion. These groupings make separate dimensions; much as you might measure a furniture cabinet by its height, length, and width, you could measure personality along dimensions. More than three dimensions are needed to measure most of the range of personality differences. For example, in addition to height, width, and length, you also want to measure the height of shadows cast by the cabinet.

Quite a few personality measures are available, but few have examined personality across the life course. An important exception is the five-dimensional model (the five-factor model) developed by Paul Costa and Robert McCrae at the National Institutes of Health, Bethesda (United States).[18] This five-factor model is made up of five independent dimensions labeled openness, conscientiousness, extraversion, agreeableness, and neuroticism. Other researchers have found comparable dimensions in their personality data, and Costa and McCrae also have found their dimensions in data from other cultures. Most relevant to the aging of emotions is that, in addition to consistent results from one group to another, Costa and McCrae found these five factors remain remarkably stable over ten years or more of adult life.

From a common-sense standpoint, this finding seems even more remarkable when we consider what may have happened to a large sample of adults over a ten-year period. Some individuals would have suffered great unhappiness and personal loss, others would have been upwardly mobile and become worry free, or yet others would have been able to stop work altogether.

Longitudinal studies on personality have answered some of the questions about the stability of personality traits with aging.[19] When personality is measured over short segments of the life course, adjacent segments agree with one another quite well. Over longer intervals, however, major differences between men and women are much less stable. Changing roles, for example, from lover to parent, from mother to widow, have profound effects on self-rating of personality. Personality develops across the life course and will continue to do so into very old age.

VIII. LIFE NARRATIVES, SELF-CONCEPT, AND POSSIBLE SELVES IN LATE LIFE

We can begin to think again about how older people talk about themselves. They do not use traits but are more likely to start on a part of their life story. Speaking in everyday terms, much as Linda did in the second vignette, they will talk about their achievements at certain times of their lives. Men who once saw military action may use that as a reference point, as a witness to their strengths and weaknesses. Friends made under fire leave lasting traces of personal worth, more so if they fell in action. Listening to older adults talk about their lives contrasts importantly with a dimensional approach to personality. These conversations can provide an enriching and enlightening exchange of experiences, values, and taboos, and may prove to be much more valuable than scoring a personality questionnaire.

The use of life narratives makes a lot of sense when trying to understand the effects of aging on our personalities. A sense of the person he or she wanted to be as a young adult can be remembered, and with encouragement, a rough sketch will fill out with detail. This self-concept does not set firm and remain unchanged but rather continues to mature during adult life. With aging, people develop their personalities, and their ambitions are altered by self-awareness of strengths and limitations and by society's ideas of what is expected and what would be "normal" for the individual. Looking back, many older adults identify episodes that prompted them to change, sometimes radically, in light of events.

A "self-concept" involves self-image, self-belief, and consideration of "possible selves." In conversation with older healthy people it sometimes is surprising how often "fate" or "luck" are mentioned when asked to explain events. Occasionally, people talk about what they did to produce a major change: "I knew I couldn't go on the way I was. I wanted something better. Living from hand to mouth wasn't just to do with being short [of money] . . . it was more than that. I knew I could do a better job, do something I could be proud of." The woman speaking here easily could be talking about her family, her job, or her finances. In fact, she was talking about her work as a textile designer when she had always wanted to be an architect.

With aging, fears about health come to the forefront, but relationships with others are never far away. It is interesting, too, to hear people talking more often, especially after age seventy, in a self-accepting fashion. Usually expressed in a positive way to stress that they always tried to be truthful and compassionate, their self-acceptance acknowledges that they have showed many different faces to the world. Each time, although different, they tried to make the "real person" show through. During this later stage of adulthood, the range of possible future selves diminishes. To keep some sort of coherent sense of self, older adults think more about the positive aspects of their past lives and current predicament than dwell on "what might have been" or on another life course that might have provided a happier future.

Reading around this topic, some research crops up more often than others. Because of the lack of data-based investigations, it is difficult to compare one theory with another or even to state with confidence that a theory holds up. This concern differs from what is known in child psychology, where child development in cognition and emotions is studied extensively using robust, highly reproducible methods.

The psychodynamic theorists led by Sigmund Freud performed a great service by providing the vocabulary to help understand emotional development. This vocabulary was so successful that many of the terms have entered popular speech. Most people can recognize excessive use of "defense mechanisms" as

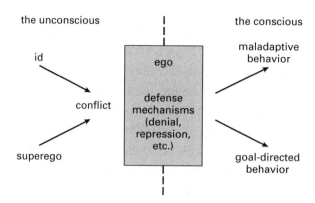

FIGURE 8.1 Schematic summary of the origins of maladaptive behavior using psychodynamic terms. Mental life is divided by the vertical dashed line into the "unconscious" and the "conscious." The ego lies on the boundary between the two so that some aspects remain in conscious awareness; others are hidden in the unconscious. The integrity of the ego is protected by defense mechanisms. These serve the ego by protecting it from conflict between the id (source of primal drives such as libido and self-preservation) and the superego (the repository of rules and behavioral controls). Overuse of defense mechanisms (like denial) impairs the efficiency of goal-directed behavior and allows maladaptive behavior to emerge.

moderators of goal-directed behavior. Figure 8.1 provides schematic of how these defense mechanisms protect the integrity of the ego. This term defines the individual's sense of self, and its content is only partly accessible in consciousness. Basic instincts are energized by the id, which is held in check by the superego, equivalent to the structure of conscience determined by parental and societal rules. Defense mechanisms protect the ego from conflict between the primal drives of the id and the control of the superego. Use of defense mechanisms comes at a cost. Goal-directed behavior is impaired to the extent that greater use of defenses allows neurotic symptoms to emerge alongside adaptive behavior.

That is a glimpse of psychodynamic theory through a very small window. Although Freud developed his original ideas extensively, he rarely wrote about aging. His views about personal biographies and clinical history taking were largely dismissive: He was quoted as saying, "all are works of fiction." and appeared to ridicule therapists who studied in detail their patients' personal histories. His focus was on the origins in infancy of the abnormal psychopathology of adulthood. He wrote little about the aging mind. In contrast, Carl Gustav Jung was much more interested in aging. He saw that young people were more outgoing ("extraverted") than the elderly. Old people were more reflective ("introverted"). He thought they were more concerned with their mortality. Jung also thought he could detect emerging trends toward greater femininity among older men, whereas women became more masculine with age. By this, Jung thought that men were allowed to explore their inner selves in later life in ways that were not open to them as young men. Evidence is hard to find that stands up to scrutiny to support Jung's theorizing. High-performing older men who invest in their personal development sometimes find common ground with these ideas. Later psychoanalysts have shown little curiosity about the older adult. Their disinterest probably derives from a presumption that older people would not be able to recall accurately their early formative experiences and that the capacity of older patients to benefit through personal growth and fresh insights is diminished through age.

Among the post-Freudians, Erik Erikson[20] did more than anyone to help care workers understand the emotional tasks to be completed over the life course. His approach is grasped easily by those who support troubled children and adolescents. Erikson's concept is that a child's developmental program is largely innate and is shaped by society's demands on an individual. Part of his appeal is that his scheme is simple and easily implemented. It can be used widely when training new staff to work in a care setting.

The sequence of stages envisaged in Erikson's eight stages are set by biological signals, which follow a prefixed order. Each stage is thought of as a struggle ("maturational task") between two opposing drives. Success at each stage is equated to control over the opposing forces with the result that competencies are established in areas relevant to the maturational task set by the opposing drives. On this basis, secure foundations are laid at each successive stage that will lead to emotional well-being in old age. Care worker trainees are asked to imagine Erikson's developmental stages as a stack of maybe a hundred pennies:

> Imagine how difficult it would be to build an entirely stable stack. Now think what would happen if you were obliged out of a hundred pennies to use a slightly bent penny at the tenth level. The column would be unstable above level ten and you would need to make allowance for it every step you took thereafter. Personality development is like that. Once you fail at a low level maturational task the greater the instability thereafter. Failure at an early task will predispose to failure for the rest of your life.

Erikson's first stage is the first task faced by an infant who must successfully establish a bond of trust with caregivers and eventually the inhabited world. This success provides lifelong lasting comfort and feelings of security. The next stage sets autonomy against shame. The infant realizes that their next task is to take responsibility for their own actions. By the time of old age, the task is to resolve the conflict between ego integrity and despair through the examination of a lifetime of successes and failures. Erikson foresaw that few old people would reach such a satisfactory conclusion at the end of their life course.

IX. SOCIAL SUPPORT, SOCIAL COHESION, AND SOCIAL PAIN

The importance of living a socially integrated lifestyle, with supportive relatives and friends, is at the core of understanding quality of life and good health in old age. Some old people feel excluded from society because of their age and disabilities. This type of exclusion triggers feelings of social pain. It is reinforced by the many visual and verbal cues old people receive. Not least among these is the stereotyping of old people as frail or weak and unable to contribute to society, upon which they place an imaginary and unwelcome burden. Patronizing remarks sting no matter how often they are heard. Failure to adapt public places to ensure ease of access and the safety of old people all contribute to the restrictions and barriers placed by society on the elderly. It is against these obstructions that many old people struggle to play a part in the wider community.

Our communities provide many social networks to engage and support older adults. These networks are mostly informal and range in size from just a few friends to larger groups led by social entrepreneurs. Leaders are mostly women; many are quite competitive and occasionally highly skilled in keeping others in line. When men play this role, they most often have very socially active wives.

These networks would not succeed without the agreement of its constituents. Although consent is actively given, particularly by women members, others make a fairly half-hearted effort to participate, preferring private arrangements with a few old friends. Conversations with members of small networks quickly identify who are the most active, but these conversations are rarely fully informative. Remarks mayseem superficial, often hearty and at pains to appear undemanding. Many old people work hard to appear as though they are coping well and are sympathetic to those who cannot cope. Their reality, in fact, can be a grim sort of social isolation with few if any confidants and even fewer people with whom to share their fears or aspirations.

These small social networks are the principal channels through which older people express social criticism. Sociologists confirm that within each social network, such criticisms often are shared, becoming well known and well rehearsed among their members. Attitudes about poor health, health care, and death are discussed freely but a great deal else remains private. Discretion is the watchword concerning personal finance and, surprisingly in view of the ease with which health issues are discussed, so are experiences of personal illness. Although it is fine to make a brief mention of an illness, it often is unacceptable to speak at length about one's own illness. To do so is considered "shopping for sympathy" and others frown upon this.

Like emotions, unless we are in a social relationship, we cannot directly observe that relationship. From observation, we can infer that a certain type of relationship is present. If a couple touch and kiss and this forms a pattern repeated over time, we can decide that the relationship is close. If an adult repeatedly hits a grandmother over a sustained period, we can state with confidence that the grandmother is in an abusive relationship. Relationships have unique properties, and every relationship has qualities not found in any other. These qualities include varying degrees of intimacy, steadfastness, or even dedication. They add another distinct layer to our relationships.

Men differ from women and, historically, many gay and lesbian older adults have been unable to overcome social obstacles (related to prejudice, social stigmatization, and, for some men, high levels of heterosexual or homosexual promiscuity) that prevented them from remaining in a lasting and loving relationship. Fortunately, many contemporary communities have well-developed networks that can befriend and support older homosexuals as they age and face the loss of a long-term partner.

A sociogram summarizes what goes on in relationships. With aging, the number of relationships declines steadily, but this is mostly among relationships that are least close. The closest relationships are sustained with aging, the effects of bereavements notwithstanding. Although relationships exist within social spheres that surround the index individual, what happens in one relationship does not happen in isolation. Relationships affect other relationships making

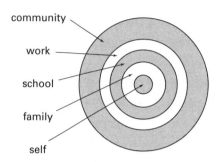 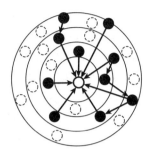

community
work
school
family
self

FIGURES 8.2A AND 8.2B This is a type of sociogram overlaid on a social landscape. Sociograms were invented by Jacob Moreno to study how groups establish preferences about group members. This figure does not distinguish between social isolates and those making "Mutual Choices." Some individuals make choices that go unreciprocated, the "One Way Choice." Cliques are groups of three or more people within a larger group who all choose each other.

up a complex grid of interconnecting lines. These are shown in Figures 8.2a and 8.2b and look a little like a plan of airline routes between major cities. Like air travel, some routes will be busier than others with more flights, more passengers, and a lot more baggage. For clarity, interconnecting relationships not involving the index individual are not shown. These secondary relationships are affected by the relationship between the index individual and the other person.

X. HOW MUCH TIME IS LEFT?

With aging comes an acute sense of time running out. The "locus of control" is the extent to which individuals believe that they can influence events through their own actions. Time is something that we can never control so, eventually, we are obliged to reach an agreement over how best to spend what time is left. We recognize, too, that factors associated with external or internal locus of control also are linked to our general health and to our sense of well-being. Factors like female sex, low social status, and low average educational attainments are associated with feelings of external control and a greater risk of dementia.

From a health perspective, we can be confident that compared with those with an external locus of control, men and women with an internal locus of control have lower mortality and a reduced risk of heart disease. They are less likely to suffer from depression or anxiety and they enjoy better self-rated health. We also know that the experience of illness will change our locus of control, so that as we become more dependent on health workers, those who previously exercised internal controls realize that controls over their health now are mostly external. The same is also true of caregivers when illnesses are disabling and slowly progressive (as are most dementias), so the locus of controls shifts from internal to external perspectives.

Locus of control applies to how we control our emotions. Emotional expertise is gathered across the life course, as each day we manage our relationships and resolve conflicts and their accompanying negative emotions. Simple exposure does not guarantee greater adult expertise, and many individuals never seem to learn from experience. Among these individuals, those who score highly on the personality trait of "neuroticism" seem to learn the least well.

Overall, social relationships and emotional health seem to benefit most from experience and a balanced sense of time perspective. Certain types of experience lead to better emotional control, and shorter time perspectives lead older adults to assign a higher priority to aspects of life that mean something to them. Old people seem sure-footed as they navigate social environments. They appear to use social controls effectively, taking care to pick their friends and companions. They become judicious in the deployment of their cognitive resource. Their thoughts turn more often to positive memories, and they will pay better attention to positive stimuli as they age. It is never possible to insulate oneself entirely from unpleasant events. The types of distress associated with illness in oneself or a close companion can greatly impair physical or mental health. In these situations, social isolation or exclusion exacerbates failing health. Toward the last years of their life course, older adults are able to function very well in social and emotional spheres. Their capabilities are usually equal to or superior to those of young adults. When faced with prolonged and unavoidable stress, however, age-related advantages are jeopardized.

In the field of social relationships, the boundary between emotional and cognitive aspects of mental life is rarely clear. In this uncertain territory, the term "social cognition" describes the reciprocal exchanges between emotions and cognitive performance. It is used to reflect a mental capacity to understand and anticipate the behavior of others. Social cognition is based on what an individual knows about the beliefs, attitudes, and intentions of other people. We use it to respond to emotional cues and to understand their significance.

Cognitive ability, motivation, and emotions all contribute to the effectiveness of social cognition.[21] Many aspects of cognitive function, including selective and sustained attention, memory, and comprehension, are each involved in social cognition. In Chapter 5, The Aging Brain, these functions were shown to be susceptible to the harmful effects of aging. In conjunction with emotional changes in aging, age-related cognitive deficits contribute to impairments of social cognition. It is, therefore, perhaps surprising that relatively little attention was paid, until recently, to the investigation of social cognition in late life. Most interest has been focused on childhood autism and victims of head injury or stroke.

Current interest includes assessment of social cognition in suspected dementia, where it may have diagnostic value. To explore social cognition in brain aging and dementia, it is useful to separate the concept into its two major subcomponents: theory of mind and empathy.[22]

Theory of mind (TOM) has several synonyms, including "mind reading," "mentalizing," and "mental state attribution." The term "affective theory of mind" refers to the ability possessed by most people to draw inferences about the emotions of others, and "cognitive theory of mind" refers to the ability to make inferences about the mental state, intentions, and motivations of others. TOM is detectable from about eighteen months old and is of diagnostic value in autism.[23]

TOM allows children to engage in social behavior and is best classified as a separate emotional domain of cognitive function. This ability improves as children mature and follows the same stepwise path in most normal children. In children with autism and Down syndrome, however, although cognitive abilities improve, TOM lags behind, showing that this ability can be dissociated from cognition.

Functional brain imaging studies show that in healthy adults the frontal cortex, limbic system, and amygdala are involved in TOM. Studies in adults with brain injury also implicate the PFC. Among the dementias, major changes are evident in the frontal lobes in patients with frontal lobe degeneration.

The capacity to communicate intentions concerning the disposal of assets after death underpins the concept of "testamentary capacity" with regard to making a will or exercising rights over property. Medical opinions about this capacity vary between societies and there are no clearly set criteria accepted by all legal systems. It is important, therefore, that physicians responsible for the care of the elderly anticipate that the grounds that guide their opinions are set out consistently and routinely in clinical records in ways that can withstand legal scrutiny. These grounds will include not only overall cognitive ability but an appraisal of sufficient emotional competence to assign property to inheritors in ways that reflect the intentions of the elderly person. TOM is relevant to making these types of appraisal.

XI. BRINGING IT ALL TOGETHER

The aging brain undergoes structural changes, some of which should increase the risk of emotional distress in old age. Rather counterintuitively, this does not appear to be a frequent cause of emotional problems, and there is no strong evidence that the presence of cognitive impairments leads consistently to emotional difficulties. In Chapter 12, the importance of interventions that improve social support of people living with dementia and their caregivers will be examined. In the presence of dementia, emotional life often is disrupted severely with negative effects on quality of life and survival. In healthy old age, however, many positive gains in emotional life are achieved even when cognition is somewhat impaired. These effects may be related to differences in temperament and character, to certain types of life experience conducive to successful aging, and to the nature of wisdom outlined in Chapter 7, The Disconnected Mind.

The capacity to share feelings, to experience love, and to remain committed to relationships continues to develop during maturity. Emotional life appears to some to be one of many personal attributes that is impaired by aging, but this is not the case. Important positive gains in feelings of self-worth, major changes in the order of priorities, and not a little pressure from a sense that "time is running out" are realized. Successful aging may provide a substantial basis for emotional gains in late adulthood, or the opposite may be true. To age successfully, it is necessary to prepare adequately for unavoidable changes linked, for example, to retirement, to loss of a loving relationship, and to the physical disabilities of old age. It is misleading to attribute emotional distress and failings to age-related changes in brain structure and cognitive performance. Furthermore, "to be old and wise" is not a matter of good fortune. These achievements develop across the life course and are built on solid foundations in social relationships, awareness of the thoughts and feelings of others, and a strong desire to care about the well-being not only of individuals who are close but also to the wider community.

Social rejection and isolation are powerful influences on mental well-being and limit the capacity to respond effectively to stresses in late adulthood. No matter how well prepared an individual might be in anticipation of aging, the task of facing adversity, especially the loss of someone who has been loved dearly, is better accomplished with the support of others.

Many make the choice to remain single, and others had no choice at all: Their relationships failed and they moved on. Life without an intimate partner is more hazardous for men than women. Compared with married men, single men tend to have poorer health and to die prematurely. Conversely (and this is not understood), single women enjoy better health and live longer than married women. Issues that arise among single adults as they age, therefore, are difficult to generalize.

9

DEMENTIA SYNDROMES

I. INTRODUCTION

Chapter 5, The Aging Brain, set out the effects of aging on brain cells and their blood supply. This account fell short of a full description of the impact of all the changes caused by brain aging. Nevertheless, the effects of aging even in summary appeared diverse and complex. They involved oxidative mechanisms, inflammatory responses, "wear and tear" of much molecular machinery inside brain cells, and critical disruption of brain blood flow and its microcirculation. To understand the syndromes of dementia, a basic understanding of some underlying mechanisms is needed. Some of this information will be specific to each dementia syndrome and some will be shared among components of brain aging.

WHAT IS A DEMENTIA SYNDROME?

Dementia is attributed to the development in an adult of impairments of memory and understanding and disturbances in temperament and social judgment (loosely termed "personality") that are sufficient alone or in combination to impair an individual's capacity for independent living. Importantly, dementia must affect several areas of higher cognitive function. It is a "global" condition and not attributed to a single type of cognitive deficit, although often one deficit (e.g., memory loss) will appear to predominate. Some clinicians prefer the use of the word "syndrome" because it emphasizes the provisional nature of attempts to subclassify the dementias. Used in this way, "syndrome" conveys the idea of a group of symptoms and signs that often occur together and implicate shared underlying pathologies.

The *Diagnostic and Statistical Manual* (DSM, http://dsm.psychiatryonline.
org/doi/book/10.1176/appi.books.9780890425596) criteria for dementia syndrome are more specific than this. To meet DSM accepted standards in support
of a dementia diagnosis the following are required:

> Multiple cognitive deficits, including memory impairment and at least one
> of the following: aphasia, apraxia, agnosia; cognitive deficits that are severe
> enough to interfere with working and/or social life; that have declined from
> previous levels and do not arise only in the presence of transient clouding of
> consciousness (as in delirium).

Most problems with these criteria concern (1) the recognition of dementia in conditions in which memory impairment is absent or slight in the first
years of dementia (this occurs typically in frontotemporal dementia [FTD] and
in Huntington's disease), and (2) the distinction from "age-related memory loss."

About thirty years ago, most Alzheimer's disease (AD) researchers expected
near-perfect agreement between valid clinical AD diagnoses and findings on
brain examination after death.[1] Gradually, this acceptance gave way to the considered view that such a level of agreement between clinical and biological
examination may be impossible and that a different way of thinking about AD
was needed. Experts widely discuss many differing views.[2] One view is that AD
is a distinct clinical entity with precisely defined criteria for diagnosis during life
as well as at autopsy. Another is that AD is a syndrome with antecedents that
are as complex as those for age-related diseases of blood vessels like ischemic
heart disease. This view has found support in the strong links that exist between
brain–blood vessel disease and AD, such that one appears to predispose to the
other (Figure 9.1).

Other views seem more pragmatic: Clinical descriptions of AD should be
divided into three phases. In the last phase, when dementia is firmly established, AD should be defined by clinical criteria strengthened with biomarker
data, such as reduced hippocampal size found on brain imaging. In an earlier
phase, during which time symptoms are insufficient in severity to meet criteria
for dementia, the syndrome of mild cognitive impairment (MCI) is attributable
to underlying brain changes resulting from AD.[3] Criteria for this phase require
clinical features similar to dementia with AD but are less severe.[4] To link MCI
to underlying AD, biomarker data are needed to support the presence of AD.

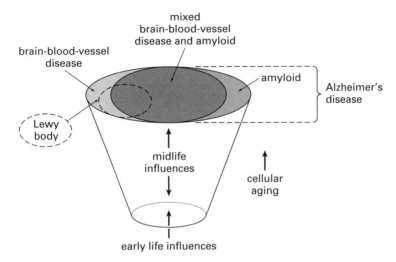

FIGURE 9.1 AD is not a simple concept. After the age of eighty years, most people living with dementia have increasing evidence of both Alzheimer-type brain changes and brain–blood vessel disease. The figure shows this as the larger area of "mixed" brain–blood vessel disease and amyloid with less common dementia types of relatively "pure" disease lying alongside. Life course factors contributing to dementia are shown as cumulative age-related contributions to dementia susceptibility. As for many types of blood vessel disease, these factors do not directly cause dementia but typically involve, for example, a predisposition to midlife high blood pressure linked to prenatal exposures, or to diabetes linked to maternal nutrition.

In the earliest preclinical phase of AD, no clinical signs or symptoms are present and only biomarker data are relevant. This considered view of AD encompasses a spectrum of AD changes that is continuous over a period of ten to twenty years.[5,6] Alongside this AD spectrum lies another syndrome that is made up of mostly brain–blood vessel changes that alone are insufficient to cause cognitive impairments but that in conjunction with developing AD changes can act additively and eventually lead to dementia.

From a life course perspective, other factors influence the clinical diagnosis of dementia. Among the "oldest-old" (older than eighty-five years of age), multiple illnesses and age-related handicaps are frequent. A set level of symptom severity is needed to meet criteria for MCI or AD that is "greater than expected for an individual's age and background." This poses a problem when trying to classify the "oldest-old" who can have at least one sensory handicap and among whom disabilities related to chronic illnesses are not uncommon.

In these circumstances, it seems unwise to impose boundaries around dementia syndromes when relationships with underlying brain AD and blood vessel changes are not straightforward and when, with increasing age, other factors confound reliable dementia ascertainment and thus many disorders appear to overlap in the same individuals. A further layer of complexity is added when the contributions of cultural differences modify symptom formation and exposure to influences that increase risk or delay the onset of dementia.

A full causal classification of dementia syndromes is not yet available: All current classification systems are provisional and require revision as causal mechanisms are attached to specific dementia syndromes. Specialist clinics have great interest in discovering methods to detect dementia subtypes using selected biomarkers and then testing how these perform alone or in combination in the prediction of dementia classification.[7,8] We shall see later how this occurred in the FTD syndromes. Ground-breaking advances in clinical classification by Tony Neary in Salford (United Kingdom) laid the foundations for clinic–genetic FTD studies with far-reaching consequences for the management and counseling of affected families.

Potentially, the level of detail required for a causal assessment of dementia syndromes will require sequential brain imaging, biological measurements in body fluids, and genotyping. It is easy to think this might become so intensive (and extensive) that large-scale clinical studies (including drug trials) would become prohibitively expensive and doubtfully feasible. In contemporary practice, a simple, pragmatic method of clinical classification of the late-life dementia syndromes currently is used to determine the number of new cases arising, how long each subtype survives from onset to death, and the benefits of different patterns of care and drug treatment. Clinical assessment of dementia, therefore, is based on what is available to most dementia service teams and what, in trained hands, can be used to measure useful properties to predict the course of dementia and compare geographic areas and different components of dementia care. Current diagnostic practice in AD reflects a dichotomy between what is needed for AD research and what is needed to plan care and services. This is further complicated, however, by finding families with early onset clinical dementia syndromes for whom all affected members carry the same genetic mutation.

Clinical teams working in dementia accept the approximate nature of schemes that aim to classify dementia, although their solutions have differed in important ways. Some do not take up the challenge of accurate subtyping of dementia, preferring not to impose subdivisions and referring instead to "dementia, not otherwise specified." Less often, subgrouping is based on the presence of specific signs or symptoms, such as memory impairment, aggressive behavior, or age at onset of dementia. In routine practice, most clinicians use a few simple criteria to place a person living with dementia into one of a small number of categories. The divisions between categories are rarely clear-cut—even at postmortem examination—and boundaries are set to take account of the effects of aging on the brain and its blood vessels. These are ubiquitous after about the age of seventy, when most dementias are first recognized.

Grounds are strong to suspect that specific kinds of brain aging of blood vessels can trigger a cascade of molecular events that leads to deposits of abnormal proteins in the brain. These processes cannot be followed as they develop. They are examined within a single time frame that reveals a mixed picture of vascular changes and the molecular pathology of a specific subtype of dementia, neither of which can as yet with confidence be regarded as "primary" changes. The current position is that among the dementia syndromes, a single group of disorders is characterized by the presence in the brain of extracellular plaques containing abundant β-amyloid and, in addition, the presence of neurofibrillary tangles (NFTs) in intracellular and extracellular deposits. Before the recent surge in molecular biology, plaques and tangles were identified by staining brain sections. Because structures stained in this way seemed in early studies to correlate with the severity of dementia before death, a strong case could be made that plaques and tangles were a "gold standard" for an AD diagnosis.

Neuropathologists like Nick Corsellis (who founded the Corsellis Brain Bank containing a repository of more than 8,600 brains held in London, United Kingdom) drew attention to the diversity of brain structures differentially affected by plaques and tangles. He commented on the wide variation in the nature and extent of clinical signs and symptoms when a single category of AD was defined only by the presence of plaques and tangles. If Corsellis's suspicions are well founded, AD will prove to be not a single disease. We might expect, therefore, stratification of the clinical AD syndrome using causal data to enhance clinical trial design and promote the chances of discovery of effective drugs. Subclassification methods include genotyping, precise measurement of clinical features, and sequential brain imaging with or without the inclusion of body-fluid biomarkers for AD. At present, age at dementia onset provides the strongest clue that AD is made up of many subtypes with additional differences caused by the presence of brain–blood vessel disease and differences in the distribution of plaques, tangles, and brain cell death. Thus far, evidence is insufficient to conclude that genetic variation contributing to longevity also is implicated among the causes of AD, although this theory will continue to be investigated.[9]

An important exception to the limitations of current clinical classifications is the identification of families multiply affected by early onset dementia in which case affected members share a genetic mutation that, when present, is proven always to cause their dementia. The mutation is said to show "complete penetrance." Such families are relatively uncommon, and major international efforts are needed to recruit sufficient unaffected members who carry one of these mutations to participate in long-term clinical trails that aim to prevent the development of dementia in these asymptomatic carriers. In addition to the ADAIN collaboration, Francisco Lopera and colleagues at the University of Antioquia (Columbia) have provided an excellent example of the large-scale scientific cooperation required to undertake this type of study. Over twenty-five years, they have recruited the largest known early onset of familial AD pedigree, containing about 5,000 individuals mostly from a mountainous area (Medellín) of Northern Columbia. With the help of U.S. researchers, they plan to test the efficacy of a humanized monoclonal antibody in family members who carry a presenilin 1 mutation.[10]

Three important issues arise in studies like this. First, if brain abnormalities suggestive of the presence of AD are detected in family members who are carriers but asymptomatic, the clinical trial is not truly preventative but rather is AD disease modifying. Second, if brain AD changes are present but asymptomatic, then (as argued) the preclinical (prodromal) phase of AD forms part of the diagnostic spectrum of AD. Third, early onset AD contributes less than 3 percent to overall AD prevalence. Causal mutations are detectable in about 40 percent of these early onset cases and in fewer than 0.03 percent of late-onset AD. The diverse age-related changes linked to brain–blood vessel disease occur in about 80 to 90 percent of late-onset AD cases and are rare in early onset cases. Causal models of late-onset AD are made up of deposits of β-amyloid and NFTs, plus varying degrees of blood vessel disease. Observations on early onset AD may prove irrelevant to the late-onset AD syndrome.

THE CLASSIFICATION OF LATE-ONSET DEMENTIAS

The names used to define subtypes of dementia are based on traditional concepts taken from observations using tissue-staining techniques and a light microscope. These techniques provided a reliable clinical classification and remain useful in the present molecular biological era. With refinements (outlined earlier), consensus criteria have reached acceptable agreement when used by experts. Molecular biological techniques have since provided "molecular signatures" for the subtypes of dementia. These subtypes do not agree exactly with existing provisional clinical classifications. When common biological changes are found to be mixed together in a single case, clinicians explain them by the high frequency of their separate incidences yielding an expected number of "mixed cases."

In Chapter 6, The Biology of the Dementias, the two main types of dementia were divided into the *tauopathies* (AD, FTD, and corticobasilar degeneration) and the *synucleinopathies* (Parkinson's disease [PD] and dementia with Lewy bodies [DLB]). We did not add to these major groups several newer types of dementia that could not be classified in this way—largely because these are rare and complex diseases. FTD illustrates these issues very well. Initially, twenty years ago, it seemed to clinicians learning about these disorders and their relationship to Pick's disease (an older term that encompasses all dementias that affect mostly the frontal lobes) that FTD would be established as a new dementia entity. Within ten years, it was clear that a better term would be frontotemporal lobar degeneration (FTLD) that encompasses a range of disorders that share some but not all molecular features.

The relationship between AD and DLB also provides a good example of the problem of classification. When AD was first described, the gold standard for diagnosis was provided by pathological examination. As clinical interest in AD gathered momentum from 1978, the AD concept widened and diagnostic boundaries blurred. Recognition of DLB produced new diagnostic criteria that also rested on gold-standard pathological findings. These methods produced clear-cut boundaries between AD and DLB using pathological data but omitted a large number of indeterminate cases lying between the two. It was customary to hear of frequent numbers of "AD with DLB features," although this never seemed satisfactory.

Studies of early onset families multiply affected by AD or FTD and rarely Creutzfelt–Jakob disease (CJD) contributed hugely to a better understanding of problems occurring at the intersections among dementia syndromes. What seemed plausible was that "molecular signatures," highly specific for each type of genetic mutation, appeared imminent. By 1990, some biochemical neuroscientists championed a "molecular classification" of the dementias, a proposal that now seems wildly premature. So many proteins are disturbed in the dementias and so many of these are common to more than one type of dementia that classification using "molecular signatures" alone remains impractical. Some experts had fallen to the temptation to deduce that the "primary" pathological process was initiated by the protein coded by the mutated gene, and this knowledge must surely simplify matters. Although this, indeed, may be the case in familial forms of dementia, sporadic forms of dementia are different. Whereas a mutated gene may cause abnormal proteins to form toxic aggregates in familial AD or FTD, a quite different set of initiating events probably generates abnormal protein aggregates in sporadic dementia. Additional problems arise when aggregates resist degradation in the living brain and undergo subtle alterations in structure so that what is examined at postmortem differs importantly from what was deposited many years before. An analogy would be the problem faced by an accident specialist called to examine the wrecks in an automobile scrap yard and asked to explain how each accident had occurred ten years earlier. The chances of success might be slim and probably would diminish when told that many autos involved in the same accidents already had been crushed and could not be examined. Brain cell death presents a similar problem when only the surviving cells remain and make up a picture of dead and dying tissue surrounded by the ghosts of dead brain cells.

Currently, it is widely accepted that relatively few pathways lead to cell death in different types of disease, and we have no reason to suspect that the dementia syndromes differ from this general rule. The molecular signatures attached to each dementia, therefore, may include some of these common elements but fail to explain the highly specific patterns of brain cell death that typify each dementia syndrome.

The current position in dementia classification is summarized in Figures 9.1 and 9.2 showing overlaps between common dementia syndromes. Blurred boundaries between dementia syndromes may reflect imprecise methods of characterizing and defining each syndrome. Figure 9.3 shows how physical boundaries between physical states can be defined very precisely when methods of measurement are reliable and accurate: in this example temperature and density

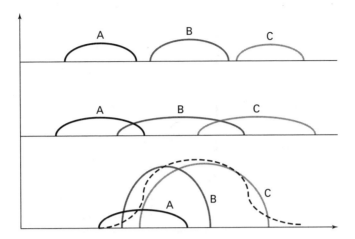

FIGURE 9.2 Three scenarios are presented in schematic form. In the top row, there are three dementia entities with clear-cut boundaries ("areas of discontinuity") between each. In the middle row, the three dementia subtypes overlap, and in the lower row, all three overlap to the extent that they form a continuum.

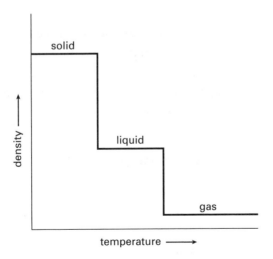

FIGURE 9.3 This shows the boundaries between three phases of a substance that passes from solid to liquid to gas as temperature rises. The point to stress is the clear-cut nature of the boundaries between each phase when temperature is plotted against density. This diagram shows nothing unusual: It simply shows a physical property of matter when the methods of measurement are precise.

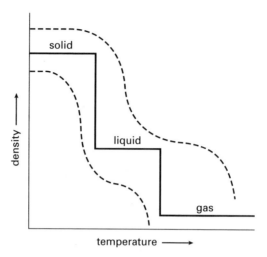

FIGURE 9.4 This shows the same substance passing through the same three phases (solid, liquid, and gas), but now the methods of measurement of both temperature and density are prone to error so the true relationship might lie anywhere within each range of error. Observations obtained in this are more likely to fail to detect boundaries and presume that passing from one phase to another occurs over a range of temperatures and densities. In short, it might be erroneously concluded that there is a continuum of phases in which mixtures of gas–liquid and liquid–solid frequently occur.

are measured in three phases gas/liquid/solid. When methods are imprecise as shown in Figure 9.4, boundaries are much more difficult to find and the three phases become indistinguishable. When methods of measurement are imprecise or critical data are missing, true boundaries between separate types of dementia may fail to be detected and boundaries will be blurred as shown in Figure 9.4.

CAUSES OF DEMENTIA

Dementia is defined as an irreversible loss of higher cognitive functions, including memory. It affects the mature brain and is not detectable at an earlier age. There is some uncertainty whether the level of consciousness is always affected by dementia. If consciousness is defined as the ability to discriminate among external stimuli, then consciousness is well preserved until dementia becomes severe. The dementias, therefore, are chronic acquired disorders divisible into those that originate in brain tissue (the primary dementias) and those that originate outside the brain (the secondary dementias).

TABLE 9.1 The Major Subdivisions of Dementia with Representative Examples

PRIMARY DEMENTIAS	SECONDARY DEMENTIAS			
	Vascular Dementias	Metabolic Disorders/ Nutritional	Toxic Dementias	Infections
Alzheimer's disease, dementia with Lewy bodies, mixed AD-vascular dementias	Cortical infarcts, subcortical dementia, vasculitis, diffuse white matter disease	Liver failure, parathyroid, thyroid diseases, vitamin deficiency, folate/B12	Alcohol, pollutants, metal poisoning, solvents, plant poisons, illicit drugs	Creutzfeldt–Jakob disease (CJD), Subacute Sclerosing Panencephalitis (SSPE), HIV
Frontotemporal dementias, tauopathies, corticobasilar degeneration				
Huntington's disease				

Primary disorders of brain tissue cause the death of neurons through the local release of toxins produced by the disturbed metabolism of naturally occurring brain molecules. Table 9.1 lists the causes of primary dementias. Secondary dementias also lead to the death of neurons, but their causes lie outside the brain. The most frequently encountered causes of secondary dementias are those diseases that affect the integrity of the brain blood supply either by direct effects on the structure of brain blood vessels, by blockage of a blood vessel, or by damage to the blood-brain barrier. Most often such blockages are caused by small blood clots (thrombi) that are displaced from their site of origin (to form an embolus) and pass through the circulation until they stop in a brain blood vessel. The ingestion of toxic substances provides another major cause of secondary dementia. These include excess alcohol, naturally occurring neurotoxins of plant origin (as in lathyrism), environmental pollutants that include heavy metal poisons and other industrial waste, and some atmospheric solvents and contaminants of illicit drug and (possibly) illicit drugs themselves.

Importantly, the harmful effects of some toxins are delayed. Harm may require long-term cumulative exposure or may not appear until the effects of brain aging are superimposed on acute brain injury and through an additive or synergistic pathway cause dementia symptoms many years after acute exposure to the toxin. Some infective causes of dementia also show delayed effects, once referred to as "slow viruses," but it was not clear whether the virus was modified in such a way as to become slowed in its development in the host or if a form of virus–host interaction was slowed. Chapter 6, The Biology of the Dementias, described a small number of infective particles known as prions. Lacking DNA, these are unlike any other type of infective agent and consist simply of a misfolded protein. This type of agent was identified in a brain disease of sheep (scrapie).

Sheep susceptibility to the infective agent required the animal to possess a gene that coded for the same misfolded protein to be affected. Alan Dickinson (MARC Neuropathogenesis Unit, Edinburgh, United Kingdom) pioneered many of the genetic studies on this group of disorders that in addition to sheep affects wild deer, mink, and cattle. In this last group, the neurological disorder is known as bovine spongioform encephalopathy (BSE or "mad cow disease"). The causal agent was passed to humans where it is called human new variant CJD (nvCJD).[11,12]

II. ALZHEIMER'S DISEASE

What features distinguish a person living with dementia from anyone else? Such a simple question emerges as quite a tricky problem when all the facts are set out. To become a "dementia case," an adult must fall below thresholds in key areas of cognitive performance and activities of daily living. They must have deteriorated from an original ("premorbid") level of mental performance. These thresholds are not set in stone and reveal wide individual variation.

People of higher socioeconomic status, who have strong family support, and who can retain sufficient mental flexibility to compensate for early deficits attributable to the presence of brain pathology will present to health services later in the course of their dementia. At presentation, these individuals often show a greater degree of brain pathology than those who do not have similar support or advantageous socioeconomic circumstances. From this, it follows that along with improvements in the material well-being of a society, when measures are in place to enhance family and community support, the number of "dementia cases" seen in clinical services will seem to be reduced. The opposite is also true: When families are smaller and geographically dislocated, older relatives find they must rely on community support services that become available only once they are defined as a "dementia case." The best solutions to this problem rely on prospective longitudinal studies with access to a wide range of data sources relevant to both hospital-treated and community support services.

Most industrial countries now have reliable estimates of the number of people living with a clinical dementia syndrome, have introduced reforms to provide health workers with useful guidelines on assessment and standards of dementia care, and have facilitated access to the few drugs with efficacy in the reduction of memory problems and improved activities of daily living. At a political level, these developments in dementia services are widely supported, but their implementation sometimes falls short of the health planner's intentions.

At one time, it seemed that "counting heads" in dementia studies would be superseded by advances in laboratory molecular genetics. Supported by identification of genetic mutations in early onset AD, claims were made that the causes of AD, irrespective of age at onset, were genetic and that these causes soon would be remediable. All that would be needed was a program of genetic tests to count the number older adults who were carriers of "susceptibility genes" not only to obtain the numbers currently affected but also to identify future trends based on gene frequencies in the general population. These hopes were never fulfilled. Carriers of genetic mutations that predispose to dementia proved far fewer than anticipated (probably less than 1 percent) and the much hoped-for molecular classification of the dementias never materialized. So far, among many putative AD genes, only *APOEε4* has remained a well-established genetic susceptibility factor for late-onset AD. Those who possess one or two copies of this allele are at increased risk of dementia. The consensus view (reached after wide consultation) is that testing for *APOEε4* is never justified in the assessment of an individual's risk of dementia. Too many old people who complete almost all of the risk period for AD (to the age of ninety-five and beyond) are *APOEε4* carriers but do not develop dementia.

FAMILY HISTORY

The best established risk factors for AD among Caucasians of European descent are increasing age, family history of dementia, and the presence of one or two copies of the *APOEε4* allele. When family pedigrees are investigated, the relatives of AD patients at greater risk are among women and risks increase further when *APOEε4* is present. Among other ethnic groups, age is the major risk factor, and *APOEε4* appears less important than among Whites. Some evidence indicates that other AD risk factors vary among ethnic groups and between geographic areas.

Family studies of patients with AD and follow-up studies of cognitive aging among identical twins reared apart or together provide the best opportunities to evaluate lifetime risk of dementia. Data from these studies help understand the roles of genetic and nongenetic factors in the course of AD. Among many reports on this topic, the work of Lindsay Farrer and his colleagues at Boston University School of Medicine (United States)[13] and Nancy Pedersen of the Karolinska Institute (Sweden) with Margaret Gatz at the University of Southern California[14] provide much of the information relevant to the evaluation of risk among first-degree relatives of AD patients.

In The Multi-Institutional Research in Alzheimer's Genetic Epidemiology program (MIRAGE), Farrer's group collected probably the largest series to date with more than 17,000 first-degree relatives of 2,339 White patients with reliable AD diagnoses. These patients were compared with 2,281 first-degree relatives of 255 Americans of African descent. The average age at dementia onset was about seventy years of age among the AD patients and was reported up to the age of eighty-five among the relatives. To examine the effects of a shared environment on individuals who were not genetically alike, spouses of AD cases also were studied.

The MIRAGE study provides the best available estimates of the risk of AD among first-degree relatives of AD patients. The cumulative risk of dementia among White relatives was about 27 percent and about 44 percent among Black relatives. These estimates did not depend on the educational level of the AD patients. Among spouses of White AD patients, the risk of dementia was about 10 percent, and among spouses of Black AD patients, it was about 19 percent. Although these risk estimates differed significantly between Blacks and Whites, the proportional increased risk of relatives (over spouse) was the same in both groups (i.e., about 2.5). The presence of a single *APOEε4* allele in the AD patient increased the risk of AD in a first-degree relative by about 1.4 in Blacks and Whites.

Genetic mutations can explain only a very small number of AD cases. Understanding how genes and environment contribute to the risk of AD is an important goal of AD research with major implications for the design of interventions that aim to prevent AD. The work of Pedersen's group in Sweden has contributed to these questions with their studies of Swedish 11,884 twin pairs.[15] A twin study is a fascinating type of "natural experiment" during which the twins are either identical (monozygotic) or nonidentical (dizygotic). When AD is present in both twins, they are said to be "concordant" for AD. If concordance is significantly greater in identical than nonidentical twins, this suggests genetic influences are present. When concordance is significantly less than 100 percent in identical twins, this also merits explanation. Discordant twins are also potentially informative in the search for biomarkers for AD. Concordant twins and monozygotic twins tend more often to volunteer for research. The Swedish twin sample showed that the age of onset of AD is affected by genetic factors, that heritability of AD lies between 60 and 80 percent, and that women are not at greater risk of AD than men. (Heritability is the proportion of observed variation in a characteristic that is due to genetic variation in the particular population studied.) Pedersen reasoned that the same genetic factors are operating in men and women. The Swedish findings on heritability of dementia are the best available and probably more soundly based than those reported in the United States, Norway, and Finland. Pedersen's investigations of dementia among twins are certainly the largest so far reported, and apart from showing the importance of genetic factors among the causes of late-onset AD, they underline the sizable contribution of environmental factors—particularly in the modification of age at AD onset.

AD AND MILD COGNITIVE IMPAIRMENT

Interest in the category of MCI has diminished somewhat as the definition of AD has extended to include preclinical AD with or without cognitive deficits. Nevertheless, many old people worry about slight degrees of impairment and seek help or reassurance. The main features are its mild nature and frequent lack of detectable impairment in domains other than memory. Several overlapping terms often are used synonymously for adults fifty years of age or more with mild memory impairment. These people lie on a continuum of cognitive impairments that are of insufficient severity to meet dementia criteria. The introduction of reliable biomarkers to detect underlying AD is likely to replace this term with another firmly within the AD spectrum. When brain–blood vessel disease is the preferred explanation of MCI, a term (or its equivalent) such as "mild vascular cognitive impairment" might be used. No matter how these mild disorders are classified, many people living with mild impairment will seek advice about further probable decline.

COURSE OF ILLNESS IN AD

After age seventy years, age-associated memory impairment becomes more frequent and troublesome. When cognitive tests show significantly greater than expected lower scores on two or more cognitive domains (one of which must concern memory), criteria for a diagnosis of MCI may be met. Among individuals with MCI, after age seventy, about one in eight each year will become more cognitively impaired and meet the criteria for AD.

Once an AD diagnosis is established, most people decline gradually to death over a period of five to twelve years, typically around eight or nine years.[16] Age at AD onset before age sixty is associated with shorter survival (around seven years) than later onset after age eighty, when life expectancy sometimes does not differ significantly from the general population at the same age but without dementia.

Currently available antidementia drugs will slow dementia progression in about 35 percent of patients for periods of up to about two years. After this point, when drugs have ceased to be effective, the rate of decline accelerates so that within three years from initiation of drug therapy, patients experience the same degree of dementia that was expected had they not received the drug. Nevertheless, this period of slowing of disease progression is highly valued by those living with dementia, and it is accepted that this class of drugs, although considered palliative by some, in fact, does have very real benefits for a minority of patients. They sometimes delay significantly the time to admission to residential care.

WHAT CAUSES DEATH IN AD?

Premature death is frequent among all patients living with dementia. Death certificates remain unreliable sources for specific causes of death. When autopsy data are available, respiratory disorders account for almost 50 percent of all dementia deaths, with bronchopneumonia as the most frequent specific cause.[17] This probably is linked to impaired ability of patients living with dementia to clear secretions from their airways, often complicated by poor coordination between in-take and swallowing food, causing aspiration of foodstuffs. Heart disease often is found in dementia as in the general population, but there is some doubt about the frequency of deaths caused by stroke: Some reports suggest these are increased in dementia, but other reports cannot support this. General frailty and severe weight loss (cachexia) increase the likelihood of falls and bone fractures.

During the last years of life, great care is needed to ensure adequate nutrition. Failure to achieve this sometimes is linked to lack of help at mealtimes. Loss of body mass and immobility predispose to the development of bedsores, which, with poorly managed urinary or fecal incontinence, can lead to most distressing circumstances at time of death.

AD, DOWN SYNDROME, AND CHROMOSOMAL INSTABILITY

Extensive data establish increasing age and, to a lesser extent, female sex as major risk factors for AD. Among a small number of diseases claimed to be associated with AD, only the association with Down syndrome is significant. At first, the link was regarded as a medical curiosity, more evidence that the person living with Down syndrome was, in the terminology of the time, "degenerate." Subsequently, a small number of careful surveys of autopsy reports in Down syndrome showed that the link was almost invariable in those who lived beyond age forty.

Down syndrome or trisomy 21 is a chromosomal disorder resulting from the presence of all or part of an extra chromosome 21.[18] It is among the most common forms of intellectual disability with a spontaneous frequency of about 1 in 700 births. Links with increased maternal age and familial recurrence were recorded from antiquity. The stigma of Down syndrome lessened when in 1959 Jerome Lejeune and his colleagues Gautier and Turpin in Paris (France) described trisomy 21 in nine cases of "mongolism." Once sufficient chromosomal data from Down patients were available from Down patients examined at autopsy after age forty, a small number of patients with partial trisomy 21 who did not develop AD and some with a "mosaic" form of trisomy 21 also were unaffected by AD.

When the β-amyloid protein was linked to the amyloid precursor protein (APP), and by reverse genetics, the gene coding for APP was identified; it was quickly located to chromosome 21.

Studies on trisomy 21 prompted new lines of enquiry in AD.[19] First, a small number of studies searched for a link between early onset AD and increased maternal age without consistent success. Investigation of chromosomal arrangements in AD were more informative. These studies showed that the loss or gain of chromosomes (aneuploidy) in cells cultured from patients with AD provided evidence of premature aging effects in AD. There was no sign of trisomy 21, however, and because these were peripheral leucocytes stimulated to divide in culture, they widely were thought to be irrelevant to brain cells that do not divide. More recently, when chromosomal arrangements within brain cells from AD patients were examined, stronger evidence was found.

First, technical advances showed that the DNA content of neurons varied, suggesting that about 10 percent of neurons were aneuploid. The significance of this mixture of aneuploid and euploid brain cells is unknown but may be linked to the selective vulnerability of aneuploid neurons to the effects of aging. Tom Arendt and colleagues in Leipzig (Germany) showed that the DNA content of surviving neurons in areas most affected by AD contain greater amounts of DNA and were able to demonstrate an "inverted U-shaped" relationship between numbers of hyperploid neurons and dementia progression from control values through preclinical, mild, to severe AD.

In a second line of enquiry, Ivan and Yuri Iourov at the Russian Academy of Medical Sciences, Moscow (Russia), directly examined the stability of individual chromosomes in brain cells from AD patients and found chromosome 21 was three to four times more likely to be triploid than any other chromosome.[20] The discovery of a technique to silence extra genetic material on chromosome 21 is a possible prevention of AD in Down syndrome. If successful in trisomy 21, this approach may prove relevant to sporadic late-onset AD.

These chromosomal data eventually may be linked to genetic mutations in AD so that, for example, a presenilin mutation could disrupt cell cycle division. Alternatively, chromosomal abnormalities may be explained by differences in rates of aging, much as differences in the length of telomeres already are linked. Other than aging, alternative pathways are implicated in chromosomal instability and include the effects of putative neurotoxins and some viral infections that together suggest lines of enquiry comparing the origins of some cancers with those of late-onset AD.

AD AND BRAIN INJURY

Head injuries are classified broadly as "penetrating" into brain tissue or "nonpenetrating." Cognitive deficits after penetrating head injury are readily attributed to localized damage to brain areas, and the symptoms and signs of focal injuries of this type can be correlated readily with known cognitive functions of damaged brain tissue. Other types of brain injury include a progressive form of dementia following stroke, dementia linked to anesthetic misadventure, acute poisoning, exposures to solvents or neurotoxins, and some infections. The critical shared components of these types of injury are deprivation of blood supply to an area of brain critical for memory functions and the effects of neurotoxins that are either locally generated or ingested. Some types of brain injury do not arise in isolation and may be linked to exposures to additional hazards. For example, a young drug user may experience brain damage at different times—after using an illicit drug (e.g., MMDA), being the victim of violent assault, and having suffered occasional nonfatal drug overdose with periods of brain hypoxia. It is, therefore, unsafe with many types of brain injury to expect all ill effects to be determined by one specific event or agent. Complex multifactorial effects frequently present in most emergency rooms and may preclude definite diagnoses.

An association between repeated nonpenetrating head injury and dementia is the well-known punch drunk syndrome or dementia pugilistica. In aggressive physical contact sports (such as professional boxing or U.S. "grid iron" football) long-term consequences of head injury are extensively documented. In these young men, the characteristic feature of increased density of NFTs is most marked in superficial layers of cortex (deeper layers are involved in AD). This association prompted many studies in experimental neuropathology to follow the molecular events that lead from acute brain injury toward AD and their possible modification by factors (e.g., *APOEε4*) already implicated in AD. Currently, many experts agree that brain injury can trigger a molecular cascade that leads to AD in some vulnerable individuals. At a population level, it is estimated that brain injury contributes about 4 percent to the overall incidence of AD. The attributable risk associated with some contact sports is accepted as significant, but its actual effect size is not yet estimated: It may prove to be substantial.

AD AND STROKE

Dementia following stroke is a topic of lively interest. There are now many well-conducted surveys, and experts on acute stroke agree that at least 10 percent of first stroke victims progress to a dementia syndrome that is irreversible and probably little modified by the reduction of exposure to factors that increase risk of recurrent stroke. Should stroke recur, about 30 percent of stroke victims will develop a dementia.

The findings from Framingham, Massachusetts (United States), provide data of general relevance to U.S. Whites. About 700,000 strokes occur each year in the United States. These strokes occur in adults who, for reasons of age alone, are already at increased risk of dementia. The question becomes What is the relative risk of dementia attributed to stroke after adjustment for the expected incidence of dementia? Phil Wolf and his colleagues at the Boston University School of Medicine (United States) have made progress on this question. In their careful study,[21] first stroke victims (19.3 percent) experienced a two times greater risk of dementia than age- and sex-matched controls (11 percent). This increased risk was uniform over the entire follow-up period, so it was safe to conclude that the observed dementia increase was not confined to the immediate poststroke period but evolved gradually over time. This time frame is compatible with the development of AD-type pathology initiated by stroke. It is also relevant that risk factors for stroke (hypertension, diabetes, smoking, and atrial fibrillation) did not affect risk of dementia.

Thus far, it is unclear following acute stroke how much of the brain pathology is classified as the AD type (loss of cortical tissue, diffuse plaques, and NFTs) and how much is caused by deprivation of blood supply to the affected brain area and any subsequent adjacent brain swelling (the penumbra).

EARLY DIAGNOSIS OF AD

Many families now living with AD regret the time taken to reach an AD diagnosis for their family member. Health care service planners agree and urge families to seek advice as soon as they recognize what might be the first symptoms of dementia. These same planners, through education of doctors and the public alike, stress that the capacity to make an early diagnosis should be a priority. Their arguments are consistent with the views of families: Once a firm diagnosis can be made, families can begin to anticipate their future care needs and ensure that whatever time is left is used as fruitfully as possible.

Faced with a new patient who can describe their problems clearly and who otherwise appears well, a clinician begins to gather as much additional information as possible to establish a firm AD diagnosis. A clinical history with corroboration from someone who has known the patient well for some years provides the starting point. Data are needed to show that the patient's current difficulties reflect reliable decreases from a previous higher level of functioning. Most doctors will attach greater significance to memory deficits than any other symptom. Mild cognitive symptoms are commonplace and cannot be explained by an underlying physical illness (as in "silent" bronchopneumonia or myocardial infarct) or by psychological distress attributed to recent stress or loss (as in depressive illness). At this stage, a clinician needs normal values against which to judge the significance of what appear to be mild symptoms but may be portents of a serious progressive dementia.

Clinical judgment at this stage relies on recognition of three components of each symptom. These are the exact nature of the symptom, its severity, and duration. The nature of the symptom is understood by comparing the symptom to its precise definition. This is not as easy as it might seem, and success depends on the doctor not only knowing the definition but also having the skill to incorporate components of this definition into clinical inquiry. Early symptoms of dementia fall into six major domains that form the basis of the first stage of clinical examination: (1) orientation for time, place, and person; (2) attention and registration of new information; (3) verbal memory after immediate and delayed recall; (4) understanding of simple sequential commands; (5) use of language; and (6) visuospatial ability. For good reasons, some individuals without early dementia perform less well on such questioning. These reasons include feeling overly anxious, fearing failure, having poor skills in literacy and numeracy, having sensory impairments (hearing or vision), and having lower than average educational attainments. With experience, most clinicians learn how to introduce topics of increasing difficulty as sensitively and comprehensively as possible.

Symptom severity relies on asking about the extent to which the presence of a symptom interferes with whatever the patient has wanted to do. This line of questioning often is helpful (especially when an informant can independently verify an account) and can reveal critical decrements in cognitive and social domains that were not obvious on preliminary inquiry. Importantly, alterations in social judgments may remain undetected until direct questions are asked of both the patient and, separately, of an informant. In FTDs, these alterations can be the first areas of difficulty that persist for many months before memory problems are either reported or detected on testing.

Symptom duration rarely is established satisfactorily. At first, what appears to be a minor problem is dismissed as trivial, "no worse than expected for his age." Only later, as problems become more frequent, are thresholds of concern crossed and perhaps more openly discussed within the family. Occasional watershed moments may be revealed in such statements such "it was around then that I began to feel unsafe when he was driving" or "after that I never left her in charge of the grandchildren; she seemed to lose it for hours at a time . . . off in her dream-world." The issues of what would be a level of forgetfulness that reasonably could be ignored given the age of the patient and what level of severity and duration must be investigated further are common causes of concern. In broad terms, a clinician can safely assume that if every patient who presents after the age of sixty with complaints of memory impairment were assiduously followed up, from about seventy years of age, about one in eight of these patients would have progressed to dementia by the age of eighty. Cautious circumspection frequently is recommended and, as a minimum, some form of oversight should be put in place that would ensure prompt review should symptoms worsen.

In many clinics, the need for review and for more precise detection of symptoms established universal referral of all patients with suspected early dementia for cognitive assessment. This assessment is rarely comprehensive and most often includes a limited range of mental tests that estimate original levels of mental ability ("premorbid intelligence"), current performance on tests of mental speed, visuospatial ability, verbal fluency and use of language, verbal memory (immediate and delayed recall), logical memory, and abstract reasoning. When early dementia is suspected, cognitive test results obtained at presentation provide an invaluable baseline against which to compare later changes and, of huge importance, against which to estimate rate of change in dementia progression. These data, in turn, are used to advise families of possible prognoses and to guide future investigations and choice of treatments.

III. BRAIN IMAGING IN EARLY DEMENTIA

When brain imaging became available, neurologists who wanted to exclude "treatable" forms of dementia requested it most often. In routine clinical practice, these examinations rarely detected more than 5 percent of patients with underlying disease, although specialist tertiary referral centers did obtain much higher rates of detection. Currently, brain imaging is used most often to distinguish between brain aging and dementia and for subtyping dementia as an aid to treatment. Magnetic resonance imaging (MRI) is preferred in some specialist centers to computed tomography (CT), largely because MRI is better at assessment of regional brain atrophy and detection of other brain changes, such as white matter hyperintensities (WMH) and bleeding in cerebral microvessels.

Distinguishing between normal aging without dementia and the brain changes typically seen in dementia is an important task in early assessment, and MRI is the investigation of choice. Despite the importance of this task, the decision to conduct this assessment is difficult to make and the hurried or careless clinician can cause much unnecessary concern. The problem arises when after a brief consultation, a nonspecialist feels confident enough to conclude prematurely that early dementia is likely and to request brain imaging. Slight degrees of cortical atrophy are reported as "consistent with early dementia" and inappropriately are conveyed to the patient and family as evidence to support an early diagnosis. By itself, the result of brain imaging does not establish a clinical diagnosis of dementia. This conclusion can be based only on the presence of symptoms that are sufficient in extent and duration to establish that an individual has progressed from a previous higher level of cognitive function to a current level that is significantly below what is expected for a person of that age and background. Findings from brain imaging are used to support a dementia diagnosis and may be confirmatory.

FREQUENT MRI FINDINGS IN THE AGING BRAIN WITHOUT DEMENTIA: NORMAL AGING

Following brain imaging, several findings are common in the aging brain without dementia. These results are considered to be a part of normal aging.

Brain Appearance
Normal loss of total brain volume does not exceed 1 percent per year, more usual around −0.5 percent per year. The hippocampus appears more vulnerable

to aging and can lose up to 1.5 percent per year. Enlarged spaces around brain blood vessels ("perivascular") are found in normal aging without dementia.

White Matter Hyperintensities

Small scattered lesions are normal; when these join up to form large lesions ("confluent") these are always abnormal.

Microbleeds

About one in five people over the age of sixty have cerebral microbleeds, which can increase slightly in the absence of dementia. Exact distribution can indicate the effects of chronic high blood pressure and requires expert evaluation.

Cerebral Infarcts

A silent cerebral infarct occurs without the signs or symptoms of a stroke. It is caused by a disruption of the supply of oxygen or nutrients to a discrete brain region caused either by a burst (hemorrhage) or block (thrombus or embolus) of a cerebral blood vessel. Loss of blood supply causes death (necrosis) of brain tissue. Silent cerebral infarcts are found on MRI in about one in five older adults. These may increase the risk of dementia.

For convenience, brain volume lost with age in the absence of dementia usually is expressed as percent total brain volume lost over time. During brain development, the capacity of the skull expands to accommodate exactly the size of the growing brain within. Total "intracranial capacity" reliably estimates the total original brain volume before the effects of aging or dementia caused brain shrinkage. This capacity helps estimate decreases over time in total white or gray matter. Although total brain volume approximates to the sum of losses in regional brain structures, it provides an informative guide as to what is happening in the aging brain. For example, a simple measure such as brain volume can be related to survival in old age, which in part, is attributable to decreased survival in dementia.[22]

The volume of the hippocampus is of greatest interest in the early detection of AD. When repeated MRI shows loss of hippocampal volume greater than 3 percent per year, this is likely to support an AD diagnosis when accompanied by other evidence of cognitive decline. Rates of hippocampal loss average about 5 percent per year in confirmed AD. The principal difficulty when using this estimate in clinical practice is the problem of defining the exact boundaries of the hippocampus. Ideally, this method manually traces the borders of the hippocampus in three-dimensional space followed by computing the volume of the irregular object contained within. The method requires technical expertise, and when used in follow-up exams, it should be accompanied by measures of reproducibility. Fortunately, automated methods of measurement are improving quickly and soon should become part of clinical practice. This development is relevant to clinical trial designs that aim to detect the benefits of disease-modifying drugs in the treatment of AD.

Most clinicians incorporate findings from clinical history and examinations based on brain imaging data in a way that acknowledges a spectrum extending from normal aging through various subtypes of dementia syndromes. Such a continuum of change forms one dimension of a multidimensional approach to brain aging and dementia. This information provides the foundation of clinical judgment about the probable causes of the patient's dementia and the course it will follow. Other components of this clinical model include functional capacity to complete activities of daily living, physical fitness[23] and general well-being, social supports needed to keep safe, cognitive effects of prescribed medications, use of alcohol and illicit drugs, and relevant aspects of lifelong personal adjustment.

By extension, this model is relevant to understanding the relationship between normal aging and the dementia. Although it is attractive to assume that such a relationship remains relatively fixed through late adulthood into extreme old age, this belief may not be justified. People living to age ninety and beyond account for the largest growing segment of the elderly population. This is found in countries as diverse as the United States, Japan, and New Zealand and, paradoxically, among rural Africans, despite their high infant mortality rate. Among the very old, the usual predictors of mortality do not operate so effectively— lower socioeconomic status, obesity, and smoking appear to be less important. When compared with the young-old, the oldest-old have more effective antioxidant defenses, better immune surveillance, and better blood glucose control. Their psychological health is also better than the young-old and, taken together, this suggests that the oldest-old are a highly selected group of tough survivors with relatively slow rates of aging and more effective ways of coping with stress and age-related diseases.

To date, clinicians assumed that the highest rates of dementia were found in the oldest-old, but this assumption was based on clinical impressions rather than on detailed assessment of the small numbers of oldest-old available for tests. In one recent study of 1,694 people living with dementia, Lautenschlager and colleagues showed that dementia rates declined after age ninety, which was consistent with reports elsewhere that far from reaching 100 percent at age one hundred, observed dementia incidence from ninety to more than one hundred years old fell in the range of 30 to 60 percent. Postmortem studies of brain tissue from centenarians provide strong grounds to accept that they are relatively resistant to the formation of aggregates of abnormal proteins in the form of plaques and NFTs and that dementia is more likely to arise when the brain also has multiple microvascular lesions. When NFTs are found in centenarians, they are less likely to form in brain areas critical for memory function, suggesting a possible genetic variation in brain region susceptibility. In turn, such genetically determined decreased brain cell susceptibility may be linked to genes that promote longevity.

As findings from clinical research in brain aging and the dementias of late life are folded into routine care and assessment of older adults with suspected early dementia, methods to detect long-term brain structural changes will be introduced. These most likely will be computer automated with centralization of validation methods to support remote locations. Although visual rating scales have provided useful preliminary data, they now appear somewhat crude and open to observer bias.

The telltale signs on brain imaging of moderate to severe AD are loss of volume that is most marked in the parietal or temporal cortical regions and that present, to a lesser degree, in frontal and occipital regions. Cortical loss tends to be symmetrical, although this is not a constant feature. The hippocampus is more affected in AD than other brain areas, sparing the major motor and sensory areas until late-stage AD. The death of so many brain cells in the cortex leads understandably to commensurate loss of much of the structure of the large white matter tracts that provide cortex-to-cortex connections. Gradually, advanced methods of statistical analysis of brain structures will be introduced from research into clinical practice. Major developments will allow for comparisons between findings from individual patients and centrally located "reference atlases" that suggest how a patient may differ from a typical comparator drawn from the general population with similar educational and occupational backgrounds. Presently, it is feasible to anticipate merging of large databases that include brain structural and functional observations as well as genetic and epigenetic data with clinical outcome data. This approach makes the best use of the information currently available but will place great demands on the limited resources for mathematical analysis in clinical assessment. Long foreseen by Alan Turing,[24] the pioneer of artificial intelligence, mathematical applications to the stratification of "big data" may herald the introduction of "supercomputers" into clinical practice, where the probabilities of specified outcomes can be estimated only by machines and not by clinical judgment (Table 9.2).

The general consensus is that effective treatments of dementia will be based on disease-modifying interventions that are tested in those at the first stages of disease or, potentially, in those at highest risk before the onset of disease processes. Three methods now are being used widely to investigate early dementia, and all are relevant to the problem of early diagnosis. The first relies on positron emission tomography (PET) to detect subtle regional differences in brain metabolism and the deposition of abnormal β-amyloid protein using tracers, such as Pittsburgh Compound B (PiB). Almost all researchers now accept that the lag time between disease initiation and symptom onset lies between ten and twenty years and that the average may be closer to twenty than to ten. The U.S. and the U.K. brain banks were established to investigate the occurrence of brain pathologies associated with dementia. The findings from these general population surveys are relatively unbiased and support a consensus that most dementias seen in late life are attributable to "mixed pathologies" with frequent observations of

TABLE 9.2 Recent Advances in the Early Diagnosis of Dementia

BRAIN AREA	TYPICAL AD	ATYPICAL AD	FRONTOTEMPORAL DEMENTIA	DEMENTIA WITH LEWY BODIES
Hippocampus	General atrophy	Usually absent	Atrophy most marked in anterior	General atrophy
Temporoparietal cortex	Atrophy more marked than elsewhere	More evident in posterior region	Relatively spared	General cortical atrophy
Asymmetry	Absent	Often present	Widening of frontal ventricles	Absent

AD-type changes (in the form of plaques and tangles) and widespread evidence of disease in the brain's small blood vessels (small vessel disease [SVD]).

In the living, observation of the brain at rest using radiolabeled (oxygen-15) water can trace blood flow through the brain. Before the onset of dementia, blood flow falls compared with those who maintain cognitive function but not in all brain areas. Some areas seem spared, whereas in others, such as the frontal areas, blood flow actually can increase. When blood flow studies are combined with methods to detect β-amyloid aggregates, the findings suggest that the blood flow changes are coupled with the earliest stages of AD and occur some years before AD symptoms are reported. This type of study has little practical value at the individual level and is useful only when comparing groups of individuals.

The strong links between AD protein deposits and disturbances of the brain blood flow, especially at a microvascular level, raise the possibility that careful analysis of brain microcirculation could contribute to the differential diagnosis of dementia. SVD may disrupt clearance of β-amyloid from the brain and could be relevant to the early (preclinical) diagnosis of AD.

Postmortem studies of large general population-based samples of old people carefully assessed for dementia symptoms in late life are rare. Carol Brayne and Fiona Matthews at the University of Cambridge (Medical Research Council Cognitive Function and Ageing Study, MRC-CFAS, United Kingdom) led one important study.[25] This study required a large, well-organized team to compare clinical and brain pathology data to explore the nature and extent of brain pathology associated with clinical dementia syndromes (not otherwise specified).

The most frequent findings in postmortem brains from old people are of generalized brain tissue loss, amyloid plaques, and NFTs, but these are not the only findings. When pathological data are used to distinguish between those with and without dementia before death, the risk of dementia is associated with greater pathology. Alzheimer pathology is made up of moderate to severe loss of brain tissue, amyloid in the walls of brain blood vessels, plaques, and NFTs. When dementia was diagnosed before death, NFTs in the cortex, especially the hippocampus, were the single most clear-cut association with dementia.

The brain pathology of people dying without dementia merits comment. These individuals often show pathologies that appear to predict the presence of dementia of which there was no evidence before death. NFTs are sparse in the cortex (4 percent, see Table 9.3), in which case a severe NFT score is absent, but multiple vascular disease (24 percent) and SVD (47 percent) are common.

TABLE 9.3 Percentage of 426 Individuals (median age 81 years at death) with Different Levels of Severity of Brain Pathology Grouped by Presence of Dementia Before Death

BRAIN PATHOLOGY	NO DEMENTIA % (*n* = 183)				DEMENTIA % (*n* = 243)			
	None	Mild	Mod	Sev	None	Mild	Mod	Sev
Hippocampus								
Diffuse plaques	56	26	15	3	28	24	35	13
Atrophy	69	19	11	1	33	22	36	8
NFTs	19	37	29	15	5	16	30	48
Cortex								
Diffuse plaques	30	24	31	15	16	18	26	39
Atrophy	57	29	14	1	25	23	41	12
NFTs	63	33	4	0	33	26	16	25
Entorhinal NFTs	15	30	43	12	3	13	42	43

Note: NFTs = neurofibrillary tangles; Mod = moderate; Sev = severe.

Source: Data adapted with permission from Matthews et al., *PLoS Med.* 6, no. 11 (2009): e1000180, table 4.

Overall, these data emphasize how difficult it is to predict the presence of a dementia syndrome using neuropathological data alone.

General population surveys of brain pathology of old people dying with or without dementia provide valuable insights into mechanisms at critical points along the pathways leading toward dementia. When the pathological data obtained by staining brain tissue followed by visual inspection using a light microscope are combined with biochemical studies to find molecular signatures for subtypes of dementia, exact boundaries between subtypes of dementia remain to be established. In fact, it seems possible that such boundaries do not exist and that the dementias represent a continuum of change from normality that disrupts dendritic structure and synaptic function and leads to the death of the most vulnerable brain cells in patterns that are specific to each dementia sub-type. Until better methods to study the cascades of molecular pathology in the living brain are developed, the limitations of current methods will fail to detect boundaries between dementias, even if such boundaries exist.

From the perspective of possible dementia treatments, the ubiquitous nature of changes in β-amyloid and tau proteins continue to encourage the search for agents that will reduce their formation or impair their toxic effects. The fact that dysfunctional proteins are formed in more than one pathway, however, widens the search for effective interventions. Any single dysfunctional protein contrib-utes only to a moderate degree to the overall risk of dementia. An effective inter-vention, therefore, could include a number of elements, each targeted upon a separate pathway toward dementia. This conclusion does not sit comfortably with the intentions of clinical trails that aim to identify dementia subtypes and test single rather than multiple therapies in dementia prevention or disease progression.

These issues are quite relevant to understanding the contribution of amy-loid pathology to late-onset dementia. Although a strong case can be made for a causal role of amyloid in early onset dementia, in which a mutation in one of several genes influences the production and removal of amyloid, this is much more contentious in late-onset dementias that appear mostly to be sporadic (i.e., nonheritable) and are associated with genetic mutations. In this case, a credible pathway can be charted from impaired neural health to the death of brain cells and subsequent deposition of aggregates of amyloid. In this scenario, amyloid pathology is seen as a consequence and not as a cause of neural death.[26]

These observations do not disprove the amyloid hypothesis of AD, which remains at the heart of many antidementia drug development programs. The role of amyloid need not necessarily be primary for antiamyloid therapies to be effective in dementia prevention. Longitudinal follow-up studies of brain amy-loid deposition became possible with the development of radiolabeled tracers that attach to brain amyloid and are localized using PET. In one such study, the Australian Imaging Biomarkers and Lifestyle Research Group described a three- to five-year follow-up study of 200 individuals (145 healthy controls, 36 with MCI and 19 with AD).[27] Using repeated cognitive tests, brain MRI, and 11C-labeled PiB, the research group described the natural history and rates of change in amyloid deposition, memory impairment, and hippocampal shrink-age that precede the appearance of the clinical phenotypes of AD. This type of study relies on establishing the validity and reliability of biomarkers of amyloid deposition and reveals the time course of underlying brain changes occurring in the preclinical period before the clinical AD syndrome is detectable. In time, data provided by these types of prospective studies will aid the clinical trials of novel interventions that specifically target amyloid deposition. These trials will be encouraged by an apparent slow rate of amyloid deposition that indicates a wide window of opportunity to decrease risk or even prevent AD. Although data from prospective 11C-PiB studies that use intraindividual comparisons support the amyloid hypothesis of AD, these do not establish amyloid deposition as the primary cause of AD. Nevertheless, these strengthen the view that an antiamyloid therapy could slow AD progression. If effective, such therapies could begin even decades before symptom onset.[28]

IV. FRONTOTEMPORAL DEMENTIAS

After AD, the FTDs are the second most common primary dementia. The FTDs are made up of a mixed group of disorders that occasionally affect more than one individual in the same family. Early studies from Sweden separated FTD from among a spectrum of related disorders previously labeled Pick's disease, but that term rarely is used. The most obvious FTD symptoms and signs are as follows: progressive coarsening of personality, antisocial actions, poor self-care, socially intrusive and impulsive behavior, and deficits in use of language and comprehension. Diagnosis of FTD was much improved through the work of David Neary and Julia Snowdon in Salford (United Kingdom) whose diagnostic criteria are now used widely.

Three FTD subtypes are recognized: (1) apathetic, (2) disinhibited, and (3) stereotypic. These seem poorly differentiated in clinical practice, as their features appear often to overlap and not to discriminate satisfactorily among the subtypes. One clinical subtype termed "semantic dementia" is a disorder of conceptual knowledge caused by bilateral dysfunction of the temporal cortex. It is linked to progressive nonfluent aphasia, which probably is caused by loss of neurons in the language areas of the left cerebral hemisphere. The underlying disease process involves deposits of abnormal tau or ubiquitin. When Parkinsonism occurs with FTD, it is much easier to accept a distinction between FTD with or without Parkinsonism because these features are recognized readily.[29]

VIGNETTE 9.1 FEATURES OF FRONTOTEMPORAL DEMENTIA

The younger unmarried sister of a church minister developed difficulties at age sixty-seven in naming objects, poor understanding of reading, and impaired ability to sing in the church choir. Her self-care diminished, she became child-like when reprimanded, and she had begun to eat with her fingers in company. After two years of slow progression, she saw an experienced neurologist who reported his suspicion that she was demented and asked a psychiatrist to assume her care. She attended his clinic accompanied by her brother and, to his discomfiture, was overfamiliar with the psychiatrist saying she would have worn a more revealing neckline had she known how handsome he was. She wore inexpertly applied facial rouge and on leaving tried to kiss the psychiatrist on his lips. Her brother (the church minister) said she had become "a victim of gluttony" and was now obstinate if prevented from one of her recently adopted routines. Brain MRI showed left temporal cortical atrophy and detailed psychological tests revealed slightly impaired memory, poor comprehension, and inattentiveness. Four years later her memory problems had not worsened noticeably, and although she seemed largely mute, her behavior was significantly more embarrassing, to the extent she was no longer welcome in her brother's congregation.

Other more complex mixtures of signs and symptoms of FTD are encountered in FTD that sometimes is accompanied by bizarre motor symptoms or features suggestive of amyotrophic lateralizing sclerosis (ALS or Lou Gehrig's disease). FTD is not rare. About 5 to 15 percent of a carefully diagnosed population-based series of dementia patients who also were examined at post-mortem were shown to have FTD. These patient samples are notoriously biased toward the inclusion of dementia cases with behavioral problems.

The causes of FTD are fairly well defined. In one Dutch series, 43 percent of cases had a family history of FTD. Causative mutations occur in about 25 percent of FTD cases. These mostly affect (1) the microtubular assembly protein tau (*MAPT*) gene or (2) the progranulin (*PGRN*) gene, although other mutations are suspected, as only about 10 percent are linked firmly to either *MAPT* or *PGRN* genes.

V. PARKINSON'S DISEASE WITH DEMENTIA

The aging brain is susceptible to damage in areas critical for the control of movement. As these age-related changes spread, they hinder fine movements, impairing smooth voluntary actions, and later they produce involuntary tremors most obvious in hands at rest ("pill-rolling" action of opposed thumb and finger as if rolling an imaginary pill). These movements are typical of PD and accompany a shuffling gait and stooped posture. They are experienced increasingly with age so much so that among the oldest-old a slight degree of abnormal movements (termed "Parkinsonism") occurs in more than 30 percent of individuals, although only about 2 percent will be diagnosed with PD. The underlying brain changes in PD are also found in PD with dementia and in DLB.

These three conditions—PD, PD with dementia, and DLB—have abnormal microscopic deposits containing α-synuclein. The synucleins are a family of small proteins found mostly in neurons. Although only discovered and investigated since 1994, much is now known about their structure and function, how they interact with other proteins, and their role in disease. In the brain, they are localized in the nucleus of neurons and in the presynaptic area. Their roles include unknown functions in mitochondria, where they may predispose some cells to selective neuronal damage. Synucleins affect the movement of vesicles toward the synapse and interact with tubulin in the cytoskeleton.

PD, PD with dementia, and DLB form a spectrum of closely related disorders that overlap with AD. Some experts interpret the close relationships between these four disorders as indicating shared causes, but others—especially clinicians—see clear differences between subgroups in their characteristic symptoms and management. Many more people are living with AD than PD, which is less common and affects around 1 million Americans compared with about 5 million people with AD. Among PD sufferers, about 50–80 percent will progress to AD. This wide estimate is related to the success of PD treatment with L-Dopa, which has significantly prolonged life after a PD diagnosis but at a cost of greater incidence of PD with dementia.

PD is a progressive brain disorder. It affects several brain areas of which the most important is the substantia nigra, controlling balance and movement. The early symptoms of PD include shaking or trembling, especially when at rest, and increased muscle tone experienced as rigidity. As PD worsens, slowing of movements increases, as do difficulties with balance and coordination of movements.

In addition to these motor signs and symptoms, there are problems in PD with thinking (cognitive functions) and feeling (emotions). Occasionally, unusual psychiatric symptoms are seen that include depression (common) and visual hallucinations (rare).

Around one in seven people with PD have a reliable family history of PD; for most cases, a complex interplay of genetic and environmental causes seems to be implicated. Familial PD is linked to mutations in a small number of specific genes, whereas in nonfamilial cases, other genes are involved that influence individual susceptibility to PD. It is largely unknown how these genes impair the functions of neurons in the substantia nigra. The main effect of impaired neural health in the substantia nigra is loss of a chemical transmitter, dopamine. This loss weakens the control of movements so that these become "jerky" and less smooth.

Impaired health of dopamine-containing neurons suggests that some genetic mutations may influence their normal working. This seems likely to involve the inefficient processing of certain intracellular proteins that causes these to accumulate and lead to neuronal death. The production of energy by the same neurons produces highly active molecules that, when unopposed, degrade critical intracellular structures. Some evidence shows that other types of genetic mutation in PD can weaken cellular defenses against this type of damage.

DEMENTIA WITH LEWY BODIES

In some U.K. surveys, AD is the most common form of late-onset dementia, and DLB is the second most frequent cause. In some case series, DLB accounts for 10 to 15 percent of all cases. Other studies suggest lower estimates for DLB, but this does not mean that one is wrong and another is right. At present, some experts argue that, within ten years, DLB could become one of the most treatable causes of late-onset dementia. This claim rests on the favorable response of DLB to cholinesterase inhibitors, which are widely available. Claims like this rarely are made for the dementias and could be overly optimistic. General agreement, however, is that DLB is important. The main point for discussion concerns the area where DLB overlaps with AD and elsewhere with PD with dementia. These form a spectrum of dementias that might be only of academic interest were it not for the clinical importance of recognizing DLB to ensure that a major class of sedative tranquillizers is avoided because of the high risk of sensitivity reactions, with occasional fatal consequences.

DLB lies on a spectrum that includes AD, PD, PD with dementia, and types of failure of the nervous system (primary autonomic failure). Within the spectrum of these diseases, abnormal aggregation of the synaptic protein α-synuclein occurs to varying extents. The correct identification of DLB patients is based on recognition of specific symptoms, signs of certain cognitive impairments, and functional disabilities. These features support the accurate distinction between DLB and other late-onset dementia syndromes that include AD, vascular cognitive impairment, and FTD. The clinical criteria for DLB can fail to detect some DLB cases with unusual (atypical) presentations, which most often are attributed to changes of DLB, SVD, and AD arising in the same patient.

DLB patients often have severe sensitivity reactions to a group of drugs (neuroleptics) that was introduced to control the symptoms of acute psychosis in young adults but that are used widely to relieve unwanted behaviors in dementia (e.g., aggression or wandering). The use of neuroleptics in DLB is firmly associated with significantly increased morbidity and mortality and must be avoided in all circumstances. Cholinesterase inhibitor treatment is well tolerated in DLB and can improve cognitive and behavioral signs and symptoms.

WHAT ARE LEWY BODIES?

Lewy bodies were named after Frederick H. Lewy, a colleague of Alois Alzheimer. Lewy first identified microscopic, spherical protein aggregations in nerve cells in old people. α-Synuclein protein is the main component of these aggregations, but the processes that lead to its deposition are largely unknown. Lewy bodies disrupt the brain's normal functioning, interrupting the action of important chemical messengers, including acetylcholine and dopamine. Lewy bodies first were associated with PD, a progressive neurological disease that affects movement. Many people who initially are diagnosed with PD later go on to develop a dementia that closely resembles DLB.

VIGNETTE 9.2. FEATURES OF DEMENTIA OF THE LEWY BODY TYPE

Winnie was seventy-four years old and until then her husband felt her slight memory problems caused little inconvenience. He became suddenly concerned early one evening when she hurried into their bedroom shouting that there was a "horse in the bath."

"Don't be ridiculous," he'd said, but this did not assure her. At psychiatric examination, she described several similar experiences mostly, she said, when looking at patterns on the carpet or wallpaper: "My imagination plays tricks on me. I thought it was silly . . . that horse was real . . . white all over and standing in the bath."

During clinical assessment, Winnie experienced sudden and severe drops in blood pressure. Closer inquiry revealed a history of unexplained falls that later were attributed to faulty control of her blood pressure. Compared with other patients in the clinic, Winnie was less affected by memory problems, but her visual problems distinguished her from other patients. She had difficulty understanding signage in the clinic area and continued to describe recurring unexplained visual experiences that were consistent with visual hallucinations. The "horse in the bath" remained fixed in her memory, and over several weeks, she repeated her description, refused to enter her bathroom, and appeared frightened by the experience. At review, some months later, her appearance had changed. She now showed many features of Parkinsonism (rigid muscles, tremor, and hunched posture). A short trial of a cholinesterase inhibitor drug was extended for almost two years. Over this time, her visual symptoms improved, but her memory slowly declined. At three years after diagnosis, her main problems did not distinguish her from other patients living with AD with signs of PD. Throughout her care, use of neuroleptic drugs was specifically excluded.

COGNITIVE IMPAIRMENTS IN PARKINSON'S DISEASE

A strong association exists between PD and cognitive impairment.[30] This association is relevant to patients and their families who will require careful advice about possible future risk of dementia and possible treatment choices. The distinction between DLB and dementia with PD can be critical and will affect drug choices in the course of dementia. In DLB, the dementia develops either before or in step with motor signs and symptoms of PD. In PD with dementia, PD is well established before the cognitive features of dementia are noticeable.

Specific criteria[31] are available for the diagnosis of MCI in PD and include an insidious decline in cognitive function that is noticed by a patient, a caregiver, or a doctor. At this point, most clinicians would complete a short cognitive assessment to be repeated at a later clinic attendance. If cognitive decline persists, then a more detailed evaluation of a wide range of cognitive functions would be appropriate. When two or more cognitive domains not affected by motor problems associated with PD are scored significantly lower than expected for the patient's age, sex, and education, then an MCI diagnosis can be supported, but this is often a difficult task, especially when PD symptoms (e.g., apathy, mental slowing) are severe.

When cognitive loss is sufficient to impair the functional capacity to remain independent, a diagnosis of PD with dementia will be more likely once other possible causes of poor cognitive performance are excluded. Expert clinicians take great care during this diagnostic process and are watchful for the copresentation of PD with depressive symptoms, which also can impair cognitive function.

The neurobiology of cognitive impairment with PD is relevant to understanding the biology of the dementias. A family history of PD suggests that genetic factors probably are relevant. Genetic association studies have identified a small number of genes that influence PD risk. Some of these genes also are linked to AD, particularly those in the assembly of microtubules in the neuronal cytoskeleton (*MAPT*[32]) and brain-derived neurotrophic factor (*BDNF*[33]). No convincing data, however, link cognitive impairment in PD to differences in apolipoprotein E gene (*APOEε4*), as discussed in Chapter 6, The Biology of the Dementias.

The challenge of understanding cognitive impairment in PD is faced in the design of PD-modifying drugs, and this is related closely to comparable problems in AD. One informative approach is to visualize all the interactions between molecular components of pathways implicated among the genetic and environmental causes of PD. This task can be broken down into two phases. The first names the components of a map of PD pathways toward PD with roles for synaptic disruption, mitochondrial dysfunction, α-synuclein biology, failures of protein degradation and clearance, neuroinflammation, and programmed cell death. The second requires plotting how the tools of information science can explore such a map to reveal where novel experimental results would fit in and to show how new drugs would modify the map's structural and functional dynamics. This is an example of how systems biology pulls together all that currently is knowable about a disease in ways that will support future drug design and new experimentation.[34] For an example of a PD map, see Fujita and colleagues' article in *Molecular Neurobiology*.[35]

VI. DEMENTIAS ASSOCIATED WITH BRAIN BLOOD VESSEL DISEASE

SUBCORTICAL DEMENTIA

Among the dementias, a small number of disorders spares the cortex but shows extensive damage to subcortical structures, excluding those involved in PD. These include Huntington's disease, rare types of spinocerebellar degeneration, and the much more frequent PD with dementia. Brain structures damaged by these dementias include the major subcortical nuclei (thalamus, basal ganglia, and other brain stem nuclei). Recognition of these syndromes was important in the history of neurology and remains valuable when grouping prominent signs and symptoms together in a preliminary phase of diagnosis. Routine clinical practice, however, now requires a more precise classification of dementias. For example, once all investigations are completed, it is usual to categorize the clinical features as consistent with a subcortical dementia but to rely on the results of investigations before placing a patient precisely within a single diagnostic group. Some tests provide clear results (e.g., genetic tests in Huntington's disease or MRI findings of loss of specific subcortical brain structures). The most frequently encountered subcortical dementias are Huntington's disease and PD with dementia.

DIFFUSE WHITE MATTER DISEASE

Chapter 5, The Aging Brain, summarizes the results of surveys of brain–blood vessel disease. These surveys showed that brain–blood vessel changes occur in the majority (about 95 percent) of people over age sixty-five. When signs of a previous stroke are investigated in older adults without a history of stroke, around 25 percent have MRI findings indicating the presence of a small stroke that was not recognized previously.

The importance of these findings lies in their contribution to the risk of development of dementia or acute stroke. In the famous Nun Study, by David Snowdon of the University of Kentucky (United States), the presence of brain–blood vessel changes was linked to a greater risk of progress to dementia. The underlying mechanisms seemed to be that brain–blood vessel changes in gray or white matter lowered the threshold at which dementia might occur and thus required less AD pathology to cause dementia. The current position in the United States and United Kingdom is that in community surveys, the coexistence of brain–blood vessel and AD pathologies provides the most frequently occurring group of brain changes in late-onset dementia.

Imaging studies of brain–blood vessel disease have applied rating scales to MRI appearances. These studies usually rely on visual inspection, but semiautomated methods are now available that provide quantitative measures of the space occupied by blood vessel changes. These studies have revealed how location and extent of these changes are associated with specific signs and symptoms. The most reliable finding is that speed of information processing is slowed in proportion to the extent of white matter disease. When these changes predominantly occur in subcortical structures, disturbances in gait, urinary incontinence, slurred speech, and episodes of elation or depression are common.

Diffuse white matter disease is now well recognized in the investigation of early dementia. It plays a central role among a complex interaction of factors. Among these factors are the deposition of abnormal protein aggregates as in AD, perturbations of the dynamics of blood flow through the brain's microcirculation, and a group of factors that may mitigate against any impairment of cognitive processes. This complexity aside, clinicians have devised strategies taken from lessons learned in the prevention of strokes to address the additional burden posed by brain–blood vessel disease that affects white matter.

HIPPOCAMPAL SCLEROSIS

Hippocampal sclerosis (HS) is difficult to distinguish from AD, and the diagnosis usually is made at postmortem. The main feature is widespread loss of neurons in the hippocampus. The clinical history is of a slowly progressive dementia with severe memory problems. In surveys of patients dying with dementia, HS makes up 2–4 percent of total cases, although it can accompany other forms of dementia in an additional 10–20 percent of cases. About 50 percent of HS cases are clearly unilateral. Thus far, there are no clear-cut causal factors. For many, it seems that some form of impaired blood supply to the hippocampus on one side (usually the right side) of the brain seems responsible. Most often, HS is linked to brain–blood vessel disease. Little is known about its molecular pathology, its treatment, or its prevention.

VII. BRINGING IT ALL TOGETHER

All classifications of the dementia syndromes are provisional and await a comprehensive causal understanding of their origins. This will be provided by studies in their neurobiology, their links with aging, developmental psychology, and the social sciences. Notwithstanding, great clinical and research value is attached to the current distinctions drawn between the dementia syndromes, not least because these can be roughly separated using clinical descriptive methods, brain structural and functional imaging, and by observational outcome studies. At a future date, some clinical presentations, as in FTD and FAD findings from molecular biology, will contribute to diagnosis so that it is reasonable to anticipate that molecular abnormalities will reveal boundaries between a significant proportion of people living with dementia and those other people with dementia who remain incompletely understood. A critical question will concern the possibility that there are multiple AD subtypes. Potentially, successful treatment and possible prevention of AD will remain undiscovered until methods to identify AD subtypes are available. If the principle is accepted that successful AD prevention will require valid methods to detect a "preclinical" stage of AD with certain progression to symptomatic dementia syndrome, then the discovery of reliable biomarkers will become an essential step toward this goal.

Among the general population, reduction of exposure to factors that increase dementia risk appears currently to provide useful prospects of reducing dementia risk in late life. Improved treatment of underlying neurovascular abnormalities (as in acute stroke when treatment may be as simple as optimizing current management), reducing the risk of head injury (through vehicle design, driver regulation, rules governing contact sports, etc.), and as shown in Chapter 12, better management of delirium in old people—when taken together could significantly reduce dementia in older adults. Nevertheless, some common concomitant clinical disorders (e.g., PD) that are associated with increased dementia risk will require a specific research program in order to discover how to reduce dementia risk. Currently, common ground between the molecular pathology of AD and PD provides the best prospect of discovery of a means to prevent PD leading to a dementia syndrome.

Clinical investigation of the dementia syndromes has progressed beyond establishing dementia diagnosis in an individual with florid signs and symptoms of dementia. Successful investigation of early onset dementias (symptomatic before age 65 years) has encouraged specialists to intensely investigate those with the earliest symptoms and signs of late-onset dementia, whereas previously their concerns may have been dismissed. In the modern era, it has become widely accepted good clinical practice to identify reversible factors that increase dementia risk and/or to advise specific life changes that revolve around better nutrition, a more active and socially engaged lifestyle, and optimum control of vascular risk factors (as set out in Chapter 12).

10

DEMENTIA RISK REDUCTION, 1: CONCEPTS, RESERVE, AND EARLY LIFE OPPORTUNITIES

Expert groups[1] have made a great deal of effort to evaluate the merits of competing claims for dementia prevention or risk reduction. The present position is that lessons learned from the prevention of heart disease and stroke may be applicable to the problem of dementia.[2] Otherwise, sufficient scientific evidence has not supported any steps that an individual could take to reduce the risk of dementia.

I. THE DEMENTIA EPIDEMIC

After World War II, the United States, like most of the industrial world, witnessed declining mortality rates with increased life expectancy. In step with longer life expectancy, the number of old people living worldwide with dementia rose sharply and will continue to do so. Among the elderly of European ancestry at age sixty-five or older, the prevalence of dementia is estimated at around 5–10 percent. On the basis of increased life expectancy, estimated future numbers of U.S. citizens living with dementia will have tripled from around 3 million in 1990 to about 10 million in 2030. Predictions of an *epidemic of dementia* rely on three key assumptions. First, that the number of old people will continue to maintain expected increases in life expectancy. Second, that increased mortality linked to obesity, sedentary lifestyles, late-onset diabetes, and substance abuse will only slightly offset increased survival. Third, there is no immediate prospect of effective interventions to slow or prevent dementia onset.

The first of these assumptions has been discussed widely. In the twentieth century, the golden generation born between 1925 and 1935 experienced the greatest improvements in healthy life expectancy ever seen in the history of humanity.[3] In large part, improved health was attributed to better health care and improved living conditions. From pregnancy to old age, social historians can chart dramatic falls in maternal and infant mortality, fewer deaths caused by infections, and easier access to more effective health care. Notwithstanding these advances, even with hindsight, it is difficult to separate sources of major gains from those of minor benefit.

On the one hand, public health measures to encourage better diets, more active lifestyles, and smoking cessation are certain sources of improvement in heart disease and stroke deaths. On the other hand, these alone do not seem sufficient. Improvements in medical education and better recognition of risk factors for heart disease together led not only to better control of high blood pressure using drugs with fewer side effects but also to the advice to lower blood fats. The international pharmaceutical industry can be credited with promoting these health gains through investment in continuing medical education and better drugs that can be tolerated safely even when taken continuously for many years. The drug industry did much to improve public understanding of science and strengthened trust in medical scientists as important contributors to our overall well-being. A healthy skepticism pervades, however, about an industry that seems to discourage cooperation between scientists and is not as fully committed as it should be to completely impartial evaluation of its claims.

These concerns give rise to several fundamental questions about interventions intended to reduce the risk of dementia. First, there is the question whether increased life expectancy is a reward or a curse if it increases the risk of living long enough to become incapacitated by age-related physical or mental impairment. For this reason, a new indicator of health status in late life was devised: the Healthy Life Expectancy Index,[4] which is defined as "the duration an individual at a specific age is expected to live without any significant morbid condition." The application of Healthy Life Expectancy indexes is fairly straightforward. When individuals born between specified birth dates (a birth cohort) are compared with an earlier or later cohort, the relationship between healthy life expectancy and life expectancy becomes critical. If life expectancy remains the same while healthy life expectancy increases, fewer years will be spent with disability or disease. This is termed "compression of morbidity." If life expectancy increases but healthy life expectancy stays the same or lags behind, greater prevalence of disability, including dementia, is reliably forecast. A second concern is related to the disease model of the dementias. To put research and development resources into antidementia drugs, industry leaders are obliged to make a case for dementia research that they can communicate effectively to their investors and their staff. Simple disease models of the dementias do not reflect the complex gene environment; interactive causal models are needed that provide a comprehensive explanatory framework of the dementias. Although many leading dementia researchers readily accept such complexity, their vision is focused tightly on what is achievable and the shortest route to reach a prespecified goal. So far, this strategy has been unsuccessful, and other dementia models, which may prove to be multifactorial, will be required on the road to drug discovery.

MULTIFACTORIAL NATURE OF ALZHEIMER'S DISEASE

Following a life course approach, this book has set out to understand how and why individuals differ in their rates of cognitive aging and why such a large proportion of the old succumbs to dementia. So far, no single cause or set of related causes can explain the sources of these differences. By extension, many experts argue that effective interventions in slowing dementia onset will be multifactorial, requiring combinations of pharmacological, behavioral, and social strategies to be tested in randomized clinical trials.[5] In scientific medicine, these trials sometimes are designed as pivotal tests of key hypotheses about the causes of disease. In dementia, they pose particular problems of design and long-term commitment but remain essential for progress.[6]

When the aim of these trials is to prevent decline to dementia in unaffected individuals, it is easy to anticipate several questions. First, which criteria will identify an unaffected person? Will these individuals be symptomatic (i.e., cognitive function unimpaired)? Or will it be critical to show that the first biological changes of dementia are absent from the brain? Second, what should be the outcome criteria? Should these be the prevention of decline that meets clinical criteria for dementia? Or will it be sufficient (and certainly less costly) to show that an intervention under testing prevents progressive cognitive decline typical of the early clinical stages of dementia but without meeting criteria for a diagnosis of dementia? If differences in rates of cognitive decline are selected, then the analysis becomes very tricky. Certain childhood advantages for cognitive development appear to carry over to late adulthood, so those with higher child-hood mental ability not only begin to decline later from a higher level of ability but also do better on repeat cognitive tests. In these individuals, the observed overall rate of decline approximates the net effects of age-related impairments and the positive effects of practice. When the effects of higher educational attainments and more complex (cognitively demanding) occupations are added, it is easy to see that a large sample of the general population unaffected by but at risk of dementia because of age will include a complex mix of individuals who, in the absence of dementia, would display a wide variety of possible trajectories of cognitive decline in late life.

Currently, there is some agreement that the only practical approach to test-ing dementia prevention strategies is to identify biomarkers in people at risk of dementia—on the grounds of age or family history or both. Biomarkers need to be highly specific for substantially increased risk of dementia and must change in a predictable way with progress to a dementia syndrome. This is the declared aim of several current clinical trials that hope to discover drugs with disease-modifying effects in dementia. These drugs contrast from those that produce symptomatic relief. This latter group includes the acetylcholinesterase inhibitors described in Chapter 6, The Biology of the Dementias.

Prevention of dementia is often divided into (1) the reduction of risk fac-tors as "primary prevention"; (2) detection of the earliest biological changes of dementia (biomarkers) during a preclinical state as "secondary prevention"; and (3) optimum care for individuals living with dementia as "tertiary prevention." Attempts to modify disease processes are subsumed under both primary and secondary prevention. Although widely used, this subdivision is not detailed in this chapter; instead, possible preventive interventions are linked to points across the life course.

The Introduction discussed the problems of reductionism. In discussions of dementia prevention, these problems are closely related to the hope that a single, universally effective silver bullet will be found that is effective in slowing processes that make the brain vulnerable to dementia, regardless of the category of clinical dementia syndrome. Interventions that might slow brain aging fall into this category and include strategies that aim to maintain the lifelong health of neurons and brain blood vessels. Contrary to reductionism is a consensus that in addition to candidate drug therapies, it will be important to test adjunctive social or psychological therapies. This approach would be comparable to the care of distressed patients in mental health therapy settings who fare much better when social–supportive therapies are used skillfully in combination with psychotropic drugs that have been selected carefully and given at an optimum dose.

Opportunities to test dementia prevention strategies present most obviously among the elderly, many of whom are already aware of the risk of dementia linked to aging. In keeping with the life course approach, however, it is helpful if opportunities for intervention are identified at various points along one's life journey. The plan is to identify the best timing of possible interventions from conception through late adulthood. A further layer of complexity is added when the social setting of a person at risk for dementia is considered to be a possible source of variation in response to intervention. The majority of elderly individuals living with mild cognitive impairment (MCI) in their own homes, for example, do not require assistance or supervision from caregivers, who often would be a spouse. The role of the family caregiver for someone with MCI at risk of dementia is not straightforward. The caregiver is at greater risk of developing stress-related somatic and psychiatric health problems. This risk becomes more acute should MCI increase progressively to a dementia syndrome. In anticipation of progressive deterioration, dementia care teams recognize that counseling and psychosocial interventions for caregivers have valuable positive effects not only on patients living with moderate to severe dementia but also for those with MCI, and these benefits extend to their caregivers.

Other conceptual issues concern the complex nature of biological systems as targets of interventions to prevent dementia. Although there is great variation among individuals in rates of cognitive decline in old age, some "successful agers" appear to be unaffected before ninety years of age and others seem to improve well into their eighties. Described in more detail in the following section, these commonplace observations have supported the idea that the brain possesses a form of reserve that is available for use when needed but not employed in routine operations.[7] This is known by many synonyms (including "brain reserve" and "cerebral reserve"), but in this chapter, it is called "cognitive reserve."

Cognitive performance in old age can be considered to be a balance between the positive effects of cognitive reserve and the negative effects of brain–blood vessel disease and the insidious, steady development of Alzheimer-type changes in the aging brain. Although quite difficult (but not impossible) to establish on the basis of precise measurements in life, this sort of balance makes good intuitive sense. Individuals can understand that if cognitive reserve is essentially "money in the bank" and if this reserve is improved by education and lifelong interests in mentally effortful pursuits, then at least this positive part of the aging process is under their personal control. The dangers of brain–blood vessel disease can be reduced by adherence to blood pressure control measures, dietary restriction, and regular check-ups for abnormal glucose metabolism. Smoking cessation and regular exercise also help and are best established as lifelong patterns of behavior not left until heart disease, stroke, and possibly dementia seem imminent.

II. COGNITIVE RESERVE

If amyloid deposition is the main cause of brain cell damage in AD, then the relationship between amyloid and clinical dementia might be straightforward. More β-amyloid-containing plaques should equate to greater mental impairment. Often, however, this is not the case. When brains are examined after death, some people show very high densities of amyloid plaque deposition in the brain, yet before death, there was no convincing evidence of dementia. Conversely, others can show few amyloid deposits yet have suffered a severe degree of dementia. These findings challenge the idea that the degree of compromised brain function should precisely predict the severity of dementia. To explain these discrepant findings, the idea discussed widely is that the brain possesses a reserve capacity to resist any age-related or dementia-type changes.

Cognitive reserve, therefore, merits separate consideration alongside other conceptual issues in dementia prevention. Reserve includes two hypothetical components: (1) a passive function of brain structures that can be mobilized when other areas lose function, and (2) an active capacity of mental processes that can be called on to devise new ways of maintaining brain work and thus compensate for loss of function.[8] These ideas have promoted a search for strategies to reduce the risk of dementia by increasing cognitive reserve, either through the enhancement of relevant brain structures or flexibility of cognitive processes. Education, job training, and active and cognitively effortful leisure pursuits all are considered to be possible methods that might improve both active and passive components of reserve and thus reduce the risk of dementia.

An alternative explanation of cognitive reserve derives from the view that critical differences in rates of cognitive aging and dementia probably originate in differences in neuronal maintenance, repair, and regeneration. The biological systems that maintain and repair brain structures have been studied extensively. Rather than introducing a hypothetical concept of brain or cognitive reserve, some experts regard well-understood brain maintenance and repair systems as the more relevant reasons why individuals vary in their rates of cognitive aging and risk of dementia.

Early education makes a large, persisting contribution to cognitive reserve. Studies on this topic have compared the length of formal education and have shown that less education increases the risk of dementia. This leads to the question: What is it about those last few years at school that add to dementia protection? The first idea about dementia protection is that those extra years ease entry into better jobs with supervisory responsibilities. Better jobs usually pay better, allowing more choices about leisure time activities and, possibly, social worlds that are more varied. It is difficult to decide how the benefits of education actually arise. A second idea about dementia protection is attributable to the fact that historically, the education of women was not tightly linked to their lifelong work pattern. Competing demands of childcare and the need to retrain to reenter the labor market can limit job opportunities even for well-educated women.

SOCIAL INFLUENCES ON THE CAUSES AND COURSES OF DEMENTIA

Chapter 8, Emotional Aging, examined how social judgments affected our emotional lives in old age. In Chapter 7, The Disconnected Mind, we saw that information processing deficits could impair efficiency of emotional regulation. In total, our perceptions of our cognitive and emotional competencies could make us feel less capable, anticipate failure, and cause us to perform less well under pressure. This finding introduces the possibility that cognitive reserve possesses some social and emotional components that can inform and direct information processing.

The concept of cognitive reserve, therefore, can be developed further and usefully compared to the idea of individual differences in resilience. This idea conveys the sense of a capacity to adjust to or face up to adversity. From a psychological perspective, resilience includes those aspects of character and temperament that predict successful coping with stressful life events, including illness, as well as many chronic stressors such as long-term unemployment or the cumulative ill effects of aging. From this viewpoint, the onset of dementia is seen as a developmental challenge that arises in the setting of aging when personal resources to cope with adversity may be diminished by social isolation or chronic disability, both difficult to counter in an aging predicament. When this line of reasoning is followed, it is noticeable that among the economically advantaged, individual differences in cognitive reserve may have their sources in personality structure; enduring aspects of temperament or character; or the ability to appraise potential stressors, learn from experience, and acquire across the life course a range of coping behaviors appropriate to challenges associated with aging lifestyles. Personality traits such as openness to experience described in Chapter 8 are relevant in this regard.

More obviously, aging may cause poor people who have experienced sustained economic hardship to cope less well with further reductions in material welfare.[9] Possible vicious circles of cause and effect seem plausible, with socioeconomic disadvantage causing ill health and social isolation, which, in turn, weaken resilience and cause further decompensation.

Our emotional and cognitive lives are intertwined; neither exists in isolation. We know that both are influenced strongly influenced by our social environment. When old people collaborate, learning from each other, finding ways of spotting an error of judgment, or finding new solutions, they can act in concert and improve the early deficits attributed to an aging brain. Resilience in the face of an aging brain is a type of self-possession held powerfully together by emotional and social forces.

We make sense of the world by making a mental effort to learn about what is happening. This is not achieved by passive acquisition and it is most effective with the help of others. With aging, we come to rely more on the help others can give. Between parents and children, this help is called "guided participation." Between older couples, help of this type often is unspoken and can be viewed as a type of collaboration, with neither party in the ascendancy. The benefits for performance of an old person, however, are comparable with a child's improvement when helped by a parent or older child.

When children are growing up, they transition from reliance on others to total self-reliance; with age-related cognitive decline, the transition is reversed. For some, reliance on others is distressing and weakens their self-esteem. When this feeling provokes lasting anxious or depressive feelings, mental performance may be impaired, not improved. Many experts are now more careful to distinguish two major components within cognitive reserve: "functional" and "structural" reserve capacities (sometimes called "active" and "passive" reserve, respectively). Although some scientific writing about reserve remains inexact, most authors no longer regard cognitive reserve as a unitary entity and instead emphasize distinctions between several underlying cognitive processes.[10–16] When a further distinction is made between "cerebrovascular reserve" and "metabolic reserve" the idea that "cognitive reserve" represents a single entity no longer remains tenable."[17–23]

HOW USEFUL IS THE CONCEPT OF COGNITIVE RESERVE?

A Novartis symposium on cognitive reserve was held in 2004.[24] No consensus was reached on a better definition of the concept or even whether the concept of cognitive reserve remained useful. At that meeting, some scientists argued that the use of an umbrella term like cognitive reserve was an indicator of ignorance about basic brain processes involved in brain maintenance and repair, learning and memory—much like the state that prevailed fifty years ago concerning the causes of heart failure in old people. Progress in understanding cardiorespiratory physiology, the effects on heart structure, and the function of disease processes and aging made general terms, such as "heart failure," redundant and, in the view of some symposium members, the same should happen to "cognitive reserve."

This view of cognitive reserve is too restrictive, and it does not meet the aim of this chapter to understand possible strategies to reduce the risk of dementia. Cognitive reserve is used to draw attention to the distinction between "active" or "passive" processes available to buffer the effects of differing forms of "burden" on brain function. We will extend the contributions made to reserve beyond these brain-centered processes to include social influences on central information processing. These considerations strengthen the position of cognitive reserve as helpful in developmental and aging brain science because they remind us that (1) over the life course, early life endowment and later enrichment of experiences influence cognitive aging; (2) many different factors provide protection from dementia; and (3) dementia-protective factors mediate individual differences in rates of cognitive aging and are relevant to these differences in addition to the development of dementia-related brain pathology.

This consideration is important because of the frequency with which some individuals believed to be at high risk of dementia within their normal life expectancy are found never to succumb to dementia because of the disease-slowing effects of neuroprotective factors currently subsumed under cognitive reserve. Thus, cognitive reserve warrants continued use in disease models of dementia and recovery after brain injury because it remains a useful proxy for a group of factors that otherwise would be labeled "unknown."

IS COGNITIVE RESERVE AN EXAMPLE OF REDUCTIONISM?

In the Introduction, "reductionism" was explained as a process that reduces a larger number of items to a smaller number of connected ideas or theories. When cognitive reserve was first used in brain science, it was thought help-ful. Its introduction followed reports that early life experiences were associated with differences in the risk of late-onset dementia.[25,26] Experiences linked to a greater risk of dementia appeared to reflect suboptimal neurodevelopment and included smaller brain size, less full-time education,[27–30] lower childhood intelli-gence,[31] and lower occupational complexity.[32] The general inference drawn from these studies was that the association between each risk factor and increased incidence of dementia was explained by their association with a lesser capacity to withstand the presence of age-related brain pathology or perhaps of normal brain aging. No firm conclusion about the precise nature and roles of cognitive reserve can be drawn. It remains a hypothetical construct, useful in the planning of cognitive experiments in the aging brain, but as yet, without a sound scientific basis.

III. BIOLOGICAL PLAUSIBILITY OF STRATEGIES TO REDUCE THE RISK OF DEMENTIA

In 1978, soon after reports of cholinergic deficits in AD were first reported, the Medical Research Council's Neurosciences Board (the U.K. government's main funding body) formally reviewed researchers' plans to build on their discovery. Proposals that aimed to test the efficacy of cholinergic replacement or stimula-tion therapies were immediately discounted. The chair of the visiting committee mocked a planned trial: "It would be like stopping a fuel tanker when your car breaks down and asking the tanker driver to pour petrol over the engine . . . surely someone has a better idea than that!" In similar fashion, another visiting committee member reproached the idea that molecular genetic studies of cholinergic neurotransmission could be fruitful. "Pointless," he declared, "only the rare early onset cases have a significant genetic component . . . the rest is simply a complication of brain aging."

In 1978, convincing funding committees that an idea might be realized as effective in the primary, secondary, or tertiary prevention of dementia was notoriously difficult. In part, this difficulty stemmed from the fact that experts on dementia risk reduction did not exist and, when asked to form a consensus, most committees supported ideas firmly connected to what they already knew about dementia. Sometimes, this was very little indeed.

Dementia risk reduction should be biologically plausible before testing in nonaffected people at risk is contemplated. The strongest grounds to establish plausibility are provided when a substantial body of data already exists that supports the effectiveness of an intervention in disease conditions that share underlying disease pathways with those that lead to dementia. Consider, for example, the "neurovascular hypothesis of Alzheimer's disease" proposed by Raj Kalaria in the University of Newcastle upon Tyne (United Kingdom).[33] In some specific circumstances, shared disease mechanisms can trigger a stroke or lead to dementia so that, overall, a sizable proportion (say, more than 10 percent) of people at risk of dementia suffer an acute stroke or progress to dementia along pathways that trigger these mechanisms. Certain events or exposures are well known: severe sepsis (infection) and surgical trauma usually linked to the use of general or regional anesthesia are probable culprits. The design and conduct of a clinical trial linked to a specific clinical event (e.g., anesthesia) would be relatively straightforward, although as yet, few interventions are considered to be likely candidates. Many lessons can be learned, however, from the successful prevention of recurrent stroke in at-risk populations, and much evidence is available to support trials in the prevention of dementia of interventions that are known to reduce the risk of heart attacks and stroke.

A small number of individuals at risk of AD are known carriers of mutations in amyloid precursor protein (*APP*) or presenilin 1 (*PSEN1*), which always lead to early dementia, making a much stronger case to test an intervention. Francisco Lopera (University of Antioquia, Columbia) and Eric Reiman (Banner Alzheimer's Institute, Phoenix, Arizona, United States) and colleagues have provided an excellent example of the large-scale scientific cooperation required to undertake this type of study. Over twenty-five years, they recruited five of the largest known pedigrees of early onset familial AD, containing about 5,000 individuals mostly from a mountainous area (Medellín) of Northern Columbia. With the help of a U.S.-based team of researchers, they plan to test the efficacy of a humanized monoclonal antibody in family members who carry a *PSEN1* mutation. The largest ever dementia prevention study is now under way. Even in its preliminary stages, the trialists have contributed substantially to resolve many critical ethical and conceptual issues arising in the prevention of AD among individuals who are at the preclinical (presymptomatic) stages of disease.[34-36]

SOCIAL EPIDEMIOLOGY

The problem of risk reduction in dementia is tied closely to the problem of distribution of dementia in space and time. From a life course perspective, the science of social epidemiology provides a far-reaching overview of the risk of dementia and its reduction. This is not to say that social epidemiology is more fundamental or important than elucidation of biochemical pathways toward and away from dementia. It is only that the type of information provided by social epidemiology illuminates those biochemical processes in ways that point to possible ("candidate") risk reduction interventions.

The life course approach to brain aging and dementia leads to quite complex causal models of dementia with many opportunities for possible causal factors to interact at multiple levels. On one hand, the temptation is obvious to oversimplify these pathways to aid understanding and communication. On the other hand, the belief is widely held that statistical treatments of multifactorial models of dementia are so often flawed that no firm conclusions concerning causality can be safely reached.

The opinion of George Kaplan from the University of Michigan at Ann Arbor (United States) is relevant. He wrote:

> The search for "independent" risk factors has been a dominant force in social epidemiology, and certainly risk factor epidemiology has come under fire from a number of quarters. As is true for much of epidemiology, claims to independent effects often conflate issues of statistical independence that are at least dependent on measurement and modeling approach with causal independence, a process not without its perils. In social epidemiology, this search for independent effects may represent more of an attempt at legitimization of

a newly evolving field—identifying "new" risk factors—than something more informative. Many analyses stop with demonstrating a statistically significant association between the risk factors and outcomes in question after adjustment for known risk factors. While it could be argued that identifying new risk factors can catalyze the search for new disease mechanisms, it can also result in a plethora of new "social" risk factors that sometimes exist in an almost miasma-like fog; we observe them to be importantly related to some outcome but cannot really identify the mechanisms that explain this association.[37]

Kaplan emphasized the need to develop coherent causal theories to guide data collection in ways that facilitate interpretation. With the exception of the role of childhood poverty and its links to higher levels of stress on brain development, it often is difficult to link observations from childhood to biological pathways that lead eventually to late-onset dementia. More frequent is the suspicion, rarely more than that, of a putative childhood risk factor playing a role in the delay of dementia. The most quoted examples are higher childhood intelligence, greater early linguistic ability (including bilingualism), higher educational achievements, and the acquisition of healthy lifestyles with cognitively demanding jobs and leisure activities.

An important strength of the life course approach is that it provides a time-based framework within which opportunities to intervene along pathways leading to dementia can be identified. The developmental framework lends itself to correctly placing biological changes in the brain that arise during the course of dementia pathology from the identification of the earliest biochemical abnormalities in a preclinical phase, to the selective loss of localized populations of neurons, and eventually to the terminal phase of dementia.

It is against this background that opportunities available in childhood to prevent or delay dementia arising in late life must be established. Epidemiology rarely translates satisfactorily into successful clinical trials and even less often into effective public health measures to reduce the morbidity of disease in a population. For example, in childhood social epidemiology, it is usual to observe associations between birthplace and risk of adult disease. Although it is straightforward to measure current physical properties (e.g., exposure to airborne pollutants) of a birthplace, reliable historical data may be unavailable. When the likely causal factor is less easy to measure (e.g., stress related to overcrowding and hostility), the task becomes much more difficult. When birthplace also relates to father's occupation, the possibilities of confounding increase with the introduction of socioeconomic status, occupational exposure to neurotoxins, and longer periods of unemployment. If birthplace also is linked to local cultural factors that differ in exclusion of specific nutrients from habitual diets, and unusual parenting practices, what seemed to be a simple relationship between birthplace and risk of disease now becomes extraordinarily difficult to disentangle. Kaplan's reasoning can be applied in the following sections. Rather than list all the factors identified in dementia epidemiology as risk factors and argue for their role in dementia prevention, the following sections stress those risk factors that can be linked plausibly to the biology of the dementias.

IV. EARLY LIFE OPPORTUNITIES TO PREVENT DEMENTIA

FAMILIAL CLUSTERING OF DEMENTIA RISK

The fact that dementia occurs more often among the close relatives of people living with dementia is the basis of genetic causal theories of dementia. Statistical models of familial occurrence of dementia consistently place estimates of genetic contribution to dementia in the range of 50 to 80 percent. Other nongenetic reasons, however, could explain why dementia should affect some families more than others. In early life, people acquire many resources (e.g., physical, psychological, cultural, material, and social comprising "social capital") that, through such behavioral adaptations as lifestyle, coping strategies, social networks, and exercise-related physique, influence current and future health. The transmission (or nontransmission) of such adaptations between consecutive generations is the subject of intergenerational continuities that are clear-cut in the study of inherited disease.

Disease patterns that recur in subsequent generations sometimes suggest, especially when associated with socioeconomic disadvantage, the involvement of nongenetic mechanisms. This point is relevant to neurodevelopmental models of dementia in which early life disadvantages can strongly influence cognitive development and, as will be discussed, probably predispose to dementia. In 1976, Michael Rutter (a psychiatrist) and Nicola Madge (a psychologist) completed a landmark systematic review[38] of evidence from anthropology, sociology, criminology, social administration, medicine, psychiatry, psychology, and social work and concluded that except in specific circumstances (inherited intellectual disabilities) the case for major genetic components was weak. An important aspect of their work, also relevant to the prevention of dementia, concerned the possible benefits of compensatory education to improve upon the disadvantaged start experienced by some children from deprived backgrounds. Political opinion remains undecided, but educationalists have reached consensus that preschool education helps many children who otherwise would begin formal education without essential social and learning skills. Later opportunities to help these children are fewer and less effective.

Intergenerational continuities in premature death, slower cognitive development, and lower psychological functioning remain important areas of scientific enquiry. Their relevance to the overall health of later-born generations is clear, and major efforts are under way to understand possible mechanisms involved and identify areas where interventions might succeed in breaking such cycles of disadvantage.

Intergenerational continuities in intellectual development can be linked to longevity in a rather unexpected way. Dick Mayeux and his Alzheimer research team at Columbia University (United States) have exhaustively studied the interplay between environmental and genetic factors in the course of dementia in carefully defined general population samples. In one such sample of 283 families recruited in the United States (from Boston, New York, and Pittsburgh) and in Denmark, they found higher cognitive function in the children of long-lived parents. The comparison group was drawn from the offspring of the spouses of children of long-lived parents.[39] Those long-lived parents were older than ninety, did not have dementia, and had at least two living brothers or sisters older than age eighty. The underlying mechanisms are unknown but seem likely to involve genetic variations in overall susceptibility to common causes of disability in old age.

Children born outside marriage are a case in point: They are recognized in many cultures as a stigmatized and more deprived group. This view was especially prevalent among grandparents and great-grandparents of current generations.[40] More so than women, men who are born illegitimate tend to be smaller as newborns and have more heart disease as adults than expected. Relatively few intergenerational studies have extended beyond two generations to examine their children and grandchildren who are born in wedlock. Studies from Sweden show that illegitimacy of children born early in the twentieth century is associated with increased risk of premature death in their children and grand-children.[41] These disadvantages persist even after adjustment for social class background. What becomes clear from these and other Scandinavian studies of this type is that a form of "social inheritance" is taking place among the offspring and their children of those born in disadvantaged circumstances. Although the underlying mechanisms remain obscure (and may yet prove to be biological), the most plausible explanation is that upward social mobility of children is hindered by disadvantaged circumstances at birth.[42]

EARLY PARENTAL DEATH: EARLY LIFE STRESS AND INCREASED RISK OF DEMENTIA

Until the early twentieth century, death of a parent before a child's fifth birthday was commonplace. Even in prosperous countries, maternal death in childbirth occurred more than 400 times for every 100,000 live births. The first account that early death of a parent was linked to late-onset dementia was reported by Ingmar Skoog's research group in Gothenberg (Sweden).[43] This Swedish study was part of a comprehensive survey of sociodemographic predictors of late-onset dementia and estimated that early loss of either parent was associated with an approximately sixfold increase in risk of dementia. Maria Norton and her colleagues in the Cache County, Utah (United States), study team[44] found support for the Swedish findings in a much larger study in which other major risk factors for dementia were included. The team surveyed 1,793 local residents and found that paternal death in childhood (before four years old) was linked to about a threefold increase in risk of prevalent dementia. This effect remained after adjustment for age, sex, years of education, and apolipoprotein E (*APOEε4*) carrier status. A 1921 Scottish birth cohort follow-up study,[45] directly comparable with the Swedish study in terms of size and birth year, and like the Cache County study included other major risk factors, also provided significant support for the original Skoog findings. A fourth research group[46] studied 9,362 men recruited to the Israeli Ischemic Heart Disease study and found increased risk of late-onset dementia was associated with an approximate threefold increase in the risk of dementia among men who had lost their father before they were six years old.

An important feature of these studies is that their findings do not appear to be influenced by the socioeconomic status of the families. Loss of a father without remarriage would be an obvious cause of hardship, but the Cache County study suggests that this does not have a major influence on increased risk of dementia. Previously, lower socioeconomic status in childhood was linked to greater risk of dementia.[47,48]

The authors of these reports share similar views about underlying processes leading from early parental death to increased risk of dementia. All agree that loss of either parent would be stressful for these families and that disruption of childcare would follow, widows would be obliged (if possible) to return to work and wean their children earlier, and children would experience a less stimulating environment with likely dilution of parental care. Importantly, many children experience constraints on socialization with adverse effects on educational attainments. Some commentators emphasize the compounding impact of nutritional privations on the developing child and proposed specific epigenetic explanations of the link between childhood disadvantage and dementia.[49]

Exposure to high levels of stress during early life affects brain structures in ways that may endure into adulthood. Like other early environmental influences, stress can impair brain growth and affect maturational development with long-lasting effects that may predispose to premature brain aging. Advances in brain imaging provide accurate measurements of brain structures and show that diseases of early life are associated with differences in brain development. For example, childhood epilepsy[50] is selectively associated with specific abnormalities in the medial temporal lobe. In addition, childhood adversity is associated with smaller anterior cingulate cortex and caudate volumes.[51] The structure of the adult brain also is associated with adverse childhood experience,[52] and this structure may impair cognitive and emotional outcomes in adulthood.[53,54] Tomalski and Johnson[55] suggested that low childhood socioeconomic status (cSES) has substantial and enduring effects on structural and functional brain development. These authors agreed that the effects of low cSES are mediated by several factors, including poor diet, low-quality parental care, and an impoverished, understimulating environment.

CHILDHOOD INTELLIGENCE

Lower childhood educational attainment[56] and poorer linguistic ability as young adults[57] were among the first indications that early life cognitive development may affect the risk of late-onset dementia. The relationship between childhood mental ability and dementia remained untested until childhood mental ability records became available for those who have experienced the risk period for dementia. Until then, the best known solution to the problem of a lack of childhood mental ability data was provided by David Snowdon of the University of Kentucky (United States) in his Nun Study. Autobiographical essays written at about age twenty-two by novitiates to the Sisterhood of the Order of Notre Dame (by then between seventy-five and ninety-five years old) were used to derive measures of linguistic ability in early life. Low scores on these measures were associated with low cognitive test scores in late life and a pathologic diagnosis of AD.

By 1998, survivors of the first national surveys of mental ability conducted by the Scottish Council for Research in Education were at increased risk of late-onset dementia. Comparisons of intelligence quotient (IQ) scores at age eleven between those who had become hospital-treated cases of dementia and those who had not been hospitalized but continued to live locally showed that average childhood IQ scores were significantly lower in the dementia cases than in the nonaffected controls.[58] As mental ability surveys conducted in diverse settings in many industrial countries matured and interest turned to late-life health outcomes and their links with early intelligence, several conclusions became widely accepted as robust. These conclusions are relevant to a possible link between childhood IQ and risk of dementia. Individual differences in childhood IQ are related to overall mortality.[59]

The proposal, advanced by our group,[60] that childhood IQ plays a major part in the acquisition of health-related behaviors was supported by later examinations of links between higher IQ and health that show associations with moderate alcohol consumption, cessation of smoking,[61] more active and socially engaged lifestyles, and better diets.[62] Studies also have shown links between childhood IQ and health-related risk factors for dementia, suggesting that the effect of IQ on risk of dementia is mediated through increased risk of some chronic age-related diseases. These diseases include high blood pressure,[63,64] cardiovascular disease,[65] and obesity.[66] "Metabolic syndrome," a recognized cluster of risk factors for midlife vascular disease (including any three of the following: abdominal obesity, high blood pressure, high blood glucose, high triglycerides, high HDL cholesterol, and high HbA1c), also is linked to dementia in some studies.[67] Marcus Richards and colleagues[68] at the MRC Unit for Lifelong Health and Ageing, London (United Kingdom), followed up the 1946 U.K. Birth Cohort and showed that childhood IQ was linked to metabolic syndrome in adulthood but that this effect was entirely lost when adjusted for education and socioeconomic status. As other birth cohort studies elsewhere mature, participants at increased risk of dementia on grounds of age will be identified and associations between childhood intelligence and dementia can be reviewed.

In light of reports of worldwide improvements on performance on IQ tests (the Flynn effect[69]), it is possible that underlying trends toward reduced dementia incidence found in Sweden,[70] England,[71] and (to a lesser extent) the United States[72] will be linked to improvements at a population level in intelligence test scores. Cautious encouragement for this proposal is found in projections drawn up by Vergard Skirbek and colleagues at the International Institute for Applied Systems Analysis in Laxenburg, Austria.[73] They used cohort-sequential data from U.K. nationally representative surveys to project observed improvements in cognitive performance and argued that these would "more than offset the corresponding age related cognitive decline." Their analyses support the tentative conclusion that as current cohorts age, they will experience less age-related cognitive decline than observed previously in earlier born cohorts at similar ages. Support for this view is found in recent surveys that suggest reductions in dementia prevalence. The largest such study was reported by Fiona Matthews[74] and colleagues from the U.K. MRC Cognitive Ageing Follow-up Studies in three areas of England. An overall reduction of dementia prevalence (Table 10.1) was interpreted as around a 20 percent reduction in dementia incidence or as between a one- and three-year delay in dementia onset.

TABLE 10.1 A Two-Decade Comparison of Prevalence of Dementia in Individuals Sixty-Five Years of Age and Older from Three Geographic Areas of England

	TIME PERIODS COMPARED USING SAME METHOD TO DETECT DEMENTIA	
	1989–94	2008–11
Men	7.4%	4.9%
Women	9.4%	7.7%
Total	8.3%	6.5%

Source: F. E. Matthews, A. Arthur, L. E. Barnes, J. Bond, C. Jagger, L. Robinson, . . . C. Brayne, "A Two-Decade Comparison of Prevalence of Dementia in Individuals Aged 65 Years and Older from Three Geographical Areas of England: Results of the Cognitive Function and Ageing Study I and II," *Lancet* 382, no. 9902 (2013): 1405–12.

GENE THERAPIES FOR LATE-ONSET DEMENTIA

Placing a commentary on gene therapy at this point in the chapter reflects the need to consider just what part genes play among the early life causes and courses of dementia syndromes. Many factors influence dementia onset, ranging from longevity to childhood intelligence and education, making it difficult to know where to begin to unravel and rank the relevance of possible points of genetic intervention. Furthermore, some critics of genetic therapeutic research point to a mismatch between the hyperbole that often follows gene discovery in major diseases and a perceived failure to identify successful treatments based on earlier genetic discoveries. Although it is true that gene therapy remains in its infancy, there is no shortage of data to support "the proof of concept" for its continuing development. Most gene therapy studies to date have examined the more common cancers, most of which, like late-onset dementia, have proven to be complex multifactorial diseases. Autosomal-dominant familial early onset Alzheimer's disease (FAD) might seem a fairly obvious target for gene therapy, but it is only over the past decade that FAD has become a feasible undertaking in clinical trials. Genetic contributions to disease mechanisms that underlie dementia involve brain differentiation and synaptic repair, brain inflammation, and immune responses. These cannot be considered in isolation. The molecular genetic components of diseases that affect metabolic and cardiovascular reserve are equally relevant. An important component of contemporary research in dementia concerns the possibility that dementia shares with metabolic syndrome abnormalities of intrauterine programming of developmental genes. If, like some other inherited genetic errors of metabolism, ill-effects of disordered metabolism harm the developing fetus, then interventions during pregnancy may be contemplated. This is not the distant prospect it might seem: studies are underway in Europe to prevent childhood obesity through implemetation of improved diet and better activity in overweight pregnant women.

In large part, the apparent tardiness in the discovery of effective gene therapies is attributable to the care needed to negotiate critical steps in drug discovery before trials can begin. These steps concern the validity of animal models of FAD in which causal genetic mutations found in FAD are inserted, most often into a mouse. Once achieved, possible gene therapies can be tested for safety and size of any therapeutic benefit. This last step informs trial design and helps determine the number of participants required for a possible trial.

The current state of gene therapy for dementia is that the safety of gene transfer approaches is established in mature adults. Ideas are clear about which groups of neurons are optimal targets for gene therapy, and difficult technical issues concerning how best to measure benefits gradually are being resolved; however, important problems remain.[75] The first concerns how best to deliver a gene therapy at or close to an intended region in the brain. Protected by a bony skull, surrounded by tough internal membranes, the brain is no simple target. Although small molecules like oxygen can pass easily into brain tissue, larger molecules like proteins cannot, unless the protecting membranes that make up the blood–brain barrier (BBB) are diseased ("leaky") or have a unique transport system for specific molecules. The BBB is unsurprisingly efficient in preventing gene transfer into the brain so most approaches have relied on direct infusions into targeted brain structures. The usual alternative approach is to inject a gene transfer vehicle into the cerebrospinal fluid to achieve wide distribution across the surface of the brain with variable degrees of passive diffusion deeper into the brain tissue.

Ideally, a gene therapy would be given orally and enter the brain from the brain blood vessels. The oral route, however, is not feasible because the genetic material would be digested in the gut. If given directly into a blood vessel in a form that is attached to a carrier molecule for which the BBB has a specific transport mechanism, satisfactory results are achievable. In animal models, techniques are under development that rely on the insertion of therapeutic genetic material into the genetic makeup of a nonpathological virus. This combination can spread through brain tissue without causing a pathological infection. This process eventually leads to the expression of the genetic material throughout the brain that can last a lifetime. In a comparable manner, therapeutic material can be inserted into cells taken from the host that then are returned directly into brain. In the case of neural stem cells, such genetically modified cells (or their progeny) will migrate within the brain to target sites and exert their therapeutic effect. These effects include the secretion of growth factors to help maintain or repair cells damaged by dementia pathology or to inhibit physiological systems (e.g., inflammation).

What almost ten years of intense research effort and a small number of clinical trials have shown is that, thus far, gene therapies for AD were introduced too late to be effective. A truly effective preventive dementia drug strategy must be started before the disease takes hold. Studies on autosomal-dominant FAD suggest that AD is detectable in the brain up to twenty-five years before symptom onset, and AD associated with Down syndrome (trisomy 21) probably is detectable from infancy. Late-onset AD is more difficult to date, although many experts accept that its time course and that of FAD are probably similar. The notion is now gathering support that intervention in the mild stage of AD aimed at clearing β-amyloid is probably too late to prevent further deterioration or to relieve symptoms.[76] A U.S.-led consortium aims to implement this recommendation in an ongoing clinical trial of a drug to prevent β-amyloid production. If such trials also fail, does this mean that unknown biochemical events precede β-amyloid deposition and these must be prevented first? Or, does it mean that β-amyloid must be addressed much earlier? Would it be ethical to test dementia prevention strategies in children at increased familial risk of AD and, if so, what would the outcome measures be if the disease was not expected to arise for fifty or more years? Some of these issues were raised by Lissy Jarvik from the University of California–Los Angeles (United States) and a wide group of clinical neuroscientists who, in 2008, foresaw many of the desirable and undesirable consequences of an increased understanding of AD for the children of parents with AD.[77] Advances in brain imaging and molecular biology used in combination already point to the probable success of methods to identify the children who are most vulnerable to AD even in the absence of mutated genes.[78]

Great care is needed in planning future prevention studies of those at greatest risk of AD. Although there is a suspicion that interventions to inhibit the biology of *APP* and the presenilins may have far-reaching effects on the developing brain, thus far, we have no grounds to believe that current gene therapies under consideration will cause undesirable effects on growth and development.[79] The case of trisomy 21 will be an important priority. Thus far, we have strong grounds to believe that almost all children born with trisomy 21 will develop premature AD, probably before fifty years of age. The abnormalities caused by the possession of an extra copy of chromosome 21 include congenital cardiac malformations, certain hematological cancers, and AD—all of which lead to premature death and thus far are unpreventable. One possible therapeutic approach has been suggested by Jun Jiang and colleagues at the Universities of Massachusetts (United States) and British Columbia (Canada).[80] Their research has pointed to methods that potentially could "silence" chromosome 21 and may prove to be an early effective approach to the prevention of AD in children with trisomy 21.

EARLY EDUCATION

Margaret Gatz and her colleagues at the University of Southern California (United States) have systematically reviewed outcome studies in dementia and have concluded that most but not all studies found a relationship between less education and higher risk of dementia in late adulthood.[81] Childhood education is completed by age seventeen in most industrial countries, an age that gradually increased from about fourteen years by one-year increments after World War II.

The reasons for a link between education and dementia are of great interest from a life course perspective. Why should a link be maintained over intervals typically greater than fifty years? Strong intergenerational continuities influence how long a child will remain at school. Parents' own education, family income, intellectual interests, and social values are important. Given the opportunity, smart children will remain in school longer, and when they do, they will learn more and may go on to more complex, better paid work. It is unsurprising, therefore, that childhood IQ is strongly associated with longer duration of formal education and higher educational attainment. Entry in better paid jobs with supervisory roles will increase lifetime earnings and improve the complex array of personal assets labeled "social capital." Income inequality, reduced investment in social capital, and overall poor health[82] with potential ill effects on the risk of dementia is well recognized in almost all cultures and is a possible source of error when estimating the strength of the effects of education on the risk of dementia.

This link between education and dementia was first explained by the observation that the less well-educated perform poorly on cognitive tests to detect dementia. David Kay in Newcastle upon Tyne (United Kingdom) raised this concern in 1964, but it was not until 1990, when Robert Katzman and colleagues identified little or absent education among old people in Shanghai, China, that low education was identified as a risk factor for dementia. Many of their survey participants had received little or no education, not because their childhood intelligence was lower than average but most often because they were women born in the late-nineteenth century when education in China was unavailable for many.

A connection between low education and disease is not specific to dementia, and education can act independently of socioeconomic status and childhood intelligence.[83,84] Some aspect of education seems to uniquely capture individual differences in disease susceptibility that are not detected by other means. Two questions recur among those who work in dementia susceptibility: (1) What aspect of individual differences is approximated by "education"? And (2) why should the power of education to predict dementia and other adult diseases be so constant over time and between cultures? These are intriguing issues and are recurring themes when factors affecting trends in dementia incidence are discussed.[85]

Education is not an independent influence on childhood development. It is, however, closely related to other powerful formative influences on the acquisition of later health-related behaviors, which are achieved through imitation and peer-group pressures to form an adult understanding and acceptance of health education. These pressures can be both positive and negative. Among the latter, social disadvantage, poor social support, and inferior social position may steer children along one of the many pathways to poor health from childhood adversity to late-onset adult disease, including dementia.

Education is affected by processes to which it is not directly linked. For example, a child from a disadvantaged background may have been poorly nourished with slowed cognitive and physical development. Some evidence for the relevance of undernourishment to the risk of dementia has been found in studies of the physical stature of adults living with dementia. Lower adult height of both men and women is reported in some studies and most often is attributed to poor nutrition in childhood. These children may arrive at school hungry, and their attention in class (and other behaviors) can be improved by a light nutritious meal. The role of nutrition in the prevention of dementia clearly extends across the life course.

Education can be judged by its duration and by its effectiveness in terms of educational goals achieved. Seen in this light, it might be presumed that twelve years of formal education would be broadly equivalent in most settings in the industrial world. This may be inaccurate: Life in a Mississippi classroom of sixty children of mixed ages and ability being taught in a rural setting in 1936 is probably quite different from schooldays spent in a class of twenty-five pupils at similar ages in Madison, Wisconsin, ten years later. The philosophical foundations of educational systems often were subject to major social and political change, and throughout the twentieth century, these changes were affected by the powerful arguments of educational theorists.[86]

Ian McDowell and Joan Lindsay at the University of Ottawa, Canada,[87] tested three hypotheses when they challenged the proposed link between education and dementia. First, (hypothesis 1) they reasoned that the link is a spurious artifact of methods of dementia ascertainment such that better educated people without dementia are more likely to take part in surveys. This is certainly true of people who remain in follow-up studies. When better educated people develop dementia, they might be more likely to remain undetected in surveys because their families have changed so substantially over time. Estimation of the risk of dementia among birth cohorts drawn from the early twentieth century, whose education differs markedly from what currently is provided following contemporary guidelines, might seem almost archaic or even purposeless, offering no fresh insights into the determinants of dementia. This topic is equally relevant to the discussion of links between trends toward rising scores on childhood intelligence tests over the past hundred years and age-related cognitive decline.[88,89] Second (hypothesis 2), shorter duration of education is a proxy for lower socioeconomic status with increased likelihood of poorly paid work in hazardous environments with less personal control over working conditions. Finally (hypothesis 3), education may be an important determinant of cognitive reserve (described earlier) and may contribute to the retention of cognitive functions in the face of age-related brain pathologies.

The Ottawa research group analyzed the Canadian Study of Health and Aging, a longitudinal population-based study of dementia.[90] Hypothesis 1, that the link is spurious, yielded a complex interaction between study retention, dementia ascertainment, and level of education but supported the conclusion that the link between low education and dementia is spurious. Hypothesis 2, that the link can be explained by other factors, including associations between education and socioeconomic status, was partly supported, but "these factors do not explain away the entire link with education." The best explanation seems to be that most of the effect of education on risk of dementia is achieved through its links with other sociodemographic factors. Finally, hypothesis 3 proved difficult to resolve satisfactorily, largely because the researchers did not have records of childhood IQ for participants in their survey.

The complex nature of education is linked in the minds of many teachers and parents to timing when specific topics or skills can be taught most effectively.

This issue was discussed in the context of *critical periods* and related to learning a musical instrument or a second language. The benefits of early learning in prescribed areas of study were seen to extend to other areas to have lasting effect. The biological mechanisms of this type of generalized adaptability are poorly understood but are believed to involve greater synaptic plasticity in the developing brain and the recruitment of additional brain structures to aid learning. The ability to learn a second language, therefore, is an important acquired skill that probably involves plastic changes in the brain. The age when a second language is learned and the linguistic proficiency attained produce corresponding increases in the density of cortical gray matter (specifically the dominant left-side inferior parietal cortex).[91]

Ellen Bialystok and her research colleagues at the University of Toronto (Canada)[92] have investigated brain structural changes in bilingual and monolingual patients with probable AD. Bilingual patients had significantly more cortical atrophy, and the atrophy was most marked in brain structures that distinguish AD patients from normal controls. Studies elsewhere[93–95] have strengthened proposed protective effects of bilingualism on brain structures and the retention of cortical functions impaired in aging and AD.[96]

The mature brain reflects diverse genetic and environmental influences. Learning a second language is just one of these influences. Typically, epidemiology identifies exposures that impair growth, and it is unusual to find experiences that affect the capacity to withstand dementia. Early life contributions that contribute to cognitive reserve include education and childhood intelligence. Research on the benefits of early bilingualism represents a major advance and suggests educational strategies with cumulative benefits extending across the life course. These benefits include greater upward social mobility, gains in several domains of cognitive function, and delayed dementia onset.

Variations in dementia incidence between cultures and within societies over time can help identify important clues to the causes of dementia and can suggest factors contributing to individual differences in the resilience of brain functions in the face of dementia pathology. Such variations can be quite large. A recent comparison among dementia surveys in England[97] suggested that a decrease of 25 percent in dementia incidence may have occurred in the final decades of the past century. Previous data from elsewhere are consistent with a decline, although not of this magnitude.[98]

ILLICIT DRUG USE IN ADOLESCENCE: LONG-TERM DAMAGE TO THE NERVOUS SYSTEM

Evidence is compelling that dementia syndromes are strongly associated with long-term use of alcohol and, in some studies, long-term use of illicit drugs.[99] Benzodiazepines and cannabis have short-term ill effects on attention and memory. When used often in adolescence, cannabis causes long-term irreversible deficits in intelligence.[100] When drugs are injected, there is also a risk of infections of the nervous system, causing a dementia syndrome (e.g., HIV causing an AIDS dementia) that is distinct from the toxic effects of the illicit drug or its contaminants (usually other substances used to "cut" the drug to improve profitability).

Two well-known examples illustrate the long-term consequences in the nervous system of illicit drug use. Since the 1980s, the long-term effects on animals and humans of illicit drug use were studied extensively. Methamphetamines (and probably all drugs of this type) and the "rave" drug ecstasy cause brain damage.[101,102] This damage is highly selective, causing specific disruption of brain systems that use serotonin to transmit messages between neurons. Serotonin is important in memory, reasoning and comprehension, mood, and impulsivity. It is unsurprising that deficits in memory often are seen in regular ecstasy users. These effects are found in young and old ecstasy users alike with no evidence (so far) of effects on risk of dementia.

In 1982, a group of young drug addicts in northern California developed severe Parkinsonism after intravenous injection of a new "synthetic heroin" (a designer drug) that was being sold on the streets at the time.[103] The toxic contaminant proved to be 1-methyl-4-phenyl-1,2,3,6-tetrahydropyridine (MPTP). This was quickly shown to selectively destroy nerve cells in an important brain structure (the substantial nigra). The resulting striatal dopamine damage produced most of the clinical features of Parkinson's disease. These findings show that illicit drugs can selectively target subgroups of brain cells sharing the same chemical compounds (neurotransmitters) to communicate between cells.

Drug addiction is linked to the biology of aging. There are many regular illicit drug users among older adults.[104] When more than 12,000 lab results collected from 1995 to 2006 compared drug addicts with general medical patients, the lab data profile of the addicts suggested accelerated aging explained by long-term disruption of the immune function in addicts.[105]

V. BRINGING IT ALL TOGETHER

Measures of overall mental ability might correctly distinguish between people as they age, identify points at which decline begins, and provide an accurate way of looking at changes in mental ability with age. The effects of aging on mental performance are the focus of aging studies in cognitive psychology. These studies show that different domains of cognitive function are affected differently by aging and the dementias. Specific patterns of cognitive decline typically are observed in aging and in specific types and stages of dementia. A major challenge in the cognitive psychology of aging (cognitive aging) is to specify the exact cognitive processing deficit that accounts for these changes. Many psychologists do not take their analysis beyond this stage, largely because the neurobiological foundation of specific cognitive domains is not yet completely secure. Nevertheless, studies on the aging brain have made substantial progress on numerous levels from the molecular through the cellular to the examination of functional connections between brain areas.

In addition to age and genetic factors, education is one of the few additional variables that is related consistently to the risk of dementia. The studies reviewed in this chapter point to the many difficulties found when trying to identify the important ingredients in this relationship. Is it protecting an individual from later dementia, or is it available as part of a cognitive reserve to compensate for the deficits caused by a dementia syndrome? Although this is not a specific association between education and AD, it may be stronger with other diseases.

One explanation for this association is that education acts as a proxy for childhood intelligence and the socioeconomic status of the birth family, but this does not remove the role of education entirely. Questions remain about which specific components of education influence the risk of dementia and, of immediate concern, what might be added to education that would reduce this risk? The answer to this last question may already be known. The past decades of the twentieth century saw the beginnings of a slight but significant decline in dementia incidence in some industrial countries[106–108] that can be linked to improvements in IQ test scores[109] and directly to better educational opportunities. Our educationalists may already have contributed to improved IQ test scores and thereby helped reduce dementia incidence. The first part of the question is more difficult, however. If we assume that the ability to recruit additional cognitive processes to support those impaired by the presence of dementia in the brain, as proposed by the cognitive reserve hypothesis, it would seem sensible to look for those learning experiences that encourage such mental adaptability and ask whether any can be linked to the risk of dementia. This is exactly what Bialystok has achieved in her work on bilingualism. By extension, we can argue that enhancement of verbal skills through education promotes lifelong learning and facilitates participation in cognitively demanding social activities. These gains may be sustained from childhood into adult life and thus reduce the risk of dementia by promoting cognitive competencies.[110]

We can be encouraged by a study of Portuguese immigrant families living in the Grand Duchy of Luxembourg. Pascale Engel de Abreu and her colleagues at the University of Luxembourg compared the abilities of monolingual and bilingual children from poor backgrounds on two types of cognitive task. They concluded that learning a second language improved scores on tests of selective attention and improved inhibition of misleading information with more efficient switching among different types of cognitive response. This work is highly relevant to understanding how bilingualism might reduce the risk of dementia because these same cognitive functions are those most affected by cognitive aging and early AD. Engel de Abreu also addressed the concern that many have about the role of bilingualism: Is it not simply a passport to gain entry into better paid work? She reduced these concerns by controlling for socioeconomic status and recruiting children from a wide range of original mental ability. Acquisition and use of a second language to varying degrees of proficiency is found widely among the general population. For some, their use is a matter of pride in their "mental flexibility" and may through practice influence adult performance on intelligence tests. Thomas Bak and colleagues at the University of Edinburgh have shown that bilinguals perform better on cognitive tests in old age even when childhood intelligence, socioeconomic status, and frequency of use are taken into account.[111] These data suggest positive effects of bilingualism comparable in importance with the negative effects of possession of the *APOEε4* allele.

Educational programs are available to improve language skills and certainly help children from disadvantaged backgrounds to participate in lessons. The studies of Bialystok and Engel de Abreu point the way toward encouraging intervention programs based on foreign language learning, and Bak's data goes further in suggesting that age of acquisition should not discourage older adults from this type of learning. These programs do not involve expensive facilities and are accepted widely as a way to improve the overall experience of schooling with widening of cultural horizons in ways that generalize to other aspects of education.

There is an obvious interpretation of the findings that longer formal education and bilingualism provide possible opportunities to delay dementia onset. Specific ethnic groups may derive greatest benefit in terms of dementia delay from interventions that aim to improve acquisition in linguistic skills. Careful studies are warranted of differences between diverse ethnic groups in risk factors for dementia. Bilingualism did not protect from cognitive decline or dementia in one study that suggested that bilinguals' higher original intelligence provided most of their observed advantage in lower dementia risk.[112] Likewise, it may be unsafe to assume that ethnic groups share the same dementia risk factors or that all ethnic groups are equally susceptible to dementia.[113,114] These are not trivial matters—and they inform the design and conduct of dementia prevention trials.[115]

Otherwise, no specific additional recommendations appear compelling from reading early life course studies on the risk of dementia. Advice about early nutrition and late-onset dementia does not differ from that already provided by expert pediatric nutritionists. It is probably worthwhile, however, to follow closely those contemporary studies that aim to correct abnormalities of intrauterine programming that aim to prevent childhood obesity. Should these be successful with benefits in prevention of late-onset metabolic syndrome/diabetes, their implications for dementia prevention will be seized upon. Early parental loss is a well-known risk factor for developmental problems in young children, and these youth are among many high-risk groups identified by social care agencies as needing additional care. This is particularly true of the young children of drug users at increased risk of early death and those who are exposed to multiple social and psychological hazards to health. One message, however, is clear: We need to talk more about intelligence. The capacity to learn, to acquire lifelong habits of good health maintenance, and to benefit from education and opportunities accruing from work are all related to individual differences in general mental ability. Understanding intelligence, knowing how and why it is measured, and developing an ability to navigate around the many pitfalls posed by intelligence are essential attributes of competency in professions as diverse as teaching, health care, and politics.[116]

11

DEMENTIA RISK REDUCTION, 2: MIDLIFE OPPORTUNITIES TO DELAY DEMENTIA ONSET

When Bengt Winblat and his research team at the Karolinska Institute (Sweden) found the prevalence of dementia appeared to have decreased on the island of Kingsholmen over the past thirty years, several explanations seemed equally likely.[1] This chapter takes recent reports of declining dementia prevalence as its starting point. When natural variations seem to arise in dementia incidence, a critical question concerns what, if any, lessons can be drawn for future dementia prevention strategies?

The report from Sweden does not stand alone. Another larger study from England suggests that dementia incidence is already falling.[2] Studies from the United States,[3,4] Holland,[5] Denmark,[6] and Spain[7] provide encouraging support; so taken together there are now grounds for "cautious optimism" that the incidence of dementia is declining. Positive influences seem to be at work in high-income countries sharing contemporary lifestyles. Trends in dementia incidence, however, remain uncertain in Asia.[8,9]

What are these positive influences? Chapter 10 examined what was happening in childhood development that might influence susceptibility to dementia. Although we could find little to encourage intervention studies in pregnancy or childhood to delay late-onset dementia, two trends caught our attention. First, we saw that educational provisions improved rapidly during the twentieth century. Many industrial countries had recognized in the aftermath of World War I that a large cheap labor source was an insufficient guarantee of industrial success. Reformers encouraged improved education of large numbers of children from whom industries could recruit apprentices and future managers. At first, all of the emphasis was on the education of boys who showed potential to benefit from training. Later, the education of girls was promoted as well, albeit for more complex reasons.

In the years after World War II, relationships between investments in education, apprenticeships, and work-related training were better understood, especially in the former West Germany. If these investments had yielded returns for individuals in improved cognitive reserve and increased social capital, subsequent rapid and far-reaching social changes might be relevant to reduced dementia incidence.

Second, we identified an upward trend throughout the twentieth century in childhood intelligence quotient (IQ) test scores. To support and extend his finding, James Flynn assembled a large body of data derived from IQ surveys conducted worldwide in industrial and developing countries.[10,11] Like many other researchers, he sought to explain why such rises occurred and identified possible drivers of the so-called Flynn effect: better education; more complex occupations; a wider range of cognitively challenging leisure pursuits; and, overall, better health education, improved nutrition, and wider provision of health services coupled with public health measures to reduce exposure to toxins and otherwise-unhealthy environments. The emerging picture seems as difficult to disentangle as the multifactorial nature of AD.

Clinical trials in dementia prevention typically take as their starting point observations from experimental neuroscience or from the epidemiology of dementia. From neuroscience, data that suggest interventions at a molecular level may modify an underlying disease process point to tests of drugs that might be effective in primary, secondary, or tertiary prevention. From epidemiology, possible interventions emerge as much less easily definable strategies. Numerous studies show that various lifestyle factors may be powerful influences on the risk of dementia or the rate of cognitive decline. When randomized clinical trials are completed successfully, these usually test a single factor, such as blood pressure lowering or homocysteine lowering. When risk factors identified in epidemiological studies are explored, these trials more often have yielded inconsistent or meager results.

The multifactorial nature of Alzheimer's disease (AD) points to another approach to dementia prevention—that is, the *multidomain approach* (discussed in Chapter 12). As an alternative to testing the cognitive benefits of a single intervention, clinical trials of interventions containing several components have been introduced. Although few trials have been completed thus far, some recent intervention trials have established the "proof of principle" that trials of interventions aiming to modify *multiple domains* not only are feasible but also may become the preferred "personalized" intervention strategy.

With these issues in mind, this chapter on midlife interventions is organized around the following themes:

1. Neurovascular
2. Inflammatory and metabolic
3. Amyloid and related therapeutic approaches
4. Stress, depression, and the role of growth factors
5. Brain activity

Themes 1–3 are discussed widely and are the focus of major research programs and will receive the most attention. Themes 4 and 5 are not so intensively studied and are described in less detail, but this should not imply that these are weaker hypotheses. Pathways toward AD do not follow independent courses, and there are many links with important interconnections. Research articles have been selected either because they are widely read or because they make a particular point relevant to understanding how interventions are chosen or tested.

I. THE NEUROVASCULAR HYPOTHESIS

Brain blood vessels make important contributions to age-related disintegration of brain structures and their connections. Their intimate involvement is seen along pathways that range from the mechanical—damage to blood vessels by sustained high blood pressure or trauma—to the intricacies of metabolic dysfunctions linked to abnormal glucose metabolism and signaling between brain cells.

In broad terms, the neurovascular hypothesis of AD proposes that the cascade of molecular events leading to selective loss of neurons in AD is triggered by biochemical changes in cells that line brain blood vessels. These cells are components of the blood–brain barrier (BBB). Their functions are compromised by prolonged exposure to high blood pressure, the formation of atheromatous plaques involving cholesterol metabolism, the effects of proinflammatory molecules (cytokines), and abnormal glucose metabolism. In more detail, the neurovascular unit includes a small arterial brain blood vessel (an arteriole), an astrocyte, a microglia, and a neuron. The lining of the arteriole is damaged readily by oxidative stress and cytokines, making the brain blood vessels sensitive to hypertension; to abnormal glucose metabolism; and, as the damaged BBB begins to leak, to the effects of toxins, including β-amyloid. Overall, brain health is jeopardized by the abnormal regulation of blood flow through the brain blood vessels, thereby increasing the chance of stroke. The brain metabolic consequences of persistent high blood pressure are complex, particularly among those who are older than eighty. These consequences include perturbations of insulin signaling in brain.

Numerous observational studies link persistent high blood pressure to cognitive impairment (or more precisely, vascular cognitive impairment [VCI]) and dementia. Initially, the association was explained by a simple mechanistic model of increased pressure waves damaging neurons nearest to arterioles. These explanations were supported by brain imaging showing a loss of brain tissue and areas of brain damage that were best explained by persistent high blood pressure. Such changes could be related to those associated with stroke, and the proposal that control of stroke-related risk factors would reduce the risk of dementia was discussed widely and now is accepted broadly by many clinicians. Not all clinicians agree, however; dissenting voices have pointed to the problem of low blood pressure and its link to dementia and the fact that some older patients suffer side effects of treatments to lower blood pressure.

Jonathan Birns and Lalit Kalra at Kings College Hospital, London (United Kingdom) summarized the current position in a review[12] of twenty-eight cross-sectional studies, twenty-two longitudinal studies, and eight clinical trials (total fifty-eight studies). Cross-sectional studies tended to suggest complex relationships between high blood pressure and cognition. These studies indicated that both low and high blood pressure were associated with cognitive impairment (a U-shaped or J-shaped association).[13] Follow-up studies were more uniform and showed high blood pressure was linked more consistently to cognitive decline. Eight randomized clinical trials gave mixed results. Birns and Kalra listed multiple reasons to account for this diversity: measurement of cognitive change was inconsistent; some patients were already cognitively impaired before trials began; and treatments differed between trials. The current position is that clinical trials of antihypertensive drugs using cognitive function tests or progression to dementia as outcomes do not yet provide a solid basis for intervention to protect cognition. Obviously, treatment of high blood pressure to avoid stroke and heart disease is unaffected by this conclusion.

The conclusions of the National Institutes of Health (NIH) expert panel[14] do not inform current clinical practice. Most clinicians agree that the prevention of dementia is one of several reasons to identify and treat hypertension in those at risk of dementia. Their views are based on a mixed bag of evidence, which, if taken together, supports an interventionist approach, although taken one by one, they are not convincing. In summary, a case is sustained in light of (1) already established benefits of blood pressure lowering for stroke and heart disease, (2) the mixed results of clinical trials of blood pressure lowering on cognitive decline and dementia, and (3) the strength of evidence from longitudinal studies.[15] Well-funded long-term trials are urgently needed to evaluate antihypertensive treatment in the prevention of dementia in older adults who are hypertensive but are without cognitive impairment. Too many trials were either poorly designed or too short, or dementia outcome was added only as an afterthought.[16–19]

Subsequently, leading experts on AD countered the NIH panel's conclusions with their own systematic reviews.[20] All of these reviews identified control of vascular risk factors in midlife as the public health strategy with the greatest potential to reduce the future incidence of dementia. By 2010, guidelines[21,22] from the American Heart Association/American Stroke Association for the protection against VCI were disseminated with adequate supporting evidence, including the following:

1. Reduction of blood pressure according to standard guidelines . . . reduces the risk of stroke, myocardial infarction, and other major cardiovascular outcomes.
2. For persons at risk of developing VCI, treatment of blood pressure is recommended according to standard guidelines to prevent VCI.
3. For persons with a history of stroke, lowering blood pressure is effective for reducing the risk of post-stroke dementia.
4. For middle-age persons and the young-elderly, lowering blood pressure can be useful for the prevention of late-life dementia.
5. For persons aged over 80 years, the usefulness of blood pressure lowering for the prevention of dementia is not well established.

THE NEUROVASCULAR HYPOTHESIS AND DELIRIUM

The neurovascular hypothesis of AD is also relevant to the timing of dementia onset. Many clinicians hear patients and their families confidently date dementia onset to a well-remembered serious incident. Most frequent are during recovery from a stroke, after a head injury, surgical anesthesia, or a hospital stay involving critical care. The history of dementia onset is described by the family as either an acute, dense pattern of cognitive deficits recognized during the immediate recovery period or as a slowly evolving dementia syndrome not recognized until weeks or even months later. These associations are important and probably account, after high blood pressure, for most of the avoidable incidences of dementia.[23] Some surveys of the proposed association between delirium and dementia estimate an increased risk of irreversible dementia following delirium to be between two and six times greater than if delirium had not occurred.

In medical textbooks, the term "delirium" is used to describe acute periods of confusion during critical care. Delirium is a severe, acute clinical syndrome that affects at least 15 percent of older adults admitted to hospital. Its main features are fluctuating levels of consciousness, profound disturbances of attention, confusion with marked disorientation, and severe amnesia. Disorders of perception are not unusual and are often visual. Delirium in hospitalized patients older than sixty-five frequently is associated with severe infections (sepsis), vascular surgery, drug dependency syndromes, some types of stroke, and cognitive impairment attributed to postanesthetic drugs. When confusion persists, or reemerges during follow-up, established risk factors include age over seventy years and a preoperative history of mental decline without dementia. It is not unusual for psychiatric assessment to identify anesthesia as the critical point in the life course that transformed normative cognitive aging into pathological decline toward dementia.

Anesthesiologists provide advice for patients and their relatives about the risk of subsequent dementia linked to general anesthesia and emphasize that postanesthetic dementia syndromes are relatively rare and sometimes associated with postoperative confusion. Unlike perioperative stroke, which is linked to the type and complexity of the surgical procedure, routine clinical psychiatric assessment typically finds only advanced age as a consistent risk factor for postoperative delirium. To date, the precise incidence of dementia syndromes arising some months after recovery from anesthesia is unknown in the United States and Europe, although recent data from a survey of hospital insurance claims in Taiwan suggest that the increased risk of dementia after anesthetic is about 30 percent.[24] The following important questions remain unanswered:

1. Does delirium increase the risk of incident dementia?
2. Among people with dementia, is a history of delirium linked to an increased burden of brain changes typical of dementia?

3. Does delirium quicken the rate of cognitive decline and lead to more severe dementia?

4. Are specific anesthetic agents more hazardous than others, or is the increased risk of dementia more strongly associated with components of postsurgical critical care?

Willem van Gool and colleagues at the Academic Medical Centre in Amsterdam (Holland) surveyed all follow-up studies on the association between dementia and delirium published between 1981 and 2010 and concluded that an episode of delirium in an older adult is associated with increased risks of death, institutionalization, and dementia.[25] These links were independent of age, sex, and the nature or severity of coexisting diseases and persisted after exclusion of preexisting dementia or institutional residence. The researchers reasoned that because treatment of dementia does not affect any of the previously noted outcomes, emphasis must be on the prevention of delirium. Importantly, because these were not short-term effects, increased risks of poor outcomes persisted long after an episode of delirium.

The city of Vantaa in Finland is the setting of the Vantaa 85+ study, which is the first to examine the potential effects of delirium history on the relationships between dementia and its neuropathological markers.[26] What makes the study interesting is that it was conducted not in a hospital setting, where the more severe cases are seen more often, but in the community and was conducted with high rates of successfully completed brain autopsies. Therefore, Vantaa 85+ describes a very unusual type of study in which more than 90 percent of the population older than eighty-four years in 1991 participated and, of these, about 50 percent donated their brains for autopsy. The Vantaa 85+ results confirmed that delirium is a major risk factor for dementia. Among those living with dementia, delirium was linked to increasing severity of dementia, worsening functional status, and higher mortality.

There are many possible biological pathways from delirium to dementia. The best current model of the pathogenesis of AD identifies a lag period between contact with one or more initiating events and the onset of dementia. In early onset AD, the molecular changes are confined largely to pathways involving β-amyloid or tau. In late-onset AD, vascular changes are much more marked so that in the oldest-old, neurovascular involvement appears to be as important, if not more so, than β-amyloid deposition or the formation of neurofibrillary tangles (NFTs). Neurovascular events trigger a sequence of molecular changes that cause selective loss of neurons, which gradually progresses until a clinical dementia syndrome is detectable. When these pathways toward dementia are considered in the context of strategies to prevent dementia at a population level, possible opportunities to prevent dementia after delirium can be identified.

At the level of practical prevention of dementia, there usually are a few alternatives among the elderly to surgical procedures with anesthesia (either general or local) that were not considered before surgery was selected. Most opportunities arise in the introduction of agents that will counter the adverse effects of the molecular events presented schematically in Figure 11.1. At present, a good case can be made for clinical trials of such agents in patients older than fifty with postoperative delirium to test the prevention of progress to dementia. The choice of agents is limited but is informed by interventions that aim to reduce inflammatory responses to acute postoperative sepsis.

In the context of dementia prevention, the aim would be to target postoperative delirium through the reduction of mitochondrial dysfunction. The broad

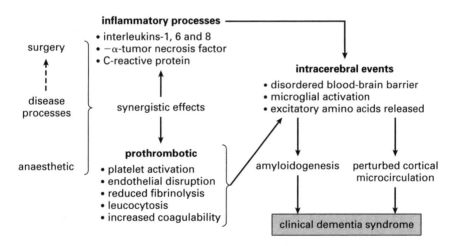

FIGURE 11.1 Candidate pathways from anesthesia to dementia: Preoperative pathophysiology, surgical trauma, and aging biology interact synergistically with anesthetic agents to trigger cascades of inflammatory and prothrombotic molecular events to induce perturbations of cerebral activity at multiple levels. These include disrupted blood-brain-barrier function, proinflammatory microglial activation, and release of potentially neurotoxic excitatory amino acids. These events can precipitate the formation of cerebral amyloid and localized disturbances of cortical microvasculature that lead to selective death of cortical neurons in brain structures critical to memory, attention, and comprehension.

aim would be to combat massive cytokine release and oxidative stress. Several antioxidant agents (e.g., melatonin, MitoQ) are available to augment endogenous antioxidant scavenging systems in mitochondria, and these might be evaluated as possible agents to reduce the risk of dementia in postoperative delirium. An unexpected cautionary note is sounded by a meta-analysis of published studies on the prevention of delirium: All types of intervention successfully reduced the incidence of delirium with the greatest benefits found in those individuals at greatest risk of dementia.[27]

Reactive oxygen species (ROS) mediate much damage to neuronal mitochondria that is greatest in those brain areas with high energy consumption—typically those affected in AD. These highly evolved brain areas contain large projection neurons that are especially susceptible to aging and environmental hazards (including anesthetics) because of their high energy requirements, reliance on axonal transport, and need for substantial trophic support. Their large cell surface areas increase exposure of these cells to potentially neurotoxic agents, which include anesthetics and perioperative drugs.

Concerns of anesthesiologists, patients, and caregivers could be addressed once a consensus is reached on the risk of dementia after anesthesia. This consensus will have medical and legal implications. When sufficient data are available to estimate reliably the risk of dementia, then current clinical guidelines must be reviewed. Wider concerns at a population level that anesthesia contributes to the risk of dementia also might lead to clinical intervention studies among the elderly who suffer a postoperative delirium. At present, there are few opportunities to test strategies to prevent dementia: A link between anesthesia and increased risk of dementia is one such opportunity and certainly merits further study and rigorous testing of methods to reduce that risk.

THE NEUROVASCULAR HYPOTHESIS AND STROKE

Stroke is linked strongly to increased risk of dementia. This is not a simple link; stroke has many causes, and several possible pathways could link stroke to dementia, especially among older adults with more than one underlying disease. For example, a new stroke can occur in patients with mild preclinical AD, in patients with significant disease of small blood vessels in the brain, and in patients with underlying brain changes caused by persistently high blood pressure. Although any of these preexisting conditions could significantly impair cognition by disruption at critical sites in brain, it is only when brain changes develop after stroke in areas that affect cognition that a dementia syndrome arises.

For more than thirty years, research on the link between stroke and dementia was confounded by a widely accepted, though unproven, assertion that multiple cortical infarcts were necessary before poststroke dementia could be diagnosed. Other diagnostic issues also have caused concern. Sarah Pendlebury and Peter Rothwell at the Stroke Prevention Research Unit in Oxford (United Kingdom) systematically evaluated studies reported from 1950 through 2009.[28] Their efforts were rewarded with clear estimates of the incidence of dementia after stroke, although some concerns remain about the quality of data they had available. They found twenty-one hospital- and six population-based studies met their criteria. A variety of types of stroke were reported (first stroke, recurrent stroke, ischemic stroke, hemorrhagic stroke, stroke in atrial fibrillation) and a range of confounders, including depressed mood and disturbances in language production poststroke, made dementia assessment particularly difficult.

When Pendlebury and Rothwell excluded patients who were without prestroke dementia and included only those in their first year after their first stroke, they estimated that dementia arose in about 5 to 10 percent of stroke victims. After recurrent stroke, dementia incidence was higher at 10 to 20 percent. Risk factors for poststroke dementia were identified, including greater age, female sex, less education, family history of dementia, preexisting cognitive impairment, atrial fibrillation, and brain imaging detection of vascular changes in brain. These factors are broadly the same as risk factors for dementia in the absence of stroke and also provide major contributions to cognitive reserve. Vascular risk factors for poststroke dementia also are as expected, including diabetes, atrial fibrillation, ischemic heart disease, high blood pressure, and smoking.

What can we make of this complex picture? First, we can see that the risk of dementia poststroke was greater than the risk prestroke when exposed to vascular risk factors. We can understand this as showing that stroke is increasing the risk of dementia and is not explained by the vascular risk factors for stroke. Second, we can accept that susceptibility to dementia poststroke is influenced by cognitive reserve. Third, recurrent stroke will increase the risk of dementia from about 10 percent following first stroke to 30 percent following recurrent stroke, and it is here that vascular risk factors play a major role in increasing dementia incidence poststroke.

From the viewpoint of preventing dementia incidence, these studies show (1) how important it is to prevent any first stroke; and (2) when stroke happens, how important it is to prevent any recurrence. It is likely that the burden of stroke placed on discrete areas of brain tissue loss poses the major cause of post-stroke dementia and steps to prevent stroke and its recurrence will contribute to reduced dementia incidence. This is just one of the factors advanced by Qiu and colleagues at the Karolinska Institute to explain observed reduction in dementia prevalence over the past two decades.[29]

THE NEUROVASCULAR HYPOTHESIS AND DIABETES

Diabetes is a common metabolic disorder caused by abnormal regulation of glucose. Some survey evidence suggests that in late adulthood, diabetes is linked to cognitive decline[30]; the evidence that this leads to higher incidence of dementia is somewhat weaker.[31] In part, this uncertainty arises because most survey data do not record the quality of diabetes care, and frequency of episodes of poor glucose control are likely confounders. Because diabetes is associated with increased mortality, high blood pressure, and impaired kidney function, many other potential sources of error occur when estimating the strength of this association. One study[32] from Dick Mayeux's Columbia University, New York (United States), group suggests it is the link between dementia and vascular risk factors (particularly poorly controlled high blood pressure) that is driving increased dementia incidence in patients with diabetes. In a large follow-up study of more than 13,000 Swedish twin pairs, Laura Fratiglioini and her colleagues[33] provided strong evidence that diabetes increases the risk of dementia by around 50 percent.

Many biologically plausible pathways could link diabetes and dementia. The life course approach identified early life risk factors shared between diabetes and dementia. These shared factors include low maternal nutritional status, parental socioeconomic position, and birth weight.[34] The Swedish twin study[35] provided an opportunity to tease apart the timing of genetic and environmental causal factors in dementia and diabetes. The team found the risk of dementia was greater among those twins whose diabetes had started by midlife rather than those who developed late-onset diabetes. Results of co-twin matched case-control analyses identified unknown familial factors (among genetic factors and early life environments) as possible contributors to the association between late-life diabetes and dementia, but they did not account for the association of midlife diabetes with dementia. Although this account of a link between diabetes and dementia is considered among the midlife opportunities to prevent dementia, these opportunities for prevention may be more appropriate among the early life risk factors for diabetes.

A few key questions remain: Is diabetes increasing the risk of dementia through a neurovascular pathway to vascular-type dementia? Is the link more directly acting to increase AD risk? Or, does low birth weight have unconnected links to both adult diabetes and cognitive function?[36] Neuroimaging studies suggest that the neurovascular pathway is the more likely pathway, with concomitant involvement of insulin resistance, oxidative stress, formation of advanced glycation end products, and proinflammatory cytokines. The Swedish research group extended the discussion of their findings to include suggestions that the molecular pathology of diabetes and AD may be influenced by the same positive factors and that these had many origins in early life experiences and obesity in midlife.

To date, no follow-up studies starting in midlife have adequately considered the influence on the diabetes–dementia relationship of poor glucose control. It is unclear, therefore, whether optimal diabetes care is associated with a lower risk of dementia in individuals with diabetes than in those who have poorly controlled glucose metabolism. Acute studies[37] on cognitive function in younger adults with diabetes suggest that this should be the case, but so far, data are insufficient to be confident.

II. THE INFLAMMATORY AND METABOLIC HYPOTHESES

The recognizable signs of inflammation are pain, warmth, and swelling. Detecting inflammation in the brain is much more of a challenge, but the underlying physiology remains the same. After an initial tissue injury, local chemical messengers are released in set order to slow the spread of injury, to draw helpful chemicals to its site, and then later to attract other chemicals that cause pain and restrict use of the injured part.

The first chemical signals constrict local blood vessels; these include thromboxane as a fast-responding vasoconstrictor and then as a mediator of prostaglandin release to trigger local vasodilation. With vasodilation comes increased permeability of the lining of blood vessels nearest the injury. Plasma leaks out of the vessel into surrounding tissues, causing local swelling. Prostaglandin release is further stimulated by local cytokines that make local pain receptors more sensitive. Many of these chemical signals are released locally from precursor molecules that make up the cellular structure of nearby cell membranes.

The chemical mix released during inflammation is intensively studied in brain aging, AD, and Parkinson's disease (PD). Historically, the brain was thought of as being isolated by the BBB from the rest of the body. It now is seen as being intimately involved with the body's immune and hormonal systems. Many of the chemicals released in the inflammatory response, in addition to their actions outside the brain, also affect brain systems, including cognitive processes. An important consequence of this bidirectional interrelationship between brain and body is that inflammation outside the brain can affect cognition. It is plausible, therefore, that many of the features of cognitive and emotional aging are influenced by inflammatory processes arising inside and outside the brain.[38]

In the aging brain, microglia (one of three types of glial cell) are activated in AD and PD. The microglia are scavengers that remove unwanted material, such as the breakdown products of cell death and abnormal materials. Much of this material is toxic to neurons, including β-amyloid. These activated microglia flood the surrounding brain tissue with chemicals and exert wide-ranging actions that include oxidation of ROS (also known as "free radicals"), stimulation of cell death ("apoptosis"), and stimulation of enzymes to digest cellular components. In broad terms, microglia are seen as beneficial. They have evolved to clear foreign material (e.g., microbes) from the brain and, in healthy individuals, complete this work very well. When highly activated, however, microglia produce chemicals that are toxic for healthy brain cells.[39] A healthy balance of microglial activation promotes but does not unnecessarily prolong the inflammatory response.

Evidence is now substantial that in brain aging and AD, inflammatory processes generate ROS and that these ROS attack neurons in the cortex (as in AD) and the basal ganglia (as in PD). Microglia are potent sources of ROS and other neurotoxic substances. These chemicals are extremely complex, and experts rely on simplified schematics to show the main families of proinflammatory chemical messengers. (These include complement proteins and complement inhibitors, pentraxins and other acute phase reactants, inflammatory cytokines, chemokines, anaphylotoxins, integrins, coagulation and fibrinolytic factors, prostaglandins, apolipoproteins and selective proteases, and protease inhibitors). Taking the cytokines as an example of one family, these include at least five subfamilies: the interleukins, the interferons, tumor necrosis factors, growth factors, and the chemokines. The inflammatory response in the AD brain includes up-regulation of acute-phase proteins, complement cytokines, and proinflammatory chemicals.

Three strands of evidence support an inflammatory hypothesis of AD. The first is finding many proinflammatory chemicals in and around the molecular pathology of AD lesions.[40] Second, epidemiology provides further support: in clinical samples, higher blood levels of some acute-phase proteins, cytokines, C-reactive protein, alpha1-antichymotrypsin, and interleukin (IL)-6 are reported risk factors for the development of cognitive decline and AD.[41,42] Additionally, surveys show an inverse relationship between the regular use of nonsteroidal anti-inflammatory drugs (NSAIDs) and the incidence of AD.[43] Additional data suggest that the cognitive benefits of NSAIDs are greater if commenced in mid- rather than in late adulthood in the presence of apolipoprotein E (*APOEε4*). Third, genetic evidence supports the associations between AD and rare variants of genes coding for proinflammatory molecules.[44–46]

These data have encouraged clinical trials of NSAIDs in the secondary prevention of AD. Thus far, these trials have not identified the benefits of NSAIDs, although they have provided helpful insights into the trajectories of cognitive decline in AD and the timing of interventions to slow such decline. Epidemiological evidence also shows that the balance between proinflammatory prostaglandins and anti-inflammatory prostaglandins is affected by the dietary composition of omega-3 and omega-6 essential fatty acids. Diets that are replete in omega-3 fatty acids can maintain a balance between anti- and proinflammatory prostaglandins. In surveys, higher dietary omega-3 intake is linked to higher cognitive function.[47] One possible explanation of this association is that inflammatory processes in the brain are held in check by anti-inflammatory prostaglandins.[48,49]

The microglia are not involved intimately in information processing but act to maintain a healthy environment for neurons. Much of their work concerns immune surveillance and the inflammatory response of brain tissue. They can identify large molecules not recognized as part of the healthy brain and can scavenge these and the products of neurons degraded by aging and by disease. Although this work is proinflammatory, the same cells contain anti-inflammatory systems that hold inflammation in check. Figure 11.2 shows schematically how these pathways function and identifies two key areas in which interventions might act to limit inflammatory processes in the brain and thus delay dementia. These are the actions of anti-inflammatory drugs and the role of fish oil containing docosahexaenoic acid and eicosapentaenoic acid and are discussed in Chapter 12.

Inflammatory pathways in AD are identified in molecular genetic studies. In addition to causal genes (mutated forms of *APP, PSEN1*, and *PSEN2*) for familial early onset AD, additional genes are involved in the immune response and in cholesterol metabolism. These are now subject to intense genetic and functional investigations and are possible targets for existing and novel drugs that may prove to be effective in dementia prevention. Cholesterol metabolism and some aspects of the immune response are implicated in AD.[50] In addition, elevated low-density lipoprotein cholesterol (LDL-C) levels are associated with raised risk of cerebrovascular disease.

In midlife, high cholesterol is linked to late-onset dementia, and statins (drugs that lower cholesterol) are reported in some studies to be associated with reduced risk of dementia. Links between cholesterol metabolism and the

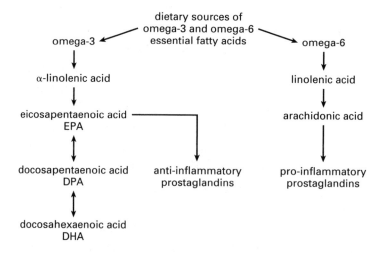

FIGURE 11.2 Dietary fish oils are rich in omega-3 fatty acids. Omega-3 acids provide precursors (EPA and DHA) of anti-inflammatory prostaglandins. Their effects are partly balanced by pro-inflammatory prostaglandins derived from the omega-6 family. Before the industrial revolution, human diets contained roughly equal amounts of omega-3 and omega-6 fatty acids. In the modern era, diets contain much more omega-6 fatty acids and are deplete in omega-3.

lipoprotein *APOE* are not well understood. On the one hand, the brain requires cholesterol to build cell membranes and relies on *APOE* to transport cholesterol from astroglia to neurons. On the other hand, it is unclear how the *APOEε4* variant affects these processes. *APOE* and *APOJ* are involved in the clearance of β-amyloid from the brain, and there seems to be a likely role for genes involved in cholesterol metabolism and β-amyloid clearance from the brain to be involved among the causes of AD. These complex interactions between cholesterol metabolism, the inflammatory response, and β-amyloid clearance from the brain are of potential importance in processes leading to AD and suggest possible, although as yet, untested interventions.

Genes involved in aspects of the immune response also are implicated in AD. Some of the proteins coded for by these genes are found among the β-amyloid plaques that are distributed densely in AD brains.[51] If these newly discovered genetic data hold up in replication studies, then grounds will be strong to consider variations in cholesterol metabolism and the inflammatory and immune response to play pivotal roles in the initiation of AD processes.[52,53] In this setting, these findings are likely to prompt the search for novel interventions to slow or prevent the cascade of molecular events (certainly involving β-amyloid) that leads to AD.

III. THE AMYLOID AND RELATED THERAPEUTIC APPROACHES HYPOTHESIS

BACKGROUND

Most anti-AD therapies have targeted the β-amyloid peptide; relatively few have pursued an antitau approach or identified as therapeutic targets links between *APOE*, cholesterol metabolism, and amyloid processing. Figures 6.2a–6.2c illustrate how processing of amyloid precursor protein (APP) by α-secretase starts a chain of biochemical events that does not produce β-amyloid (i.e., it is non-amyloidogenic). β-amyloid is produced by cleavage of APP first by β-secretase and then by γ-secretase. These processes are well organized within sections of the neuronal membrane that also contain cholesterol. β-amyloid production and aggregation may be increased when there is too much cholesterol in the neuronal membrane. *APOE* is the major cholesterol transporter into the brain, and one genetic isoform of *APOE* (*APOEɛ4*) promotes β-amyloid production and aggregation.

Opportunities to test antiamyloid therapy are present in four target conditions: (1) when criteria for mild to moderate AD are met; (2) when older adults complain of progressive age-related memory impairment (amnestic mild cognitive impairment); (3) when younger adults are without symptoms suggestive of AD and have a family history of early onset AD, consistent with autosomal transmission (FAD); or (4) have two copies of the *APOEɛ4* gene. Currently, all four groups are under intensive investigation but so far without benefit. The conduct of clinical trials in FAD is informed by the detection of the carrier status of FAD genes and *APOE* genotype and the knowledge of age at onset of affected first-degree relatives. Examples of antiamyloid therapies are given in Table 11.1.

Most of the recent and current trials in AD are based on the amyloid cascade hypothesis. This focus of intent predicates on the aggregation of β-amyloid as a critical early step in AD progression. Some experts feel that the close spatial relationship in the AD brain between tau-based NFT formation and selective loss of neurons has not been considered carefully enough. The fact that tau plays such an important role in neuronal loss in the tauopathies indicates that when acting without β-amyloid aggregation they are highly toxic. Bruno Bulic and

TABLE 11.1 Drugs in Development for Treatment and Possible Prevention

AIM	DRUG TYPE	EXAMPLES	ACTION
Inhibition of amyloid production	β- and γ-secretase inhibitors	Semagacestat	Reduces Aβ synthesis
Enhanced amyloid clearance	β-amyloid immunotherapy	Bapineuzumab Solanezumab	Binds to Aβ and increases microglial clearance
Inhibition of amyloid aggregation	Various	Scyllo-inosito Epigallocatechin (from green tea) CLR01	Inhibits or modifies amyloid aggregation
Antiphosphorylation of tau	Kinase inhibitors	Methylene blue	Multiple on tau, amyloid, and cholinergic function

colleagues at the Humboldt-University of Berlin (Germany) recently reviewed ongoing AD disease-modifying AD trials and found that of seventeen trials only one was testing an antitau aggregation drug; all others were either targeting β-amyloid production (ten trials) or its aggregation (six trials).[54] Inhibition of tau aggregation is regarded as particularly challenging by pharmacologists as the aggregation of large protein molecules takes place between relatively large flat surfaces, whereas most drugs are small molecules that interact with cellular sites that also are relatively small. In one view, however, both β-amyloid and tau aggregation must be prevented simultaneously to prevent AD. If this proves to be the case, then a combined approach that relies on two complementary therapeutic strategies (antiamyloid and antitau) will be necessary; neither will be effective used alone.

CHOLESTEROL, *APOE*, AND AMYLOID PROCESSING

The human brain contains about 25 percent of all body cholesterol; of this, about 70 percent is found in myelin, about 20 percent is present in glial cells (especially astrocytes and microglia), and about 10 percent is found in neurons. Brain cholesterol metabolism is involved in the abnormal processing of APP and tau formation and, therefore, may have a critical role among pathways toward AD. These ideas have encouraged the development of pharmacological approaches to AD based on (1) disruption of *APOEε4* function by increasing *APOEε3*; (2) blocking *APOEε4* interaction with β-amyloid; and (3) modulation of interactions between *APOE* and its receptors.[55] Future developments based on links between *APOE* metabolism, cholesterol, and amyloid processing are likely to consider some preliminary data from a clinical trial on the drug dimebon.[56]

IV. STRESS, DEPRESSION, AND THE ROLE OF GROWTH FACTORS

Some studies show that stress and depression are associated with the loss of neurons and glia. These losses contribute to the loss of tissue in brain regions that control mood and involve the prefrontal cortex and hippocampus—the brain areas also involved in brain aging and dementia. These findings suggested the actions of antidepressant treatments (e.g., drugs, electroconvulsive therapy, transcranial magnetic stimulation) on synaptic plasticity are possibly mediated through the expression and functions of nerve growth factors.[57] A smaller size of the hippocampus and other forebrain regions in subsets of depressed patients support a current hypothesis for depression. This hypothesis involves reductions in neurotrophic factors, which are neurodevelopmentally expressed growth factors that also regulate synaptic plasticity in the adult brain. The actions of growth factors include the formation of new synapses (synaptogenesis) and the activation of excitatory neurotransmitters and release of nerve growth factors.

Epidemiological studies have suggested a link between a history of depressive illness in adulthood and increased risk of later dementia. Critics argue that the link could be explained by the observation that early in the course of dementia, it is not unusual for patients to feel depressed and for doctors to recognize this and recommend treatment with a conventional oral antidepressant. Sometimes, when depressed, when self-esteem is low, individuals become more self-critical and exaggerate any minor error of memory or judgment into a self-diagnosis of dementia. It is also plausible, however, that factors predisposing to depression also predispose to dementia.

In VCI (discussed earlier), depressive symptoms are frequent, and complex interactions between depressed emotional states and the reactivity of brain blood vessels are proposed to contribute to progressive neuronal loss leading to vascular dementia. A single causal pathway is plausible that leads from stress to loss of hippocampal neurons. This pathway was first proposed in 1986 by Bruce McEwen and Robert Sapolsky at the Rockefeller University, New York (United States), as the "Glucocorticoid Cascade Hypothesis of Aging."[58] The hypothesis does not explain aging, but could explain an increased susceptibility to dementia among those who have suffered a depressive illness of moderate severity in adult life.

In a review of the reported association between depression and dementia, Amy Byers and Kristine Yaffe[59] considered that the link was stronger when depressive disorders arose in early and not late life, and that the treatment of depression may prove relevant to the prevention of dementia. They listed as plausible many of the pathways to AD considered in this chapter, including functional alterations in the neurovascular unit, increased glucocorticoids, growth factor deficiencies, inflammatory processes, and aggregation of β-amyloid. Byers and Yaffe supported the exploration of current and novel antidepressant therapies in the treatment and prevention of AD. All conclusions about the role of depressive illness among the causes of AD, nevertheless, remain provisional. Because the overlap is sufficient between estimates of onset of AD at around age forty and the occurrence of so many midlife depressive illnesses after this age, it remains reasonable to regard the link between depression and dementias as a result of presymptomatic dementia, thus lowering the threshold for depression.

ESTROGEN

In addition to its role in female reproduction, estrogen (E2) also acts as a nerve growth factor and contributes to the health of neurons. In neurodevelopment, E2 promotes selective enhancement of the growth and differentiation of axons and dendrites. In the adult brain, E2 is colocalized with receptors for neurotrophic receptors (tyrosine kinase), where its actions appear to include regulation of many aspects of gene expression in brain development relevant to sexual dimorphism. Epidemiological surveys show a slight increase in the risk of dementia among women compared with men, and this prompted the hypothesis that postmenopausal E2 deficiency could be responsible. Potentially, E2 hormone replacement therapy could prevent dementia.

Victor Henderson of Stanford University, California (United States), has systematically reviewed this question and found nine relevant clinical trials.[60] Among these trials, he found none that specifically reported a dementia prevention study. He concluded that in postmenopausal women with symptomatic AD, E2 replacement did not improve cognitive performance. He encouraged future studies on the possibility that a "therapeutic window" is present in the perimenopausal period, during which time, if E2 therapy is started, AD might be prevented. He also commented on the potential value of E2 receptor modulators and certain dietary phytoestrogens derived from plants that require investigation. Eva Hogervorst and colleagues at Loughborough University (United Kingdom) reviewed the relationships between cognitive function and blood testosterone concentrations in elderly men and noted that low free testosterone may precede AD and that this also merited further study.[61]

Most experts agree that grounds are insufficient to recommend hormone replacement therapy for the prevention or treatment of age-related cognitive decline or dementia in women or men. Some believe that advantages for cardiovascular health include a degree of protection against brain–blood vessel disease, but this opinion has not been tested in clinical trials. There are no relevant data on testosterone supplementation in men.

INSULIN

Much confusion surrounds the role of insulin among the causes of dementia. Insulin is a pancreatic hormone with important effects in the brain. It is implicated among the causes of AD and vascular dementia, and evidence indicates that perturbations of insulin control are present before the symptomatic onset of a clinical dementia syndrome. These observations raise the possibility that measurements of circulating insulin may contribute to the prediction of dementia. Importantly, they also may provide a point of intervention before dementia onset that could protect in midlife against late-onset dementia. In large part, confusion arises because of conflicting reports that insulin is too high or too low in people at greatest risk of dementia. Other factors, including sex and *APOE* gene variability, appear relevant.

Elina Ronnemaa and colleagues at Uppsala University (Sweden) completed a follow-up study[62] of 1,125 men age seventy-one without dementia or diabetes and showed that an initial low insulin response to oral glucose was predictive of AD ($n = 81$) but not vascular dementia ($n = 26$) twelve years later. This link was detected only in those without *APOEε4*, and their data supported further studies on the value of insulin in the prediction of dementia. Suzanne Craft at the University of Washington, Seattle (United States), has provided valuable insights into the complex relationship between the role of insulin and risk of dementia in older adults without diabetes. She recognized that age-related cognitive decline is linked in some studies to higher insulin levels; in others, however, that link weakens with lower insulin concentrations. She proposed that a "U-shaped" relationship exists between insulin levels and the risk of dementia, such that both low and high levels increase risk. *APOEε4*, sex, and dementia subtype all seemed relevant.[63]

Abnormalities of insulin metabolism in AD have encouraged scientists to investigate further the roles played by insulin in neural differentiation and in maintaining neural survival at different developmental stages. Evidence suggests that prolonged exposure to high insulin concentrations causes neuronal death in the human brain and that this may be relevant to AD.[64] Insulin also may play an essential role in the fetal origins of obesity, diabetes, and increased blood pressure.[65] High blood pressure is relevant to the life course approach to brain aging and dementia with implications for brain development and maturation of physiologic responses to stressors. Clinical trials are under way to find out if, when applied during pregnancy and in the newborn, these will correct alterations in programming that predispose the developing child and mature adult to obesity, hypertension, diabetes, vascular disease and, possibly, dementia.

V. THE BRAIN ACTIVITY HYPOTHESIS

The activity of neurons promotes the construction and maintenance of efficient neural networks. As described previously, early sensory experiences, particularly during critical periods, contribute to "activity-dependent self-organization of the cortex." Although the representation of sensory experience in the cortex is a dynamic remodeling process occurring throughout the life course, it is most important during early life and continues to a lesser extent into adulthood when essential cellular processes remain unchanged. To a limited extent in adulthood, some neural networks retain the capacity to add new neurons (called "regeneration"), and this appears to depend on maintenance of brain activity in affected neural networks.

Underlying cellular mechanisms depend on synaptic plasticity to store memories as traces of experience. Many brain structures are involved in memory formation, including the medial temporal lobe and hippocampus, which are most important in the earliest stages of encoding memories. Later stages in which memories are consolidated and transferred to long-term storage involve other parts of cortex.

Retrieval of memories is achieved in ways, thus far unknown, of constructive remembering. These processes allow memories to endure unchanged in storage and to be recalled usefully in their appropriate context with many of their related meanings.

Synaptic networks provide the biological basis of memories. In the context of dementia prevention, a high level of interest arises because from a practical standpoint, interventions that could improve memory performance possess the potential to delay or even prevent dementia onset. This line of thinking underpins the logical argument behind "brain training," which proposes to maintain powers of memory and reduce the risk of dementia.

Advanced imaging techniques can detect differences in mental flexibility by revealing the use of alternate brain structures (or, more precisely, neural networks) that are not used routinely for a specific mental task. Studies are designed to show differences between young and old people; to suggest differences in efficiency and capacity or flexibility; and to address the possibility that when alternative networks are used, cognitive performance is greater than expected.

In epidemiological studies, retention of cognitive reserve is associated with following an active and cognitively engaged lifestyle. For many people, age-related physical disabilities would pose a considerable barrier to their lifestyle of choice. By itself, the association could be an example of "reverse causality." Midlife studies with follow-up into old age have shown that an "active and engaged lifestyle" in midlife is associated with better cognitive function in old age. Nancy Petersen has provided evidence in support of this proposal in her follow-up studies of Swedish twins, examined first in midlife and then again after age eighty. When twins were discordant in old age for cognitive function, the twin who maintained cognitively demanding interests fared better than their co-twin, who did not perform as well on cognitive tests and had a higher incidence of dementia.

These observations provide the basis for clinical trials of physical and cognitive training in old age that aim to improve mental function and well-being and potentially reduce the risk of dementia. Cognitive training aims to improve cognitive performance on day-to-day tasks unrelated to the training materials. This improvement is achieved either directly or by teaching improved strategies to compensate for dementia-related deficits. Reminiscence therapy is the least intense and exposes older adults living with mild or moderate degrees of memory loss to objects that were familiar in their early life but were seen less often in the present. Even considering only the best studies, any claimed modest benefit was lost.[66]

Trials of physical training appear more often to produce positive results.[67] One well-designed trial, however, failed to find any cognitive benefit that could be linked specifically to either cognitive or physical training and attributed most gains to the effect of practice at taking cognitive tests.[68] Whatever benefits can be linked to physical training could be attributed to increased brain blood flow and exercise-induced reductions in blood pressure and, by extension, to reduced risk of cardiovascular disease. Because cardiovascular risk factors are known to increase the risk of dementia, lower risk may slow progression of established dementia. Most current interest in activity-based prevention strategies in dementia is focused on their incorporation in multidomain approaches, which are discussed in the next section and in Chapter 12.

VI. BRINGING IT ALL TOGETHER

When dementia researchers gather to discuss what might be immediately available to prevent dementia, they have no trouble agreeing on two things. The first, very well-argued by Ron Brookmeyer and coworkers[69] at Johns Hopkins Medical College, Baltimore (United States), proposed that if the onset of dementia could be delayed by as few as six months, within twenty years, the total number of people living with dementia in the United States should fall by almost 30 percent. If an even more effective intervention could be found, a delay of dementia onset by five years would reduce total dementia prevalence by more than 50 percent. In this context, encouraging studies from Sweden[70] and England[71] support falling dementia prevalence and suggest that some modest gains already might be realized. The second point of agreement is that the likely causes of reduced dementia prevalence achieved so far include the improved cardiovascular health of our aging population, with accompanying falls in deaths resulting from stroke and heart disease.[72,73] From a wider social perspective, what has been achieved thus far also might be attributable, at least in part, to major cultural shifts in high-income societies.[74] Over the past fifty years, these societies have witnessed huge betterments in material wealth, although great disparities remain; more complex occupations; higher educational standards; and, as demanded by a better informed populace, improved health care with superior outcomes. These considerations lead to questioning—most often by social scientists—that the risk of dementia and exposure to cardiovascular risk factors are not linked causally, as many suppose, but rather are driven by a shared association with socioeconomic hardship.[75] Studies on the effects of hardship in early life on the aging brain lend support to this proposal.[76]

Meanwhile, laboratory science follows a well-trodden path from molecular classification within a range of related syndromes to open byways leading toward well-delineated, single diseases. Thus far, the best examples are provided by Huntington's disease and, most recently, the frontotemporal dementias. Although the breakthrough in the molecular genetics of Huntington's disease was made in 1993, no therapy based on this breakthrough has been discovered. The time needed to understand fully what is happening in these and related dementias is not caused by lack of effort (or intelligence), but rather that enormously difficult problems must be overcome within the basic biology of the human brain.

Once potential disease-modifying drugs become available, highly demanding, long-term, and costly clinical trials will be needed to establish that a preventive intervention based on molecular genetic knowledge about one dementia subtype (e.g., *PSEN1* mutation carriers as in the Columbian pedigrees) can safely and effectively be applied to the most frequent causes of late-onset dementia.

At the heart of the challenge of risk reduction in late-onset dementia lie three complex and intricately entwined problems. These concern the overlapping contributions of Alzheimer-type brain pathology, the ubiquitous nature of brain–blood vessel disease, and an uncertain causal role for brain aging. Some experts argue that the only way ahead is to advance on a broad front with the means to slow the progress of dementia across multiple domains. These will include steps to prevent neurovascular disease promoting Alzheimer-type pathology and steps to hinder the aggregation of abnormal proteins and enhance their clearance from the brain. Because the structure of the brain is influenced so strongly by its activity, it seems sensible to take additional steps that will maintain the general health of neurons and their connections. A broad approach also should take account of observations from the epidemiology of dementia and incorporate proposals to encourage retention of brain function in the face of aging through a physically active, cognitively effortful, and socially engaged life style. If we add the importance of a well-balanced nutrient-rich but energy-light dietary habit, it is possible to pull together all the ingredients of a healthy lifestyle.

This chapter was not based on a comprehensive review of the enormous number of relevant scientific reports. Instead, it relies on one man's reading habits, papers cited by colleagues, the work of our multidisciplinary research group, and papers sent for review by the editors of scientific journals. Many much more systematic reviews already are well-regarded in the current literature or posted on the Internet. Among these reviews is a recent example commissioned by the U.S. Department of Health and Human Services.[77] This report remains outstanding for its comprehensive nature and clarity, although it is disappointing for the scant encouragement it provides for most dementia preventive interventions the authors and coworkers had scrutinized. Examined alone as tests of single evaluations, few interventions proved effective in dementia prevention. Improved management of clinical conditions that increase the risk of dementia (e.g., diabetes, stroke, and cardiovascular disease) seem to be promising targets, but not all clinical trials showed efficacy.

Some experts emphasize that in the absence of reliable diagnostic criteria based on a causal classification of the dementias, clinical trialists are in much the same position as those treating infectious diseases before Louis Pasteur identified which microbacteria caused specific infections. It is not unusual to hear calls for a "molecular classification" of the dementias that discourage expectations of significant advances until this is done. The work of Joseph White (http://www.independent.co.uk/news/people/obituary-professor-joseph-morrison-white-1429913.html) provides a good example of what might be possible albeit with a much simpler disease model. White was among the first physicians to translate molecular advances in the laboratory to the bedside and then into the wider community. As a recent medical graduate in Oxford he identified abnormalities in the structure of hemoglobin associated with common inherited blood diseases (the hemoglobinopathies). His success only came once Max Perutz and John Kendrew in Cambridge UK had discovered the molecular structure of hemoglobin. Within ten years, White was working with communities desperately affected by these inherited blood diseases in the Arabian Peninsula and aiming for prevention of new cases. Notwithstanding, no one expects the dementia syndromes to yield to investigations of a single large molecule but the argument is well-known: causal classification is a vital first step in dementia control.

Lack of evidence revealed in the NIH Report, of course, does not mean absence of evidence. In addition to inadequacies of current diagnostic criteria, what could the problems be? The first must be timing. If late-onset dementia first causes symptoms in the eighth or ninth decades of life and the "lag period" leading up to dementia is as long as twenty years or more (as it seems to be), then there may be an early "window of opportunity" to prevent dementia early in its course. After that, the window closes and the opportunity is lost. Most clinical trials were conducted with people at increased risk of dementia solely on the grounds of age. For them, presymptomatic dementia already may be irreversible and no treatment can be effective.

A second problem concerns the length of time an intervention must be used to be effective. Many trials are of short duration: a year is not uncommon, and five years is very rare. Can this ever be effective when the outcome is to pass a "tipping point" at which dementia pathology exceeds an individual's capacity to compensate or buffer its effects on cognition? Longer studies are tricky to interpret when data are confounded by illness or even the death of many participants.

When those recruited to a clinical trial fail to complete for whatever reason and then are added up, it is easy to see how large initial numbers of trial participants are needed at the outset to ensure that sufficient numbers remain to the end of the trial to give interpretable results. Measurement of outcome in a presymptomatic population is by no means easy. These issues are fully discussed by Tim Salthouse in a landmark review.[78]

A solution preferred by some major dementia research centers is to draw together all that appears to hold the promise of prevention into a package that addresses the multifactorial nature of dementia. Some elements by themselves would be very difficult to test as single interventions in a standalone trial. These include optimization of clinical care for chronic diseases of old age, dietary patterns, and overall lifestyle. When elements include opportunities for enhanced retention in a trial, it is possible to envisage a package containing social components that will reduce dropout and possibly reduce the number of participants required for final analysis. Earlier, we identified additional opportunities to test dementia prevention strategies. These have focused on the recruitment of individuals at exceptionally high dementia risk. Examples were found in the dementia linked to Down Syndrome, individuals who carry two copies of the *APOEε4* gene, and the unaffected offspring of families with an autosomal pattern of AD inheritance (http://www.dian-info.org/). Major research programs are already under way to discover novel genetic tests in dementia and to evaluate well-chosen interventions. Their results are eagerly anticipated (http://alois.med.upenn.edu/adgc/info/NIA_funded_genetics_awards.pdf).

12

DEMENTIA RISK REDUCTION, 3: MULTIDOMAIN APPROACHES

I. BACKGROUND

This chapter introduces multidomain approaches to the reduction of dementia risk. Complex approaches to trial design are under way at European centers. These trials choose among types of intervention to test how various combinations might reduce the risk of dementia. Choices are based on results from observational studies and some preliminary data. The chapter touches on advanced statistical methods to "mine" data from life course research to identify the best timing as well as potentially modifiable causal factors. These highly technical methods include structural equation modeling, path analysis, and estimation of contextual covariables in latent growth curve analysis of longitudinal data.

A few words of caution are necessary to introduce the issues surrounding clinical trial design and multidomain approaches to dementia risk reduction. These cautions serve as a "health warning" about the accuracy of recommendations based on trial results.[1] First, late-onset dementias are quite unusual illnesses. Alzheimer's disease (AD) does not become symptomatic until AD changes are present in the brain for many decades. This means that opportunities to prevent dementia may not be available once the disease is established, maybe at around age fifty years. To date, except for interventions among carriers of known AD genes in early onset FAD, no clinical trial has started before this point. Second, because the dementias are chronic, slowly evolving illnesses, clinical trials that begin only after dementias become symptomatic may begin too late and provide too brief of an exposure to the intervention to be effective. In fact, it is quite difficult to find any dementia prevention study that has lasted for more than one year; durations of three to six months are normal.[2,3]

A third issue concerns measurement of response to intervention. Progression from age-related cognitive decline to dementia may take several years as cognitive impairments accumulate and increase in severity. Conversion to frank dementia, although easy to identify, is an all too infrequent event that occurs over brief periods of time in small samples. Thus, either large samples are needed, which can be prohibitively expensive, or a measure of differences in rates of cognitive decline needs to be devised. Patterns of decline are subject to sizable differences between individuals, and these patterns require use of a wide range of cognitive tests to be detected reliably. In turn, as testing becomes more demanding of participants, dropouts from clinical trials increase. This high dropout rate partly is due to the efforts needed to retain the interest and motivation of those at risk of dementia as well as due to issues of physical frailty and sensory loss, which can hinder participation in a long-term trial.

As data from clinical trials in dementia prevention accumulate, critics scrutinize results and ask what, if anything, was learned. The answer, too often, is very little. Although observational studies identify candidate interventions, these studies rarely show positive results in a dementia risk reduction trial. With hindsight, the reasons seem obvious: (1) these trials were too short to detect an effect; (2) relatively insensitive cognitive tests were used to detect change; (3) interventions were introduced too late (e.g., recruits to trials all had sufficient symptoms to meet criteria for early AD); and (4) methods used to analyze data from trials failed to distinguish between the effects of practice on repeated cognitive tests or to adjust appropriately for baseline differences that affect cognitive aging. Not surprisingly, the same experts view the overall contribution of dementia risk reduction trials to date with some skepticism.

The preferred solutions to these problems are twofold: (1) to recruit participants to prevention trials only from a specific high-risk group who share characteristics or exposures that could be modified by the intervention, and (2) to use inclusion criteria that define a biomarker that is related to the actions of the intervention as in the high-risk group. The best known example of implementation of these improvements to trial design include ongoing clinical trials in familial early onset dementia in large pedigrees sharing the same genetic mutation.

One approach that may accelerate reported decreases in current dementia incidence is to accept that risk factors identified in observational studies are tested as combined (multidomain) interventions. Although it is unsafe to presume multidomain interventions will prove to be more effective than monodomain approaches, it is a reasonable, medium-term position to take on dementia prevention when many decades might be needed to reduce the public health burden of dementia.

RANDOMIZED CONTROLLED CLINICAL TRIALS

The gold standard of efficacy of a treatment in medicine is to compare individuals with the same chance of responding to a treatment after exposure to the proposed treatment at the same level of efficacy. The "randomized" part of a randomized controlled clinical trial (RCT) refers to the random allocation of participants to a drug or its comparator. This definition assumes that the likelihood of any benefit is distributed randomly among participants. Most available RCT data on dementia prevention are summarized in Table 12.1. Currently, these data provide a satisfactory guide to personal decisions and also inform public health policy. The data shown in this table are summarized from a comprehensive 2010 report by the U.S. Department of Health and Human Services.[4]

Some study results appear more compelling than others. A European clinical trial of blood pressure reduction provides strong grounds to recommend blood pressure control in the elderly and dismisses concern that high blood pressure helps maintain brain blood flow in old age.[5] Reductions in the risk of heart disease and stroke are so widely accepted as benefits of blood pressure control that few would contest the added possible benefit of reduced risk of dementia. This raises the question why should RCTs be inconclusive? This finding likely is a consequence of many trials being too short and too late as indicated previously.

At a glance, the data summarized in Table 12.1 look like a jumble of ill-assorted facts. The life course approach can begin to make some sense of Table 12.1, as shown in Figure 12.1.

Risk factors for dementia do not act in isolation to the detriment of the aging brain. Their effects arise along many paths, some of which reflect differences that appear understandable only in a culture-specific social framework. An historical context helps place an individual's experience of aging among at least three levels[6]:

1. A personal history as seen in the work of developmental psychologists
2. A social history of the communities through which an individual passes along the life course
3. The much wider perspective that shows how societies respond to the challenge of aging

TABLE 12.1 Overview of Health Care, Environmental, Genetic, and Social Factors Shown in Observational Studies and Randomized Controlled Trials to Influence Dementia Risk in Populations Mostly of European Ancestry

	INSUFFICIENT EVIDENCE	NO EFFECT ON AD	DECREASED RISK OF DEMENTIA	INCREASED RISK OF DEMENTIA
Medical History	Posttraumatic stress disorders, anxiety disorders	Anticholinergic drugs, antihypertensives		Diabetes, traumatic brain injury, depressive illness, conjugated estrogens
Metals and Toxins	Aluminum, solvents	Lead, trace metals		Pesticides, tobacco smoking
Drugs	Intravenous drug use, cannabis, ecstasy, nonsteroidal anti-inflammatories	Cholinesterase inhibitors, memantine, hormone replacement therapy (HRT), Ginkgo Biloba	Statins, antihypertensives, light alcohol intake	High alcohol intake
Nutrition	High calorie intake, saturated fats	Vitamin B12, Vitamin E, omega 3, antioxidants, fruit and vegetables	Folic acid, Mediterranean diet	
Social Factors	Occupational complexity		Physical activity, cognitively effortful lifestyle, education after age 16 years	Loneliness, early parental death
Genetic Factors			APP variant (rare)	APP, PSEN1 and PSEN2 (rare) mutations; APOEε4 variant (common), genes linked to synaptic repair, immune and inflammatory responses (frequent)

Note: Other major population groups in Africa and Asia may differ.
Source: Data are from ref.[4]

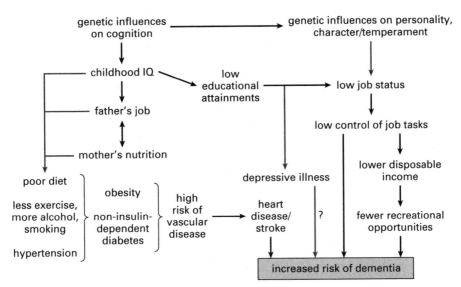

FIGURE 12.1 Major influences acting during the life course that increase the risk of dementia. There are major pathways from childhood IQ toward educational attainments and also to the acquisition of lifestyle habits of a balanced diet, exercise, moderate alcohol, and not smoking. Other childhood influences on IQ are too complex to be illustrated here; these include early fetal nutrition, childhood illnesses, "positive parenting," and other components of early education. Important hazardous influences on the risk of dementia include an acquired predisposition to develop diabetes, heart disease, and stroke and the influence of family factors and childhood IQ on job success and the lifelong accumulation of social and material capital. Protection against dementia is inferred as influences that oppose these hazards. Interventions that improve early childhood education, enable higher linguistic abilities, and help establish adult patterns of healthy eating, exercise, tobacco avoidance, and reduced alcohol and illicit drug use are largely matters of public policy and individual judgment. Management of health promotion to reduce risks of vascular disease is one of the great medical successes of the late-twentieth century and where gains seem likely to be maintained.

When social neuroscientists consider the origins of behaviors of an individual living with dementia, they observe a gradient of changes from successful adaptations through to early awareness of cognitive and social deficits and the evolution of dementia syndromes. Boundaries between syndromes are not needed until the severe end of this spectrum is reached, when with hindsight, the origins of each dementia subtype can be traced to their molecular beginnings. Considered in isolation, each of the behaviors that typify a single dementia syndrome is understood not as a simple consequence of neural loss but rather as adaptive and compensatory adjustments to an aging social context. Chapter 8,

Emotional Aging, emphasized the relevance of a complex interaction between emotional and cognitive aging. This interplay creates the medium in which the deficits and unwanted behaviors of dementia are expressed and one in which some of the risk and protective factors of dementia exert their greatest influence. These multiple levels are shown schematically in Figure 12.2. This diagram underpins understanding multidomain approaches to dementia prevention.

Social processes are available to buffer the presence of neuropathology in a brain in which neural health is compromised. Some resources were stockpiled from infancy, and others were amassed during adult development often during critical periods or at times of recovery after extreme stress. Opportunities to add to these resources arise in step with normal development to maturity, but it is

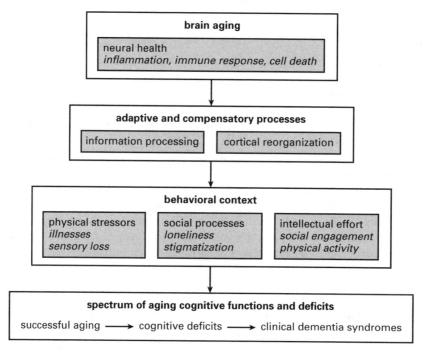

FIGURE 12.2 The schematic sets out interconnections between the neurobiology of compromised neural health that impairs neural connectivity and efficiency of information processing and places demands on cortical reorganization. The features of cognitive aging and the clinical dementia syndromes emerge alongside the forces imposed by age-related physical illness and sensory deficits. Emergent behaviors are modified by concurrent social processes specific to the aging condition and individual capacity to make good any deficits or initiate appropriate adaptive behaviors. The complete range of possible outcomes is shown as a spectrum that ranges from success to cognitive failure.

unknown how much scope is retained in later adulthood to boost resources in ways that delay the onset of dementia. The life course approach identifies some environmental influences on development and aging. The *fetal origins of adult disease hypothesis* emphasizes the potential harm of intrauterine and maternal adverse factors on the adult health of offspring with major hazards linked to adult obesity, diabetes, and high blood pressure. Chapter 2, The Life Course Approach, set out the biological mechanisms that appear to link fetal life to adult health, highlighting the importance of *critical periods* in the acquisition of behaviors and the growth and maturation of intelligence. Uncovering the neurobiology of critical periods is an opportunity to reinstate a time of great cortical plasticity and to augment resources to cope with cognitive aging and dementia.

II. NUTRIENTS AND DEMENTIA

Interest is huge in the possible role of nutrients in the prevention of dementia.[7] An international industry based on dietary supplements has developed with substantial sales of specific micronutrients that claim to have benefits for the retention of cognitive functions in the face of aging. Sometimes, these commercially available micronutrients are labeled and sold as "nutriceuticals" with an aura of specific pharmacological actions against dementia. It is difficult to accept many of the claimed benefits of any nutritional intervention. For some clinical nutritionists, the only effective nutritional intervention in the prevention of dementia is reduction in cardiovascular risk that follows weight reduction through caloric restriction among the overweight and the obese.

The conduct of clinical trials with specific nutrients is bedeviled. The first pitfall concerns the nutritional status of those whose diets are to be enriched in some nutritious way. Nutritional status of older adults is complex and great care is necessary when recruiting those at greater risk of dementia to nutritional trials. Most obviously, minor degrees of cognitive impairment can preclude the planning and preparation of some meals. This gives rise to a type of *reverse causality* in which malnourishment is presumed to be the cause of cognitive decline, whereas the opposite is possibly just as true. Second, a reliable dietary history requires retention of many cognitive functions. This is especially true of meals that are prepared by others without regard to the preferences of the older adult. Failure to record food wastage in this context often invalidates a dietary history. When the consumption of dietary supplements is high (around 30 percent of older Europeans take a dietary supplement), failure to record this raises additional problems. It may seem more straightforward to measure the concentration of micronutrients in blood and to disregard as unreliable any dietary history data. Although this approach has its merits, it sometimes is compromised by the lack of data to establish the validity of measurement of these nutritional biomarkers.

In the face of these two limitations, it is worth emphasizing that nutritional studies warrant careful evaluation. Micronutrients with a plausible link to dementia risk can be identified. Their inclusion in a healthy balanced diet tailored to the needs of old people can be set in the context of the multidomain approach. This represents a shift in emphasis away from the notion that a poor diet can be made healthy through dietary supplements and toward dietary habits that are conducive to good general health, including claimed protection against dementia. This highlights the difficulty of assessing the possible impact on dementia incidence of any single lifestyle improvement. Levels of physical activity, leisure time pursuits, social interactions, smoking, alcohol intake, smoking, level of education, and occupational complexity are all associated. These complex interrelationships hinder exact prediction of the effects of lifestyle improvements.

VITAMIN B12 AND FOLATE WITH HOMOCYSTEINE REDUCTION

Deficits in availability of folate are major risk factors for spina bifida.[8] Many studies now show folate is important in the aging brain. Initially, it seemed that patients living with dementia had poor diets, and this could explain their low folate levels in blood. Homocysteine is a risk factor for heart disease, stroke, AD, and cognitive aging. Figure 12.3 shows how homocysteine is converted to methionine and provides methyl groups for the methylation of DNA.

The *homocysteine hypothesis of AD* is based on the observation that increased blood concentrations of homocysteine are associated with age-related cognitive decline[9] and AD,[10] but not all studies support this idea. Although seven studies to date proved positive, one prospective study from Dick Mayeux's group at Columbia University, New York (United States), showed that elevated homocysteine concentrations did not predict progress to dementia. Adequate tests of the homocysteine hypothesis require blood homocysteine concentrations to

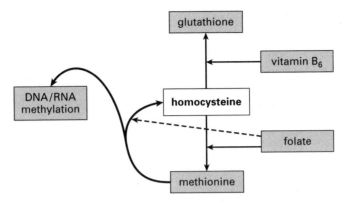

FIGURE 12.3 Homocysteine is a naturally occurring compound that is converted to methionine, and methionine has a vital role in the expression of DNA. Deficiencies of folate and vitamins B6 and B12 cause homocysteine to build up to toxic levels in tissues where it promotes damage particularly to neurons and to the linings of blood vessels.

be measured before dementia onset and statistical adjustment to be made for the contributions to established risk factors for dementia. These include gender, education, family history of dementia, and apolipoprotein E (*APOE*) genotype, and, in light of a proposed association between homocysteine and vascular disease, a history of heart disease or hypertension. Figure 12.4 shows results from an observational study in Aberdeen, Scotland.[11] Blood homocysteine levels were measured at age sixty-seven in the absence of dementia. Over the next ten years new dementia cases were observed in the study sample. The figure shows that those with highest homocysteine at age sixty-seven were at greatest risk of progression to dementia by age seventy-seven. This association remained after controlling for the effects of female sex, socioeconomic status, baseline intelligence, and many other variables.

Dietary habits are strongly influenced by low socioeconomic status (SES), which, in turn, is linked to the risk of dementia. Inconsistencies in reporting associations between homocysteine and AD, therefore, might be attributable to SES variation among studies. It is also relevant that folic acid reinforcement in cereals is permitted in the United States but not in Europe.

Plasma homocysteine concentrations are influenced substantially by intake of dietary folate and vitamins B6 and B12 as part of a healthy balanced diet. Although hyperhomocysteinaemia can be attributed to deficient intake

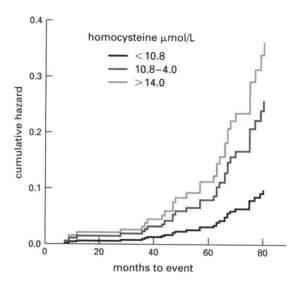

FIGURE 12.4 The risk of dementia was increased by about three times in fifty-five people with the highest homocysteine concentrations. All 199 participants were born in Aberdeen in 1921, were without dementia at the age of seventy-eight and followed up for eight years. The incidence curves are adjusted for education, childhood IQ, socioeconomic status, and blood antioxidant concentrations.

of folate/B12, hyperhomocysteinaemia need not arise in isolation and can reflect overall poor nutritional status.

Thus far, data from twelve homocysteine-lowering trials are available. The most satisfactory trial to date was reported by Jane Durga and colleagues from Wageningen University in Holland who described a three-year trial with supplementation by folic acid.[12] Their 818 participants ranged in age from fifty to seventy and had homocysteine concentrations in the range 13–26 micromoles (μmol). The previous ten trials were summarized by the Wageningen group who thought that these gave equivocal results. Their trial showed that folic acid significantly reduced blood homocysteine concentrations and that this reduction was associated with improvements in global cognitive function, memory, and information processing speed. Importantly, although revealing cognitive benefits in older adults, their study does not demonstrate protection against progress to dementia.

This question was addressed in later studies by David Smith's team at Oxford University, United Kingdom,[13] in the OPTIMA project, which found preliminary evidence that folic acid slowed cognitive decline and age-related brain atrophy. Their studies await completion and replication, but if supported, will identify a reversible cause of dementia holding significant public health importance.

FISH OILS (OMEGA-3 FATTY ACIDS)

The fatty acid composition of cell membranes determines many of their physical properties. Fatty acids also provide precursors of compounds that become critical to the health and normal function of neurons. These compounds are involved in cell-to-cell signaling, neuronal differentiation, maintenance, and repair and in modulation of responses to stress. Essential fatty acids (EFAs) must be obtained from diet (and are therefore "essential"). It is of some public health concern that increasing urbanization is associated with decreased consumption of the omega-3 EFAs and increased omega-6 EFAs. The resulting imbalance is believed to cause long-term mild omega-3 deficiency.[14] The main omega-3 EFA is docosahexaenoic acid (DHA) present in higher concentrations in the brain than in other tissues. This observation has supported a possible role of DHA in brain functions in health and disease. Studies in older adults show that dietary intake of DHA and relative amounts in accessible cell membranes (e.g., red blood cells) is associated with better retention of cognitive abilities and, possibly, a lower risk of AD.

All of the caveats about nutritional studies linking specific nutrients and cognitive aging and progress to dementia apply as well to studies on omega-3 EFAs in the diets of older people. Some doubts remain, therefore, about the possible role of omega-3 EFA supplementation in the prevention of dementia. These concerns are countered, however, by evidence that DHA plays a key role in memory formation and in the modulation of brain reactions to stress.

The typical Western diet is relatively nutrient poor and energy dense. This diet began to change substantially with migrations of the rural poor to towns and cities during the Industrial Revolution. Diets were modified as workers could no longer rely on self-grown produce. Since World War II, as food manufacture has grown to dominate markets, the balance between omega-3 and omega-6 EFAs has shifted from about 1:1 to about 1:20 in favor of omega-6. Currently, intake of DHA depends largely on intake of oily fish, particularly tuna, mackerel, sardines, and herring.

The health consequences of an imbalance between omega-3 and omega-6 EFAs are widely debated. All things considered, there appears some agreement that mild omega-3 deficiency predisposes to inflammatory processes, and these include the development of atherosclerosis (hardening of arteries). This agreement does not extend to the possible role of omega-3 EFA intake in retention of cognitive ability in late life[15,16] or to the prevention of dementia. RCTs of omega-3 supplementation are inconclusive. Participants too often are recruited from among the elderly with mild cognitive impairment who may be at too late a stage in a dementia to benefit. Furthermore, the nutritional status of participants is too often uncertain with the possibility that some are generally poorly nourished or consuming more omega-6 EFAs than expected. Some experts also suspect that the overall metabolic status associated with an imbalance of omega-3 to omega-6 EFAs is too poorly understood to permit reliable randomization into a trial.

ANTIOXIDANTS AND DEMENTIA

Antioxidants form part of the body's defenses against unwanted oxidation of tissue. These processes are more likely at locations called mitochondria that lie within cells that release energy stored in glucose. There are two classes of antioxidant. The first class forms part of the body's intrinsic defense systems, featuring membrane-bound enzymes, such as superoxide dismutase and glutathione and uric acid made naturally in the body. The second class includes dietary micronutrients with powerful antioxidant properties. Examples include micronutrients like vitamins A, C, and E, but many others are available. The total antioxidant capacity of blood plasma can be measured fairly easily using standard laboratory agents. It is, however, much more difficult to determine how much each dietary antioxidant contributes to total antioxidant capacity.

Interest in the role of antioxidants in aging and age-related disease derives from the work of Denham Harman who set out the free radical theory of aging.[17] A free radical is an atom or molecule with a single unpaired electron in its outer shell, most of which react readily with other chemicals and are highly reactive in biological tissues. These reactions lead to oxidative damage in which molecules receive an electron from a free radical. Antioxidants therefore are called "electron receivers."

Harman's free radical theory first was proposed as a hypothesis, but now with substantial experimental data available to support the idea, the proposition sits on a much stronger footing. Key experiments show that antioxidants of varying types can extend life span in single-celled organisms and that synthetic antioxidants of great potency can extend life span in the round worm (*C. elegans*).[18] As relevant data accumulated, Harman modified his original proposal to focus on the role of mitochondria (now known as the mitochondrial theory of aging).

In this version of the theory, which is well-regarded, mitochondria produce free radicals within the cell. Some of these free radicals are water molecules charged by electrons that have escaped from the electron chain inside the mitochondrion. These free radicals then cause local damage, interfere with DNA metabolism, and cause mutations. Their overall effect is to lower intrinsic antioxidant defenses and thus aggravate oxidative damage. Many chronic diseases of aging are now linked to oxidative damage. These include damage to the walls of blood vessels, initiation of amyloid processing, and damage to connective tissues. Smoking and excess alcohol can increase oxidative stress in a systemwide fashion, probably through their damage to large bioregulatory molecules. Calorie restriction is claimed to extend life span, possibly through reductions in metabolism and lower production of free radicals.

ANTIOXIDANTS AND DEMENTIA RISK REDUCTION

The free radical theory of aging predicts that antioxidants will counter the effects of free radicals and extend life span. The antioxidants present in fresh fruit and vegetables often are given as examples of natural sources of antioxidants, such as vitamins A, C, and E and the carotenes. So far no dietary intervention (with the possible exception of caloric restriction) is known to increase life span in humans.

Public interest is high in the possible health benefits of antioxidant supplements in late adulthood, with a specific focus on the prevention of aging. Many studies have explored this topic without reaching an agreement.[19-22] Contributions to this research made by studies in the United States and Europe were inconclusive.

Nutritional interventions have relied on preparations made up of a single nutrient, or combinations of micronutrients sometimes with added fatty acids. The overall impression is that no consistent benefit has been detected. Current interest has shifted away from clinical trials of nutrients toward more complex interventions based on changes in dietary habits with opportunities to add programs of cognitive or physical activity. These "multidomain approaches" are described in a later section.

OVERVIEW OF NUTRITIONAL STUDIES

Some observational studies indicate that omega-3 EFAs with or without additional antioxidant nutrients are associated with lower rates of cognitive aging, but the majority of trials do not support this finding.[23] If there are positive effects of omega-3 fatty acids, these may be detectable only in specific cognitive domains (such as information processing speed) in cognitively normal nondemented participants or in specific subgroups (e.g., only in the absence of *APOEε4*[24]). In this regard, the aim could be to decelerate the rate of age-related slowing of processing speed after prolonged administration of omega-3 EFA. These effects may be undetectable once AD brain changes are established.

A similar degree of inconsistency is evident among trial centers in tests of homocysyteine lowering by folic acid and B vitamins. This intervention may slow the rate of brain atrophy in older individuals with slight cognitive impairment, but so far without any benefits in slowing the rates of cognitive decline or progress to dementia. No benefit have been identified for cognitive function in trials of antioxidants, including vitamins A, C, and E and β-carotene either alone or in combination.

Absence of an effect of nutritional interventions in RCTs does not invalidate the many observational studies that have suggested benefits of specific nutrients. In part, this discrepancy could be explained by the differences between what is present in an overall nutritious diet that may be absent from a mix of an individual's regular diet and the food supplement under test. A little evidence already suggests this might be the case and that a well-balanced diet (as suggested for the Mediterranean diet) is more beneficial than the addition of a food supplement, no matter how well judged. Another point raised by nutritional biochemists arises from the potent antioxidant properties of β-amyloid before it is assembled into β-pleated fibrils. These actions are poorly understood, but some experts claim that administration of antioxidants in doses far in excess of normal requirements may promote production of β-amyloid with increased damage to neurons. Their point is well-taken: Great caution is needed when interfering in poorly understood metabolic processes using compounds with much higher levels of activity than encountered in healthy bodily function.

Dietary supplement use in late adulthood has become a common habit in industrial countries. Estimates vary, but it appears that about 30 percent of people aged older than fifty-five regularly consume a food supplement. Supplements often are taken for the seasonal relief of specific symptoms (such as winter joint pains), although increasingly they are taken because of their claims to enhance cognitive protection. The evidence is weak that these benefits are achievable. The "health food paradox" is that supplements more often are taken by those in lifelong good health and frequently of higher SES.[25] Put simply, supplements are more easily afforded by the affluent.

III. THE MULTIDOMAIN HYPOTHESIS

Attempts to develop a causal model of AD that takes account of social, psychological, and neurobiological data suggest that a systems approach to understanding the causes of AD might be useful. Interventions based on such an approach would integrate the many domains compromised in AD. This hypothesis places the loss of system integrity as the primary event that leads to impairments in multiple domains. Not surprisingly, discouraged by so many prior studies with negative outcomes, researchers have turned to interventions with multiple components.

This is not an unexpected outcome: Comparable multidomain approaches already have been undertaken to prevent heart disease and osteoporosis. Clinical trials of multidomain approaches to dementia prevention are ongoing in the United States and Europe and, although no results are available, the choice of components to be tested helps inform current clinical practice. For example, it is not uncommon to be asked in the clinic what steps can be taken, in light of the best available evidence, to reduce risk of cognitive decline toward dementia. Most clinicians agree that the best advice is prefixed by the fact that no intervention is effective. All things considered, however, and taking account of findings from a clinical history, several "personalized" suggestions often are made. These suggestions are divided between nutritional advice, control of vascular risk factors, avoidance of hazards (alcohol, smoking), increased physical and social activity, and the pursuit of mentally stimulating or effortful leisure pursuits.

When pressed, the clinician might argue that at least three disease processes appear to be driving dementia and that each process contains at least one possible target for risk reduction. It is unclear, however, how stages of each process relate to one another; which stage occurs first in the cascade of events toward dementia; and if priority of one over others could be established, whether this means it is a better target than another for dementia prevention.

The problem is something like trying to improve passenger journey times on the London Tube. A map of the system shows all available routes but cannot show passenger volumes, time of day effects, or locations where passenger deadlocks arise. Transport strategists might solve the problem of deadlocks by first monitoring flow across the network in ways that managers feel reflects precisely what is happening in daily use. Raising traffic volume in their model might exactly replicate what happens when a gridlock occurs. The effects of various interventions then are studied in the model before introducing preferred solutions. Unexpected consequences of a "solution" arising elsewhere in the Tube system are anticipated in this way and alternative solutions sought.

This approach is not yet feasible in dementia prevention partly because understanding the factors that determine the healthy survival of neurons is incomplete. The biology of dementia is not yet at a point that shows how the flow of proteins in brain cells can be disrupted in ways that cause overproduction, reduced clearance, or abnormal aggregation of specific proteins that when present in excess can harm the health of brain cells. The Introduction discussed a systems approach to enhance our understanding of how neurons fail in dementia. Much remains to be achieved by showing how all the individual components of neuronal systems interact in time and space to determine the functioning of the nervous system as a whole. Success depends on the intelligent organization of large amounts of data from molecular biology and genomic research, integrated with an understanding of physiology to model the complex function of the nervous system.

A multidomain approach to dementia prevention derives from an acceptance that although knowledge about the causes of dementia is incomplete, this should not prevent testing interventions that have been chosen carefully for their potential efficacy and administered in acceptable combinations. In large part, multidomain approaches mirror what is happening currently in many memory clinics. The advice given to an individual concerned that they are at increased risk of dementia is as follows: Initial advice is based on a clinical assessment to identify reversible risk factors for heart disease and stroke. Their reduction is justified easily on the grounds that this will decrease the risk of these conditions

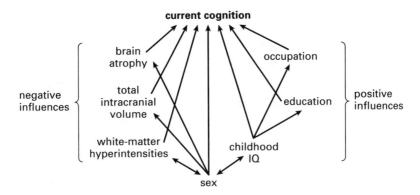

FIGURE 12.5 Major influences on current cognitive performance in late adulthood in 248 adults age sixty-eight years and without dementia. Negative influences include (1) brain shrinkage attributed in part to presymptomatic Alzheimer's disease; (2) smaller brain size in early adulthood; and (3) white matter hyperintensities attributed to brain–blood vessel disease and exacerbated by high blood pressure and abnormal glucose metabolism. Positive influences include (1) higher original intelligence from childhood; (2) longer duration of formal education; and (3) more complex occupations often with supervisory responsibilities. These relationships are shown as bold arrows, and combined, they account for about 50 percent of influences on current cognition.

regardless of any possible reduction in the risk of dementia. Lowering high blood pressure and blood cholesterol, weight loss to achieve a body mass index in the range 22–24, cessation of smoking, and abstinence from alcohol are encouraged.

Figure 12.5 is based on data from long-term follow-up studies on brain health and aging in Aberdeen from 1998 to 2013 among local people born in 1936 and without dementia. If this type of study were performed on people who were living with dementia, the contribution of "brain atrophy" would be much greater, and white matter hyperintensities also increased but less so. When the intention is to prevent dementia, complementary strategies are implemented in the multidomain approach. The aim is to lessen the contribution of the negative influences through the reduction of vascular risk factors and by taking steps that might promote retention of healthy neurons and their connections. As noted, reduction of vascular risk is the initial response to a request for advice. For the most part, these aims are met through public health measures, health education, and systematic regular review by personal physicians. Most experts agree that, at present, the strongest foundation for offering advice on the prevention of dementia is based on observational studies pointing to changes in lifestyle.

Improvements to neuronal health are much more difficult to achieve. If the aim is to repair neural circuitry by stimulating synaptic plasticity, it already is known in animals that this can be achieved in three ways: physical exercise, environmental enrichment, and chronic antidepressant treatment. Synaptic plasticity is increased by stimulating the growth of new synapses and (more rarely) new neurons. These synapses are created in numbers in excess of requirements and are allowed to compete for survival. This process of activity-driven and experience-dependent selection of circuits chosen to survive creates networks that meet the demands of the environment and are not predetermined in structure or function. Conceptually, it is possible to envisage therapeutic interventions to prevent dementia that are based on current clinical practice in the treatment of stroke, in which case the activation of adult synaptic plasticity appears desirable.[26,27] Antidepressant drugs are tested in animal models of brain injury and AD and show desirable effects on synaptic remodeling[28] and behavior.[29] This antidepressant–rehabilitation model is comparable to the routine combination of oral antidepressants with cognitive-behavioral therapy or psychotherapy in the care of depressive disorders, in which case the practice is more effective than administration of either therapy alone.

Grounds are strong to accept that work with learning and memory in rodents is directly relevant to older adults. These studies show that physical activity has the potential to enhance retention of neural networks and to enhance the richness of the connections between neurons and other networks. These positive benefits may extend to the formation of new neurons ("neuroregeneration"); however, while detectable in rodents, the extent to which this can make good any age-related cognitive deficits in the adult brain is largely unknown. The death of neurons creates spaces in the brain that are hostile to the formation of new neurons. These spaces are invaded by processes extended from the glial cells, and growth factors needed to promote the differentiation and survival of new neurons are not expressed in the damaged area. Setting these obstacles apart, experience in stroke rehabilitation has encouraged practitioners to continue prolonged activity-based therapies in stroke victims sometimes to striking effect. The evidence base for this approach to neural loss in the brain following stroke is quite firm and, by extension, has fostered the opinion that "brain training" or "cognitive training" can help retain cognitive performance of older adults.

These ideas fit well with observational studies that show that more education, more complex occupations, and an active and cognitively challenging lifestyle are associated with lower risk of dementia. "Cognitive training" used alone can help older adults without dementia improve performance on cognitive tests used in the training program and to sustain these benefits for some years. These benefits do not seem to include better performance at activities of daily living. Nevertheless, these findings have not discouraged the addition of cognitive training to multidomain approaches to dementia prevention.

Perhaps the best known multidomain approach to dementia prevention is adherence to a Mediterranean diet. This forms a component of the Finnish study described in the following paragraphs and has been scrutinized in a variety of disease prevention programs, often with encouraging results. The Mediterranean diet was followed by those living in maritime regions of Southern Europe, principally in Greece, Italy, France, and the Iberian peninsula. The diet depends on ready availability of fruits and vegetables; whole grain cereals, but with few saturated fats; lower reliance on meat and dairy products; frequent meals of seafood; and moderate intake of red wine, coffee, and olive oil.

Nine observational studies and one RCT taken together suggest that following a Mediterranean diet can mitigate age-related cognitive decline and may delay dementia. Expert reviews[30] on this topic conclude that more RCTs are needed before the diet can be widely recommended. Despite this conclusion, the health benefits of the diet extend beyond dementia and seem particularly well supported in the prevention of heart disease. In one of three analyses of the same cohort studied by Dick Mayeux's research group at Columbia University (United States),[31,32] the closer individuals kept to a Mediterranean diet, the lower their risk of dementia. This relationship was strengthened by higher levels of physical activity and is replicated in diverse populations.[33,34]

Taken together, these results point to the benefits of a particular lifestyle characterized by dietary preferences that reduce risks of vascular disease and include levels of physical activity that also promote cardiorespiratory fitness. Although some nutritionists interested in dementia have sought to find the ingredients of the Mediterranean diet that convey these health benefits, the diet is more often given as an example of a multidomain approach.

The combination of the Mediterranean diet and physical activity is particularly relevant to the example of a Finnish dementia prevention program.[35] This is an RCT of older adults believed to be at increased risk of dementia but functioning at a level only slightly below that expected for their age. The design and conduct of the study is especially helpful to this account of multidomain approaches because it draws together many of the issues described thus far with pointers to steps that might be followed at a personal level.

The duration of the trial is two years and requires much from its participants. The active arm of the trial combines nutritional guidance, physical exercise, cognitive training, social activities, and clinical management of metabolic and vascular risk factors. The passive arm of the trial involves regular health advice. The trial is notable four reasons. First, all participants receive the same health advice before randomization. Second, all four interventions are made in everyone in the active treatment group. Third, the nature, extent, and duration of each component of this multidomain approach is recorded and measured with care. For example, promotion of a healthy diet is effected through regular meetings with a nutritionist and group sessions. A healthy diet is precisely defined by a high intake of fruit and vegetables, at least two portions of fish per week, and restrictions of consumption of dairy products. Those unable to eat fish are advised to take fish oil supplements and all active participants are encouraged to take vitamin D supplements. Fourth, the use of systematic clinical review at set points over the two years is a critical means of study retention and monitoring compliance. These meetings motivate participants to make lifestyle changes as necessary and provide a point of data exchange between the study clinical staff and participants' own physicians.

Study physiotherapists supervise physical training and follow structured training programs tailored to the needs of the individual participants. Likewise, a psychologist supervises the cognitive training program, taking care to reduce boredom and to retain trialists in the study. Outcome measures include changes over time in cognitive performance, progression to dementia over seven years' follow-up, and related clinical and mortality data. Additionally, many other measures related to each component of the multidomain approach are included.

How does a study like this translate into guidance for older adults not involved in a clinical trial? The importance for the individual lies in the informed choice of components of the multidomain active arm of the study. If an individual were to make strenuous efforts to find providers, each of these components are available in large cities in most high-income countries. It certainly would be difficult, however, for most people to find all of the components as easily accessed in many less favored locations. An important part of the multidomain approach is maintaining good records of involvement and progress. Well-person clinics and some fitness gyms can help with clinical measures of cardiorespiratory fitness, but for most else, older adults will rely on their own records of dietary compliance, measures of weight, and exercise performance.

IV. BRINGING IT ALL TOGETHER

Individual approaches to the task of avoiding dementia are energized by the knowledge that delayed onset of dementia is a viable proposition and that the components of the multidomain approach listed in this chapter have the potential to achieve modest but useful delays to onset. Figure 12.6 makes this point in graphic terms.

A delay of five years has the potential to reduce the overall prevalence of dementia by 50 percent. Although none of the interventions listed has ever yielded delays of this duration, shorter delays certainly will have public health significance and likely will be welcomed by all older adults. At present, investments in multidomain approaches appear to be well reasoned and, pending confirmation that a novel antidementia strategy is both effective and safe in long-term use, seem likely to be established as the only practical alternative to the sense of therapeutic nihilism that has pervaded dementia care up to the present.

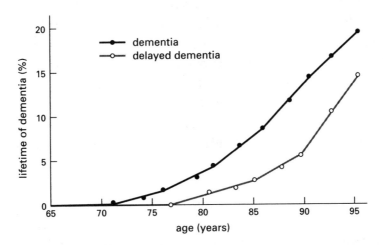

FIGURE 12.6 The effect on dementia onset of interventions that delay dementia onset by five years. These produce an overall reduction of dementia prevalence by 50 percent. At an individual level, the opportunity to postpone dementia onset optimizes quality of life in old age and discourages a fatalistic view that dementia is always inevitable.

EPILOGUE

What will people think of dementia in 2050? Today, in the absence of effective treatments, we see the dementias as largely unpreventable. These are chronic, progressive disabling diseases that affect about 5 percent of the population over the age of seventy and at least 40 percent of those who are older than ninety years of age. The underlying causes of the most common types of dementia remain unknown, and a cure seems a far distant prospect.

A deficient understanding of the causes of dementia arises because the aging brain shows so many faces. Looking toward 2050, will the dementia landscape appear much different from what we see today? We certainly hope that amid our other high expectations, the health of our descendants will improve further. If we accept that both genetic and environmental factors are important influences on dementia causes, will either have changed in a way that reduces the burden of dementia? We lack the confidence to manage our environment in ways that are conducive to successful aging. We know what needs to be done but fail: we neglect to improve nutritional standards, to optimize health in old age, and to secure the continuing social and mental engagement of the aged.

By 2050, we are sure to know whether the role that is played by amyloid biology is as central in late-onset dementia as it appears to be for some early onset dementias. If their position is secured, antiamyloid treatments will slow dementia progress and perhaps also be used to prevent dementia onset. If not, the next step probably will involve strategies that prevent neurofibrillary tangle formation used alone or in combination with antiamyloid drugs. A third research strand to promote neurogenesis and the repair of neural networks could prove to be a critical addition to dementia therapy and enter wide use as an adjunctive.

Research on the biology of aging and its links with dementia could provide new ways of thinking about the origins of dementia—why some people are vulnerable, whereas others succumb. A gambler in 2016 would consider an outside bet that Alzheimer's disease (AD) proves unique to humans and that the explanation of this uniqueness is found in genetic sources of difference in the aging brains of higher primates and humans. The possibility remains tantalizingly attractive that selective loss of cortical neurons in AD arises for reasons that distinguish these highly evolved human brain structures from nonhuman primates.

Until the beginning of the twenty-first century, comprehensive simultaneous approaches to understanding all relevant pathways to AD were not attractive and were beyond the capacity of most research collaborations. So-called Big Science is now feasible and encourages large international collaborations to pool diverse resources in biomedical investigations with shared access to patients and their families. We know that systems biology will yield data sets of a magnitude requiring advanced computing platforms and sophisticated analytical methods. By 2050, we can be sure that continuous improvements in data handling, visualization, and analysis will have adequately charted the trajectories of brain aging and pathways to dementia in very large numbers of people. These research developments on a near-industrial scale, utilizing dedicated clinical research facilities, could be among the most expensive ever in biomedical science, but they easily could be justified by the prospect of more than 100 million people worldwide living with dementia by 2050.

These prospects will not close the door on smaller research efforts. One of the most exciting scientific developments in the past decade brought together several strands of research that showed how the environment can alter the structure of DNA. Epigenetic DNA modifications produce long-lasting effects, including increasing the likelihood of some diseases like diabetes. As epigenetic mechanisms of late-onset diseases are better understood, effective means to prevent or remove these epigenetic markers surely will be found. For some scientists, this research direction holds greater promise of a successful means to prevent AD than any other.

Other life course effects of the environment do not require major scientific efforts to discover, as they already are appreciated. We will know by 2050 the answers to several important questions, such as the following: Will the widespread current use of recreational drugs like ecstasy and "designer drugs" related to amphetamine and opiates dramatically increase numbers of people with late-onset dementia? We already know that these drugs and their typical contaminants are linked to lasting deficits in cognitive ability and, therefore, should lower the capacity of the aging individual to withstand the effects of dementia. What about the late consequences of "mad cow" disease, during which slowly developing neurodegenerative dementia emerges as a major public health problem perhaps thirty or forty years after contaminated meat products entered the human food chain? Advances in understanding the biology of a range of diseases linked to abnormal protein folding remain our best hope of developing effective treatments of this type of dementia. To this we can add a major public concern about the effects of repeated traumatic brain injury among professional athletes. Enforcement of soundly based regulations that govern the exposure of athletes to head injury—and prevent those who suffer concussion from repeated exposure to injury—is widely accepted as a sound and practical measure with immediate benefits.

In the medium term, management of stroke increasingly will include prevention of progression to poststroke dementia. Likewise, if the contribution is accepted of increased homocysteine concentrations to the increased risk of dementia, public health nutritionists will direct policy makers to obtain the best scientific advice to guide physicians in the care of the elderly.

These are good reasons for early optimism. On the basis of current knowledge, we know that improved recognition, prevention, and management of in-patient delirium, prevention of brain injury and poststroke dementia, and nutritional interventions are our best opportunities to prevent the greatest number of dementia patients. When the dementia landscape is viewed from 2050, success or failure may be judged not by the success of Big Science but rather by the political will to shape societies that are committed to the health of the developing brain and the principles of social justice in dementia care. As always, education of future generations will remain at the core of how this ambition is achieved. To this we can add mitigation of the harm caused by drug use, improved care of delirium, stroke, and reduced exposure to head injuries. Looking forward to 2050, we can begin to think about how health systems will have developed to make reliable early diagnoses of dementia, to introduce preventive measures at the first opportunity and, where prevention fails, to devise methods of care that take account of the paramount need to protect the most vulnerable and not to assume that society will always have sufficient available and able-bodied caregivers. The proposal that most dementia syndromes merge with deficits of normative cognitive aging will certainly be tested by then. If, as seems possible, no clear boundary exists between normative and pathological brain aging then all of us may prove susceptible to a continuously distributed "dementia risk phenomenon" with huge public health implications as populations age with increased survival. We can also expect that the influence on dementia of early life exposures to harmful stressors and/or nutritional deficiencies will have been thoroughly explored and, where possible, effective protective strategies implemented. This may include the wide use of prediagnostic tests on the developing fetus and child with interventions tailored to the long-term personal health needs of each child. Timing may prove crucial such that opportunities to reduce dementia risk may be greater during specific critical periods of brain morphogenesis (e.g., through weeks 9 and 10 or 14 and 15 of gestation). Current national genetic screening programs are already in place with the possibility of up to sixty-seven detectable genetic errors of metabolism in newborn infants. Measures to prevent cognitive impairments caused by certain genetic errors are not always successful, with the probability that some damage has already been caused in the womb. If this also is proved to occur when genes are present that are associated with dementia (as with autosomal dominant familial AD), then public debate should be encouraged concerning the development of methodologies and interventions to prevent late-life dementia in the offspring of multiply affected families. Clinical trials to prevent childhood obesity that use methods aimed at correction of programming abnormalities are currently underway in Europe (http://www.project-earlynutrition.eu/) and led by Lucilla Poston in the United Kingdom (http://www.kcl.ac.uk/lsm/research/divisions/wh/clinical/open/upbeat.aspx).

Later life course opportunities may be detected if in late adulthood the fine neuroanatomical structure of the cortex is shown to underlie "normative" cognitive aging. These steps will not be taken, however, in isolation from appraisals of the general health of older adults especially where the environment plays a major role (e.g., social disadvantage, substance abuse, or concurrent age-related physical illness and disabilities). Many ethical and legal concerns will arise and these will demand informed debate and relevant legislation before appropriate clinical trials can begin. New regulatory frameworks will be needed in anticipation of dementia prevention strategies and may include exposing children to procedures with predicted lifelong benefits, but with unknown adverse effects and to which they are unable to consent.

One of the many blessings of having children is that your journey into an aging landscape is guided by their intrusive insistence that lifestyles are a matter of choice. It is never too late, they say, to improve diet, exercise, and take on new challenges. If we have any sense, we treat the advice of our children very well. Although we are their forerunners into aging, they are our guides in touch with the wider world, hoovering up all the information available to them. It is up to the older generation to make sense of this information and add balance that makes it fit for purpose. Adhering to a life-course narrative, this book sampled from a smörgåsbord of facts, theories, and experiences to show how the human brain changes with aging. Ultimately, it is the reader's choice to decide whether and how this information has relevance.

NOTES

PROLOGUE

1. G. Kolata, "Sharing of Data Leads to Progress on Alzheimer's," *New York Times*, August 12, 2010.
2. G. Kolata, "Years Later, No Magic Bullet Against Alzheimer's Disease," *New York Times*, August 28, 2010.
3. M. L. Daviglus, C. C. Bell, W. Berrettini, P. E. Bowen, E. S. Connolly, N. J. Cox, J. M. Dunbar-Jacob, E. C. Granieri, G. Hunt, K. McGarry, D. Patel, A. L. Potosky, E. Sanders-Bush, D. Silberberg, and M. Trevisan, "National Institutes of Health State-of-the-Science Conference Statement: Preventing Alzheimer's Disease and Cognitive Decline." *NIH Consens State Sci Statements.* 2010 Apr 26–28;27(4): 1–27.
4. Medical Research Council (UK) Vitamin Study Research Group, "Prevention of Neural Tube Defects: Results of the Medical Research Council Vitamin Study," *Lancet* 338 (1991): 131–136.
5. E. Jablonka, and A. Zeligowski, "Interacting Dimensions—Genetic and Epigenetic Systems," in *Evolution in Four Dimensions: Genetic, Epigenetic, Behavioral, and Symbolic Variation in the History of Life* (Cambridge: MIT Press, 2005), 245–283.

1. INTRODUCTION

1. National Research Council, *The Aging Mind: Opportunities in Cognitive Research, Committee on Future Directions for Cognitive Research on Aging*, eds. Paul C. Stern and Laura L. Carstensen, Commission on Behavioral and Social Sciences and Education (Washington, DC: National Academies Press, 2000).
2. T. A. Salthouse, *Major Issues in Cognitive Aging* (London: Oxford University Press, 2010).
3. I. J. Deary, L. J. Whalley, and J. M. Starr, *A Lifetime of Intelligence: Following up the Scottish Mental Surveys of 1932 and 1947* (Washington DC: American Psychological Association, 2009).
4. D. Halpern, *Social Capital* (Cambridge: Polity Press, 2005), 1–16.
5. B. Fine, *Social Capital Versus Social Theory: Political Economy at the Turn of the Millennium* (London: Routledge, 2001), 40–52.
6. S. T. Charles, and L. L. Carstensen, "Social and Emotional Aging," *Annual Reviews in Psychology* 61 (2010): 383–409.

7. J. P. Shonkoff, W. T. Boyce, and B. S. McEwen, "Special Communication: Neuroscience, Molecular Biology, and the Childhood Roots of Health Disparities. Building a New Framework for Health Promotion and Disease Prevention," *Journal of the American Medical Association,* 301 (2009): 2252–2259.

8. T. Alfred, Y. Ben-Shlomo, R. Cooper, R. Hardy, I. J. Deary, J. Elliott, . . . I. N. Day; the HALCyon Study Team, "Genetic Variants Influencing Biomarkers of Nutrition Are Not Associated with Cognitive Capability in Middle-Aged and Older Adults," *Journal of Nutrition* 143, no. 5 (2013): 606–612.

9. J. Joseph, M. K. Vitti, S. A. Cho, A. Tishkoff, and P. C. Sabeti, "Human Evolutionary Genomics: Ethical and Interpretive Issues," *Trends in Genetics* 28, no. 3 (2012): 137–145.

10. S. Rose, *Lifelines: Biology, Freedom, Determinism* (New York: Oxford University Press, 2003).

11. L. Tommasi, C. Chiandetti, T. Pecchi, V. A. Sovrano, and G. Vallortigara, "From Natural Geometry to Spatial Cognition," *Neuroscience and Biobehavioral Reviews* 36 (2012): 799–824.

12. E. Jablonka and A. Zeligowski, *Evolution in Four Dimensions: Genetic, Epigenetic, Behavioral, and Symbolic Variation in the History of Life* (Cambridge: MIT Press, 2005).

13. A. R. Joyce and B. Ø. Palsson, "The Model Organism as a System: Integrating 'Omics' Data Sets," *Nature Reviews Molecular Cell Biology* 7 (2006): 198–210.

14. N. Gehlenborg, S. I. O'Donoghue, N. S. Baliga, A. Goesmann, M. A. Hibbs, H. Kitano, . . . A. C. Gavin, "Visualization of Omics Data for Systems Biology," *Nature Methods* 7, Supplement (2010): S56–68.

15. H. Kitano, "Towards a Theory of Biological Robustness," *Molecular Systems Biology* 3 (2007): 137.

16. S. H. Jain, M. Rosenblatt, and J. Duke, "Is Big Data the New Frontier for Academic-Industry Collaboration?" *Journal of the American Medical Association* 311, no. 21 (2014): 2171–2172.

17. Royal Academy of Engineering, "Systems Biology: A Vision for Engineering and Medicine," a report from the Academy of Medical Sciences and the Royal Academy of Engineering, accessed April 10, 2012, http://www.acmedsci.ac.uk/images/pressRelease/1170256174.pdf, cited by P. O'Shea, "Future Medicine Shaped by an Interdisciplinary New Biology," *Lancet* 378 (2012): 544–550.

18. O. Ybarra, "When First Impressions Don't Last: The Role of Isolation and Adaptation Processes in the Revision of Evaluative Impressions," *Social Cognition* 19 (2001): 491–520.

19. K. S. Kosik, "Beyond Phrenology," *Nature Reviews Neuroscience* 4 (2003): 234–239.

2. THE LIFE COURSE APPROACH

1. G. D. Smith, C. Hart, D. Blane, and D. Hole, "Adverse Socioeconomic Conditions in Childhood and Cause Specific Adult Mortality: Prospective Observational Study," *British Medical Journal* 316 (1998): 1631–1635.

2. B. Galobardes, J. W. Lynch, and G. D. Smith, "Childhood Socioeconomic Circumstances and Cause-Specific Mortality in Adulthood: Systematic Review and Interpretation," *Epidemiological Reviews* 26 (2004): 7–21.

3. G. J. Duncan, K. M. Ziol, and A. Kalil, "Early-Childhood Poverty and Adult Attainment, Behavior, and Health," *Child Development* 81, no. 1 (2010): 306–325.

4. J. E. Schwartz, H. S. Friedman, J. S. Tucker, C. Tomlinson-Keasey, D. L. Wingard, and M. H. Criqui, "Sociodemographic and Psychosocial Factors in Childhood as Predictors of Adult Mortality," *American Journal of Public Health* 85, no. 9 (1995): 1237–1245.

5. J. Lynch and G. D. Smith, "A Life Course Approach to Chronic Disease Epidemiology," *Annual Review of Public Health* 26 (2005): 1–35.

6. J. W. Lynch, G. A. Kaplan, and J. T. Salonen, "Why Do Poor People Behave Poorly? Variation in Adult Health Behaviours and Psychosocial Characteristics by Stages of the Socioeconomic Lifecourse," *Social Science and Medicine* 44 (1997): 809–819.

7. I. C. McMillen and J. S. Robinson, "Developmental Origins of the Metabolic Syndrome: Prediction, Plasticity, and Programming," *Physiological Reviews* 85 (2005): 571–633.

8. P. D. Gluckman, M. A. Hanson, and A. S. Beedle, "Early Life Events and Their Consequences for Later Disease: A Life History and Evolutionary Perspective," *American Journal of Human Biology* 19 (2007): 1–19.

9. R. Ruth Leys, "Types of One: Adolf Meyer's Life Chart and the Representation of Individuality," *Representations* 34 (1991): 1–28.

10. L. M. Terman, "A New Approach to the Study of Genius," *Psychological Review* 29 (1922): 310–318.

11. I. J. Deary, J. M. Starr, and L. J. Whalley, *A Lifetime of Intelligence: Following up the Scottish Mental Surveys of 1932 and 1947* (Washington DC: American Psychological Association, 2009).

12. M. Forgeard, E. Winner, A. Norton, and G. Schlaug, "Practicing a Musical Instrument in Childhood Is Associated with Enhanced Verbal Ability and Nonverbal Reasoning," *PLoS ONE* 10 (October 3, 2008): e3566.

13. J. Bardin, "Feature: Neurodevelopment: Unlocking the Brain," *Nature* 487, no. 7405 (2012): 24–26.

14. G. Horn, "Visual Imprinting and the Neural Mechanisms of Recognition Memory," *Trends in Neuroscience* 21, no. 7 (1998): 300–305.

15. K. Lorenz, *Evolution and the Modification of Behavior* (Chicago: University of Chicago Press, 1965).

16. D. H. Hubel, and T. N. Wiesel, "The Period of Susceptibility to the Physiological Effects of Unilateral Eye Closure in Kittens," *Journal of Physiology* 206, no. 2 (1970): 419–436.

17. H. Morishita, J. M. Miwa, N. Heintz, and T. K. Hensch, "Lynx1, a Cholinergic Brake, Limits Plasticity in Adult Visual Cortex," *Science* 330, no. 6008 (2010): 1238–1240.

18. E. Jablonka and M. J. Lamb, *Evolution in Four Dimensions: Genetic, Epigenetic, Behavioral, and Symbolic Variation in the History of Life* (Cambridge: MIT Press, 2005), 101–102.

19. D. J. Barker, "The Fetal and Infant Origins of Adult Disease," *British Medical Journal* 301, no. 6761 (1990): 1111.

20. D. J. Barker, C. Osmond, T. J. Forsen, E. Kajantie, and J. G. Eriksson, "Trajectories of Growth Among Children Who Have Coronary Events as Adults," *New England Journal of Medicine* 353, no. 17 (2005): 1802–1809.

21. C. E. Finch and E. Crimmins, "Inflammatory exposure and historical changes in human life spans," *Science* 305, no. 5691 (2004): 1736–1739.

22. E. P. Davis and C. A. Sandman, "The Timing of Prenatal Exposure to Maternal Cortisol and Psychosocial Stress Is Associated with Human Infant Cognitive Development," *Child Development* 81, no. 1 (2010): 131–148.

23. L. A. Welberg and J. R. Seckl, "Prenatal Stress, Glucocorticoids and the Programming of the Brain," *Journal of Neuroendocrinology* 13, no. 2 (2001): 113–128.

24. G. D. Smith, C. Hart, D. Blane, and D. Hole, "Adverse Socioeconomic Conditions in Childhood and Cause Specific Adult Mortality: Prospective Observational Study," *British Medical Journal* 316 (1998): 1631–1635.

25. I. D. Johnston, D. P. Strachan, and H. R. Anderson, "Effect of Pneumonia and Whooping Cough in Childhood on Adult Lung Function," *New England Journal of Medicine* 338, no. 9 (1998): 581–587.

26. R. A. Cohen, S. Grieve, F. Karin K. F. Hoth, R. H. Paul, L. Sweet . . . C. R. Clark, "Early Life Stress and Morphometry of the Adult Anterior Cingulate Cortex," *Biological Psychiatry* 59, no. 10 (2005): 975–982.

27. M. W. Gillman and J. W. Rich-Edward, "The Fetal Origins of Adult Disease: From Sceptic to Convert," *Paediatric and Perinatal Epidemiology* 14, no. 3 (2000): 192–193.

28. B. Galobardes, J. W. Lynch, and G. D. Smith, "Childhood Socioeconomic Circumstances and Cause-Specific Mortality in Adulthood: Systematic Review and Interpretation," *Epidemiological Reviews* 26 (2004): 7–21.

29. M. E. Wadsworth, S. L. Butterworth, R. J. Hardy, D. J. Kuh, M. Richards,. . . M. Connor, "The Life-Course Prospective Design: An Example of the Benefits and Problems Associated with Study Longevity," *Social Science and Medicine* 57, no. 11 (2003): 2193–2205.

30. D. Kuh and Y. Ben-Shlomo, *A Life Course Approach to Chronic Disease Epidemiology*, 2nd ed. (New York: Oxford University Press, 2004).

31. Y. Ben-Shlomo and D. Kuh, "A Life Course Approach to Chronic Disease Epidemiology: Conceptual Models, Empirical Challenges and Interdisciplinary Perspectives," *International Journal of Epidemiology* 31, no. 2 (2002): 285–293.

32. J. C. Cavanaugh and F. Blanchard-Fields, *Adult Development and Aging*, 4th ed. (Belmont, CA: Wadsworth, 2002).

33. H. R. Schaffer, *Introducing Child Psychology* (Oxford: Blackwell Scientific, 2004).

3. THE WELL-CONNECTED BRAIN

1. M. Gazzinaga, M. Ivry, G. Mangun, and M. S. Steven, *Cognitive Neuroscience: The Biology of the Mind* (New York and London: WW Norton & Co., 2009).

2. C. Crossman and D. Neary, *Neuroanatomy: An Illustrated Colour Text* (Edinburgh and New York: Churchill Livingston, 2004).

3. M. Brett, I. S. Johnsrude, and M. S. Owen, "The Problem of Functional Localisation in the Human Brain," *Nature Reviews Neuroscience* 3, no. 3 (2002): 243–249.

4. K. Kosik, "Beyond Phrenology, At Last," *Nature Reviews Neuroscience* 4, no. 3 (2003): 234–239.

5. G. Kempermann, L. Wiskott, and F. Gage, "Functional Significance of Adult Neurogenesis," *Current Opinion in Neurobiology* 14, no. 2 (2004): 186–191.

6. E. Gould, "Opinion: How Widespread Is Adult Neurogenesis in Mammals?" *Nature Reviews Neuroscience* 8, no. 6 (2007): 481–488.

7. J. DeFelipe, "Brain Plasticity and Mental Processes: Cajal Again," *Nature Reviews Neuroscience* 7, no. 10 (2006): 811–817.

8. A. E. Green, M. R. Munafò, C. G. DeYoung, J. A. Fossella, J. Fan, and J. R. Gray, "Using Genetic Data in Cognitive Neuroscience: From Growing Pains to Genuine Insights," *Nature Reviews Neuroscience* 9, no. 9 (2008): 710–720.

9. N. Y. Harel and S. M. Strittmatter, "Can Regenerating Axons Recapitulate Guidance During Recovery from Spinal Cord Injury?" *Nature Reviews Neuroscience* 7, no. 8 (2006): 613–616.

10. S. M. Kosslyn, G. Ganis, and W. L. Thompson, "Neural Foundations of Imagery," *Nature Reviews Neuroscience* 2, no. 9 (2001): 635–642.

11. N. Mustafa, T. S. Ahearn, G. D. Waiter, A. D. Murray, L. J. Whalley, and R. T. Staff, "Brain Structural Complexity and Life Course Cognitive Change," *NeuroImage* 61, no. 3 (2012): 694–701.

12. C. Sagan, "Can We Know the Universe? Reflections on a Grain of Salt," in *Broca's Brain* (New York: Ballantine Books, 1974), 15–21.

13. J. W. Lichtman, J. Livet, and J. R. Sanes, "A Technicolour Approach to the Connectome," *Nature Reviews Neuroscience* 9, no. 6 (2008): 417–422.

14. W. W. Seeley, R. K. Crawford, J. Zhou, B. L. Miller, and M. D. Greicius, "Neurodegenerative Diseases Target Large-Scale Human Brain Networks," *Neuron* 62, no. 1 (2009): 42–52.

15. S. van Veluw, E. Sawyer, L. Clover, H. Cousijn, C. De Jager, M. M. Esiri, and S. A. Chance, "Prefrontal Cortex Cytoarchitecture in Normal Aging and Alzheimer's Disease: A Relationship with IQ," *Brain Structure and Function* 217, no. 4 (2012): 797–808.

16. J. Nithianantharajah, and A. J. Hannan, "Enriched Environments, Experience Dependent Plasticity and Disorders of the Nervous System," *Nature Neuroscience Reviews* 7, no. 9 (2006): 697–709.

17. R. T. Staff, A. D. Murray, T. S. Ahearn, N. Mustafa, H. C. Fox, and L. J. Whalley, "Childhood Socioeconomic Status and Adult Brain Size: Childhood Socioeconomic Status Influences Adult Hippocampal Size," *Annals of Neurology* 71, no. 5 (2012): 653–660.

18. K. G. Noble, S. M. Houston, E. Kan, and E. R. Sowell, "Neural Correlates of Socioeconomic Status in the Developing Human Brain," *Developmental Science* 15, no. 4 (2012): 516–527.

19. A. W. Toga, and P. M. Thompson, "Mapping Brain Asymmetry," *Nature Reviews Neuroscience* 4, no. 1 (2003): 37–48.

20. A. W. Toga, P. M. Thompson, S. Mori, K. Amunts, and K. Zilles, "Towards Multimodal Atlases of the Human Brain," *Nature Reviews Neuroscience* 7, no. 12 (2006): 952–966.

21. G. D. Waiter, H. C. Fox, A. D. Murray, J. M. Starr, R. T. Staff, V. J. Bourne, . . . I. J. Deary, "Is Retaining the Youthful Functional Anatomy Underlying Speed of Information Processing a Signature of Successful Cognitive Ageing? An Event-Related fMRI Study of Inspection Time Performance," *Neuroimage* 41, no. 2 (2008): 581–595.

22. D. Benson, D. Colman, and G. Huntly, "Molecules, Maps and Synsptic Specificity," *Nature Reviews Neuroscience* 2, no. 12 (2001): 899–909.

23. P. Rakic, "Specification of Cerebral Cortical Areas," *Science* 241, no. 4682 (1988): 170–176.

24. J. Stiles, and T. L. Jernigan, "The Basics of Brain Development," *Neuropsychological Reviews* 20, no. 4 (2010): 327–348.

25. K. Zilles, and K. Amunts, "Centenary of Brodmann's Map-Conception and Fate," *Nature Reviews Neuroscience* 11, no. 2 (2010): 139–145.

26. D. J. Price, A. P. Jarman, J. O. Mason, and P. C. Kind, *Building Brains* (London: Wiley-Backwell, 2011).

4. EVOLUTION, AGING, AND DEMENTIA

1. T. Dobzhansky, "Nothing in Biology Makes Sense Except in Light of Evolution," *American Biology Teacher* 35 (1973): 125–129, retrieved January 2011 from http://www.2think.org/dobzhansky.stml.

2. E. Jablonka, and I. A. Zeligowsk, *Evolution in Four Dimensions: Genetic, Epigenetic, Behavioral, and Symbolic Variation in the History of Life* (MIT Press: 2005).

3. C. N. Hales, and D. J. Barker, "The Thrifty Phenotype Hypothesis," *British Medical Bulletin* 60 (2001): 5–20.

4. P. D. Gluckman, and M. A. Hanson, "Living with the Past: Evolution, Development and Patterns of Disease," *Science* 305, no. 5691 (2004): 1733–1736.

5. C. Vitora, L. Adair, C. Fall, P. C. Hallal, R. Martorell, and L. Richter, "Maternal and Child Undernutrition 2: Consequences for Adult Health and Human Capital," *Lancet* 371, no. 9609 (2008): 340–357.

6. P. R. Hof, and J. H. Morrison, "The Aging Brain: Morphomolecular Senescence of Cortical Circuits," *Trends in Neuroscience* 27, no. 10 (2004): 607–613.

7. M. Vandevelde, R. J. Higgins, and A. Oevermann, *Veterinary Neuropathology* (London: Wiley-Blackwell, 2012).

8. C. Sherwood, A. D. Gordon, J. S. Allen, K. A. Phillips, J. M. Erwin, . . . W. D. Hopkins, "Aging of Cerebral Cortex Differs Between Humans and Chimpanzees," *Proceedings of the National Academy of Sciences* 108, no. 32 (2011): 13029–13034.

9. R. F. Rosen, A. S. Farberg, M. Gearing, J. Dooyema, P. M. Long, D. C. Anderson, and L. C. Walker, "Tauopathy with Paired Helical Filaments in an Aged Chimpanzee," *Journal of Comparative Neurology* 509, no. 3 (2008): 259–270.

10. R. F. Rosen, L. C. Walker, and H. LeVine III, "PIB Binding in Aged Primate Brain: Enrichment of High-Affinity Sites in Humans with Alzheimer's Disease," *Neurobiology of Aging* 32, no. 2 (2011): 223–234.

11. C. R. Gamble, "The Peopling of Europe, 700,000—40,000 Years Before the Present" in *The Oxford Illustrated History of Prehistoric Europe*, ed. Barry Cunliffe (Oxford, England: Oxford University Press, 1994), 5–41.

12. H. Braak, and E. Braak, "Staging of Alzheimer's Disease-Related Neurofibrillary Changes," *Neurobiology of Aging* 16, no. 3 (1995): 271–278.

13. P. D. Evans, J. R. Anderson, E. J. Vallender, S. S. Choi, and B. T. Lahn, "Reconstructing the Evolutionary History of *Microcephalin*, a Gene Controlling Human Brain Size," *Human Molecular Genetics* 13, no. 11 (2004): 1139–1145.

14. S. Dorus, E. J. Vallender, P. D. Evans, J. R. Anderson, S. L. Gilbert, M. Mahowald, . . . B. T. Lahn, "Accelerated Evolution of Nervous System Genes in the Origin of Homo Sapiens," *Cell* 119, no. 7 (2004): 1027–1040.

15. S. Dorus, J. R. Anderson, E. J. Vallender, S. L. Gilbert, L. Zhang, L. G. Chemnick, . . . B. T. Lahn, "Sonic Hedgehog, a Key Gene, Experienced Intensified Molecular Evolution in Primates," *Human Molecular Genetics* 15, no. 13 (2006): 2031–2037.

16. S. L. Gilbert, W. B. Dobyns, and B. T. Lahn, "Opinion: Genetic Links Between Brain Development and Brain Evolution," *Nature Reviews Genetics* 6, no. 7 (2005): 581–590.

17. D. Bouchard, "Exaptation and Linguistic Explanation," *Lingua* 115, no. 12 (2005): 1685–1696.

18. H. Braak, and E. Braak, "Staging of Alzheimer's Disease-Related Neurofibrillary Changes," *Neurobiology of Aging* 16, no. 3 (1995): 271–278.

19. D. Neill, "Should Alzheimer's Disease Be Equated with Human Brain Ageing? A Maladaptive Interaction Between Brain Evolution and Senescence," *Ageing Research Reviews* 11, no. 1 (2012): 104–122.

20. S. I. Rapoport, "Integrated Phylogeny of the Primate Brain, with Special Reference to Humans and Their Diseases," *Brain Research Reviews* 15, no. 3 (1990): 267–294.

21. H. Braak, and E. Braak, "Staging of Alzheimer's Disease-Related Neurofibrillary Changes," *Neurobiology of Aging* 16, no. 3 (1995): 271–278.

22. D. Neil, "Should Alzheimer's Disease Be Equated with Human Brain Ageing? A Maladaptive Interaction Between Brain Evolution and Senescence," *Ageing Research Reviews* 11, no. 1 (2012): 104–122.

23. T. Arendt, "Alzheimer's Disease as a Disorder of Mechanisms Underlying Structural Brain Self-Organization," *Neuroscience* 102, no. 4 (2001): 723–765.

24. L. Partridge, "The New Biology of Ageing," *Philosophical Transactions of the Royal Society (London)*, B 365, no. 1537 (2009): 147–154.

25. L. Partridge, and D. Gems, "Mechanisms of Ageing: Public or Private?" *Nature Reviews Genetics* 3, no. 3 (2010): 165–175.

26. L. Partridge, and D. Gems, "Beyond the Evolutionary Theory of Ageing, from Functional Genomics to Evo-gero," *Trends in Ecology and Evolution* 21, no. 6 (2006): 334–340.

27. L. Partridge, and D. Gems, "Beyond the Evolutionary Theory of Ageing, from Functional Genomics to Evo-gero," *Trends in Ecology and Evolution* 21, no. 6 (2006): 334–340.

28. S. B. Carroll, "Evo-devo and an Expanding Evolutionary Synthesis: A Genetic Theory of Morphological Evolution," *Cell* 134, no. 1 (2008): 25–36.

29. C. H. Waddington, *The Strategy of the Genes* (London: George Allen, 1957).

30. T. Domazet-Los, and D. Tautz, "A Phylogenetically Based Transcriptome Age Index Mirrors Ontogenetic Divergence Patterns," *Nature* 468, no. 7325 (2010): 815–818.

31. G. F. Striedter, "Precis of Principles of Brain Evolution," *Behavioral and Brain Sciences* 29, no. 1 (2006): 1–12; discussion 12–36.

32. E. J. Vallender, N. Mekel-Bobrov, and B. T. Lahn, "Genetic Basis of Human Brain Evolution," *Trends in Neuroscience* 31, no. 12 (2008): 637–644.

33. C. H. Waddington, *The Strategy of the Genes* (London: George Allen, 1957).

34. M. J. Hawrylycz, E. S. Lein, A. L. Guillozet-Bongaarts, E. H. Shen, L. Ng, J. A. Miller, . . . A. R. Jones, "An Anatomically Comprehensive Atlas of the Adult Human Brain Transcriptome," *Nature* 489, no. 7416 (2012): 391–399.

35. S. Dorus, J. R. Anderson, E. J. Vallender, S. L. Gilbert, L. Zhang, L. G. Chemnick, . . . B. T. Lahn, "Sonic Hedgehog, a Key Development Gene, Experienced Intensified Molecular Evolution in Primates," *Human Molecular Genetics* 15, no. 13 (2006): 2031–2037.

36. J. M. Zahn, S. Poosala, A. B. Owen, D. K. Ingram, A. Lustig, A. Carter, K. G. Becker, "AGEMAP: A Gene Expression Database for Aging in Mice," *PLoS Genet* 3, no. 11 (2007): e201. Epub 2007 Oct 2.

37. T. Domazet-Los, and D. Tautz, "A Phylogenetically Based Transcriptome Age Index Mirrors Ontogenetic Divergence Patterns," *Nature* 468 (2010): 815–818.

38. S. Stoppini, A. Andreola, G. Foresti, and V. Bellotti, "Neurodegenerative Diseases Caused by Protein Aggregation: A Phenomenon at the Borderline Between Molecular Evolution and Ageing," *Pharmacological Research* 50, no. 4 (2004): 419–431.

39. J. Ghika, "Paleoneurology: Neurodegenerative Diseases Are Age-Related Diseases of Specific Brain Regions Recently Developed by Homo Sapiens," *Medical Hypotheses* 71, no. 5 (2008): 788–801.

40. M. Stoppini, A. Andreola, G. Foresti, and V. Bellotti, "Neurodegenerative Diseases Caused by Protein Aggregation: A Phenomenon at the Borderline Between Molecular Evolution and Ageing," *Pharmacological Research* 50, no. 4 (2009): 419–431.

41. H. C. Hendrie, K. S. Hall, N. Pillay, D. Rodgers, C. Prince, J. Norton, . . . J. Kaufert, "Alzheimer's Disease Is Rare in Cree," *International Psychogeriatrics* 5, no. 1 (1993): 5–14.

42. A. L. Benedet, F. Clayton, C. F. Moraes, E. F. Camargos, C. Córdova, R. W. Pereira, and O. T. Nóbrega, "Amerindian Genetic Ancestry Protects Against Alzheimer's Disease," *Dementia and Geriatric Cognitive Disorders* 33, no. 5 (2012): 311–317.

43. A. P. Quayle, and S. Bullock, "Modelling the Evolution of Genetic Regulatory Networks," *Journal of Theoretical Biology* 238, no. 9 (2006): 737–775.

44. R. I. M. Dunbar, and S. Schulz, "Evolution in the Social Brain," *Science* 1344, no. 5843 (2007): 1344–1347.

45. G. Konopka, T. Friedrich, J. Davis-Turak, K. Winden, M. C. Oldham, F. Gao, . . . D. H. Geschwind, "Human-Specific Transcriptional Networks in the Brain," *Neuron* 75, no. 4 (2012): 601–617.

46. H. Zeng, E. H. Shen, J. G. Hohmann, S. W. Oh, A. Bernard, J. J. Royall, . . . A. R. Jones, "Large-Scale Cellular-Resolution Gene Profiling in Human Neocortex Reveals Species-Specific Molecular Signatures," *Cell* 149, no. 2 (2012): 483–496.

47. M. Venoux, K. Delmouly, O. Milhavet, S. Vidal-Eychenie, D. Giorgi, and S. Rouquier, "Gene Organization, Evolution and Expression of the Microtubule-Associated Protein ASAP(MAP9)," *BMC Genomics* 9 (2008): 408.

48. H. Kitano, "A Robustness-Based Approach to Systems-Oriented Drug Design," *Nature Reviews Drug Discovery* 6(2007): 202–210.

49. R. Plomin, and I. J. Deary, "Genetics and Intelligence Differences: Five Special Findings," *Molecular Psychiatry* (2014), doi: 10.1038/mp.2014.105 (epub ahead of print).

5. THE AGING BRAIN

1. R. Peto, and R. Doll, "There Is No Such Thing as Ageing: Old Age Is Associated with Disease but Does Not Cause It," *British Medical Journal* 315, no. 7115 (1997): 1030–1032.

2. S. Hunter, T. Arendt, and C. Brayne, "The Senescence Hypothesis of Disease Progression in Alzheimer Disease: An Integrated Matrix of Disease Pathways for FAD and SAD," *Molecular Neurobiology* 48, no. 3 (2013): 556–570.

3. T. Kirkwood, "Understanding the Odd Science of Aging," *Cell* 120, no. 4 (2005): 437–447.

4. H. Christensen, A. J. MacKinnon, A. Korten, and A. F. Jorm, "The 'Common Cause Hypothesis' of Cognitive Aging: Evidence Not for Only a Common Factor but also Specific Associations of Age with Vision and Grip Strength in a Cross-Sectional Study," *Psychology and Aging* 16, no. 4 (2001): 588–599.

5. T. A. Salthouse, "Neuroanatomical Substrates of Age-Related Cognitive Decline," *Psychological Bulletin* 137, no. 5 (2011): 753–784.

6. J. L. Price, D. W. McKeel Jr., V. D. Buckles, C. M. Roe, C. Xiong, M. Grundman, . . . J. C. Morris, "Neuropathology of the Non-Demented Elderly: Presumptive Evidence of Pre-Clinical Alzheimer's Disease," *Neurobiology of Aging* 30, no. 7 (2009): 1026–1036.

7. J. Morrison, and P. R. Hof, "Life and Death of Aging Neurons in the Aging Brain," *Science* 278, no. 5337 (1997): 412–419.

8. J. A. Schneider, N. T. Aggarwal, L. Barnes, P. Boyle, and D. A. Bennet, "The Neuropathology of Older Persons with and without Dementia from Community Versus Clinic Cohorts," *Journal of Alzheimer's Disease* 18, no. 3 (2009): 691–701.

9. W. E. Klunk, H. Engler, A. Nordberg, Y. Wang, G. Blomqvist, D. P. Holt, . . . B. Långström, "Imaging Brain Amyloid in Alzheimer's Disease with Pittsburgh Compound B," *Annals of Neurology* 55, no. 3 (2004): 306–319.

10. C. C. Rowe, P. Bourgeat, K. A. Ellis, B. Brown, Y. Y. Lim, R. Mulligan, . . . V. L. Villemagne, "Predicting Alzheimer's Disease with β-Amyloid Imaging: Results from the Australian Imaging, Biomarkers and Lifestyle Study of Ageing," *Annals of Neurology* 74, no. 6 (2013): 905–913.

11. W. E. Klunk, H. Engler, A. Nordberg, Y. Wang, G. Blomqvist, D. P. Holt, . . . B. Långström, "Imaging Brain Amyloid in Alzheimer's Disease with Pittsburgh Compound B," *Annals of Neurology* 55, no. 3 (2004): 306–319.

12. N. Raz, and K. M. Rodrigue, "Differential Aging of the Brain: Patterns, Cognitive Correlates and Modifiers," *Neuroscience and Biobehavioral Reviews* 30, no. 9 (2006): 730–748.

13. S. J. van Veluw, E. K. Sawyer, L. Clover, H. Cousijn, C. De Jager, M. M. Esiri, and S. A. Chance, "Prefrontal Cortex Cytoarchitecture in Normal Aging and Alzheimer's Disease: A Relationship with IQ," *Brain Structure and Function* (February 2012): 1–12.

14. H. Brody, "The Organization of the Cerebral Cortex. III A study of Aging in the Human Cerebral Cortex," *Journal of Comparative Neurology* 102, no. 4 (1955): 511–556.

15. J. Morrison, and P. R. Hof, "Life and Death of Aging Neurons in the Aging Brain," *Science* 278, no. 5337 (1997): 412–419.

16. A. Pollack, "Judah Folkman Researcher dies at 74," *New York Times*, January 16, 2008 (accessed April 11, 2010 at http://www.nytimes.com/2008/01/16/us/16folkman.html).

17. R. Kalaria, "Linking Cerebrovascular Defense Mechanisms in Brain Ageing with Alzheimer's Disease," *Neurobiology of Aging* 30 (2009): 1512–1514.

18. A. D. Murray, C. J. McNeil, S. Salarirad, L. J. Whalley, and R. T. Staff, "Early Life Socioeconomic Circumstance and Late Life Brain Hyperintensities—A Population Based Cohort Study," *PLoS One* 9, no. 2 (2014): e88969.

19. L. T. Westlye, K. B. Walhovd, A. M. Dale, A. Bjørnerud, P. Due-Tønnessen, A. Engvig, . . . A. M. Fjell, "Differentiating Maturational and Aging-Related Changes of the Cerebral Cortex by Use of Thickness and Signal Intensity," *NeuroImage* 52, no. 1 (2010): 172–185.

20. J. R. Wozniak, and K. O. Lim, "Advances in White Matter Imaging: A Review of In Vivo Magnetic Resonance Methodologies and Their Applicability to the Study of Development and Aging," *Neuroscience Biobehavioral Reviews* 30, no. 9 (2006): 762–774.

21. L. White, H. Petrovitch, G. W. Ross, K. H. Masaki, R. D. Abbott, E. L. Teng, B. L. Rodriguez, P. L. Blanchette, R. J. Havlik, G. Wergowske, D. Chiu, D. J. Foley, C. Murdaugh, and J. D. Curb., "Prevalence of Dementia in Older Japanese-American Men in Hawaii: The Honolulu-Asia Aging Study," *Journal of the American Medical Association* 276 (1996): 955–960.

22. W. B. Grant, "Trends in Diet and Alzheimer's Disease During the Nutrition Transition in Japan and Developing Countries," *Journal of Alzheimer's Disease* 38 (2014): 611–620.

23. D. J. L. Kuh, and C. Cooper, "Physical Activity at 36 Years—Patterns and Childhood Predictors in a Longitudinal Study," *Journal of Epidemiology and Community Health* 46, no. 2 (1992): 114–119.

24. M. P. Mattson, S. L. Chan, and W. Duan, "Modification of Brain Aging and Neurodegenerative Disorders by Genes, Diet, and Behavior," *Physiological Reviews* 82, no. 3 (2002): 637–672.

25. N. A. Bishop, T. Lu, and B. A. Yankner, "Neural Mechanisms of Ageing and Cognitive Decline," *Nature* 464, no. 7288 (2010): 529–535.

26. R. M. Sapolsky, "Glucocorticoids, Stress, and Their Adverse Neurological Effects: Relevance to Aging," *Experimental Gerontology* 34, no. 6 (1999): 721–732.

27. J. R. Andrews-Hanna, A. Z. Snyder, J. L. Vincent, C. Lustig, D. Head, M. E. Raichle, and R. L. Buckner, "Disruption of Large-Scale Brain Systems in Advanced Aging," *Neuron* 56, no. 5 (2007): 924–935.

28. M. P. Mattson, and T. Magnus, "Ageing and Neuronal Vulnerability," *Nature Reviews Neuroscience* 7, no. 4 (2006): 278–294.

29. M. C. Haigis, and B. A. Yankner, "The Aging Stress Response," *Molecular Cell* 40, no. 2 (2010): 333–344.

30. I. Martin, and M. S. Grotewiel, "Oxidative Damage and Age-Related Functional Declines," *Mechanisms of Ageing and Development* 127, no. 5 (2006): 411–423.

31. R. M. Sapolsky, "Glucocorticoids, Stress, and Their Adverse Neurological Effects: Relevance to Aging," *Experimental Gerontology* 34, no. 6 (1999): 721–732.

6. THE BIOLOGY OF THE DEMENTIAS

1. M. Roth, "The Natural History of Mental Disorder in Old Age," *British Journal of Psychiatry* 101, no. 4 (1955): 281–301.

2. G. Blessed, B. E. Tomlinson, and M. Roth, "The Association Between Quantitative Measures of Dementia and of Senile Change in the Cerebral Grey Matter of Elderly Subjects," *British Journal of Psychiatry* 114, no. 6 (1968): 797–811.

3. O. Hornykiewicz, "The Discovery of Dopamine Deficiency in the Parkinsonian Brain," *Journal of Neural Transmission* 70, Suppl. (2006): 9–15.

4. W. Birkmayer, and O. Hornykiewicz, "Der L-Dioxyphenylalanin (L-DOPA)-Effekt bei der Parkinson-Akinese," *Wiener Klinische Wochenschrift* 73, no. 8 (1961): 787–788.

5. R. Katzman, "The Prevalence and Malignancy of Alzheimer's Disease: A Major Killer," *Archives of Neurology* 33, no. 1 (1976): 217–218.

6. P. Davies, and A. Maloney, "Selective Loss of Central Cholinergic Neurons in Alzheimer's Disease," *Lancet* ii (1976): 1403.

7. K. L. Davis, and R. C. Mohs, "Enhancement of Memory Processes in Alzheimer's Disease with Multiple-Dose Intravenous Physostigmine," *American Journal of Psychiatry* 139, no. 11 (1982): 1421–1424.

8. J. Christie, A. Shering, J. Ferguson, and A. I. M. Glen, "Physostigmine and Arecoline: Effects of Intravenous Infusions in Alzheimer Presenile Dementia," *British Journal of Psychiatry* 138 (1981): 46–50.

9. L. L. Heston, "Alzheimer's Disease and Down's Syndrome: Genetic Evidence Suggesting an Association," *Annals of the New York Academy of Sciences* 396 (1982): 29–37.

10. A. Brun, and E. Englund, "Regional Pattern of Degeneration in Alzheimer's Disease: Neuronal Loss and Histopathological Grading," *Histopathology* 5 (1981): 459–564.

11. A. Brun, and E. Englund, "Regional Pattern of Degeneration in Alzheimer's Disease: Neuronal Loss and Histopathological Grading," *Histopathology* 5 (1981): 459–564.

12. K. A. Johnson, S. Minoshima, N. I. Bohnen, K. J. Donohoe, N. L. Foster, P. Herscovitch, . . . W. H. Thies, "Appropriate Use Criteria for Amyloid PET: A Report of the Amyloid Imaging Task Force, the Society of Nuclear Medicine and Molecular Imaging, and the Alzheimer's Association," *Journal of Nuclear Medicine* 54, no. 3 (2013): 476–490.

13. S. S. Stewart, and S. H. Appel, "Trophic Factors in Neurologic Disease," *Annual Review of Medicine* 39, no. 2 (1988): 193–201.

14. G. G. Glenner, and C. W. Wong, "Alzheimer's Disease: Initial Report of the Purification and Characterization of a Novel Cebrebrovascular Amyloid Protein, *Biochemical and Biophysical Research Communications* 16, no. 6 (1984): 885–890.

15. C. L. Masters, G. Simms, N. A. Weinman, G. Multhaup, B. L. McDonald, and K. Beyreuther, "Amyloid Plaque Core Protein in Alzheimer Disease and Down Syndrome," *Proceedings of the National Academy of Sciences USA* 82, no. 12 (1985): 4245–4249.

16. J. Kang, H. G. Lemaire, A. Unterbeck, J. M. Salbaum, C. L. Masters, K. H. Grzeschik, . . . B. Müller-Hill, "The Precursor of Alzheimer's Disease Amyloid A4 Protein Resembles a Cell-Surface Receptor," *Nature* 325, no. 6106 (1987): 733–736.

17. A. Weidemann, G. Konig, D. Bunke, P. Fischer, J. M. Salbaum, C. L. Masters, and K. Beyreuther, "Identification, Biogenesis, and Localization of Precursors of Alzheimer's Disease A4 Amyloid Protein," *Cell* 57, no. 1 (1989): 115–126.

18. R. E. Tanzi, and L. Bertram, "Twenty Years of the Alzheimer's Disease Amyloid Hypothesis: A Genetic Perspective," *Cell* 120, no. 4 (2005): 545–555.

19. A. Goate, M. C. Chartier-Harlin, M. Mullan, J. Brown, F. Crawford, L. Fidani, . . . L. James, "Segregation of a Missense Mutation in the Amyloid Precursor Protein Gene with Familial Alzheimer's Disease," *Nature* 349, no. 6311 (1991): 704–706.

20. E. I. Rogaev, R. Sherrington, E. A. Rogaeva, G. Levesque, M. Ikeda, Y. Liang, . . . T. Tsuda, "Familial Alzheimer's Disease in Kindreds with Missense Mutations in a Gene on Chromosome 1 Related to the Alzheimer's Disease Type 3 Gene," *Nature* 376, no. 6543 (1995): 775–778.

21. R. Sherrington, E. I. Rogaev, Y. Liang, E. A. Rogaeva, G. Levesque, M. Ikeda, . . . P. H. St. George-Hyslop, "Cloning of a Gene Bearing Missense Mutations in Early-Onset Familial Alzheimer's Disease," *Nature* 375, no. 6534 (1995): 754–760.

22. E. Levy-Lahad, E. M. Wijsman, E. Nemens, L. Anderson, K. A. Goddard, J. L. Weber, . . . G. D. Schellenberg, "A Familial Alzheimer's Disease Locus on Chromosome 1," *Science* 269, no. 5226 (1995): 970–973.

23. G. D. Schellenberg, "Genetic Dissection of Alzheimer Disease, a Heterogeneous Disorder," *Proceedings of the National Academy of Sciences USA* 92, no. 19 (1995): 8552–8559.

24. R. Sherrington, S. Froelich, S. Sorbi, D. Campion, H. Chi, E. A. Rogaeva, . . . P. H. St. George-Hyslop, "Alzheimer's Disease Associated with Mutations in Presenilin 2 Is Rare and Variably Penetrant," *Human Molecular Genetics* 5, no. 7 (1996): 985–988.

25. E. H. Corder, A. M. Saunders, W. J. Strittmatter, D. E. Schmechel, P.C. Gaskell, . . . M. A. Pericak-Vance, "Gene Dose of Apolipoprotein E Type 4 Allele and the Risk of Alzheimer's Disease in Late Onset Families," *Science* 261, no. 5123 (1993): 921–923.

26. R. E. Tanzi, and L. Bertram, "Twenty Years of the Alzheimer's Disease Amyloid Hypothesis: A Genetic Perspective," *Cell* 120, no. 4 (2005): 545–555.

27. E. Callaway, "Alzheimer's Drugs Take a New Tack," *Nature* 489, no. 7414 (2012): 13–14.

28. W. Bondareff, C. W. Mountjoy, and M. Roth, "Age and Histopathological Heterogeneity in Alzheimer's Disease: Evidence for Sub-types," *Archives of General Psychiatry* 44, no. 3 (1987): 412–417.

29. R. J. Bateman, C. Xiong, T. L. S. Benzinger, A. M. Fagan, A. Goate, N. C. Fox, . . . J. C. Morris, Dominantly Inherited Alzheimer Network, "Clinical and Biomarker Changes in Dominantly Inherited Alzheimer's Disease," *New England Journal of Medicine* 367, no. 9 (2012): 795–804.

30. A. Chandra, A. Johri, M. F. Beal, "Neuroprotective Therapies in Prodromal Huntington's Disease, *Movement Disorders* 29 (2014): 285–293.

31. A. M. Fagan, C. Xiong, M. S. Jasielec, R. J. Bateman, A. M. Goate, T. L. Benzinger, . . . D. M. Holtzman, Dominantly Inherited Alzheimer Network, "Longitudinal Change in CSF Biomarkers in Autosomal Dominant Alzheimer's Disease," *Science Translational Medicine* 6, no. 226 (2014): 226ra30.

32. L. Partridge, and D. Gems, "Beyond the Evolutionary Theory of Ageing, from Functional Genomics to Evo-Gero," *Trends in Ecology and Evolution* 21 (2006): 334–340, box 2.

33. R. A. Honea, R. H. Swerdlow, E. D. Vidoni, J. Goodwin, and J. M. Burns, "Reduced Gray Matter Volume in Normal Adults with a Maternal Family History of Alzheimer Disease," *Neurology* 74, no. 2 (2010): 113–120.

34. L. Mosconi, W. Tsui, J. Murray, P. McHugh, Y. Li, S. Williams, E. Pirraglia, . . . M. J. de Leon, "Maternal Age Affects Brain Metabolism in Adult Children of Mothers Affected by Alzheimer's Disease," *Neurobiology of Aging* 33, no. 2 (2012): 624.e1–9.

35. L. Mosconi, J. O. Rinne, W. H. Tsui, V. Berti, Y. Li, H. Wang, J. Murray, . . . M. J. de Leon, "Increased Fibrillar Amyloid-β Burden in Normal Individuals with a Family History of Late-Onset Alzheimer's," *Proceedings of the National Academy of Sciences USA* 107, no. 13 (2010): 5949–5954.

36. J. M. Silverman, G. Ciresi, C. J. Smith, J. Schmeidler, R. C. Mohs, and K. L. Davis, "Patterns of Risk in First Degree Relatives of Patients with Alzheimer's Disease," *Archives of General Psychiatry* 51, no. 7 (1994): 577–586.

37. M. Gatz, C. Reynolds, L. Fratiglioni, B. Johansson, J. A. Mortimer, S. Berg, . . . N. L. Pedersen, "Role of Genes and Environments for Explaining Alzheimer Disease," *Archives of General Psychiatry* 63, no. 2 (2006): 168–174.

38. L. A. Farrer, L. A. Cupples, J. L. Haines, B. Hyman, W. A. Kukull, R. Mayeux, . . . C. M. van Duijn, "Effects of Age, Sex, and Ethnicity on the Association Between Apolipoprotein E Genotype and Alzheimer Disease. A Meta-Analysis. APOE and Alzheimer Disease Meta Analysis Consortium," *Journal of the American Medical Association* 278, no. 16 (1997): 1349–1356.

39. J. C. Breitner, B. W. Wyse, J. C. Anthony, K. A. Welsh-Bohmer, D. C. Steffens, M. C. Norton, J. T. Tschanz, . . . A. Khachaturian, "APOE-ε4 Count Predicts Age When Prevalence of AD Increases, Then Declines: The Cache County Study," *Neurology* 53, no. 2 (1999): 321–331.

40. C. A. Martins, A. Oulhaj, C. A. de Jager, and J. H. Williams, "APOE Alleles Predict the Rate of Cognitive Decline in Alzheimer Disease: A Nonlinear Model," *Neurology* 65, no. 12 (2005): 1888–1893.

41. J. C. Breitner, B. W. Wyse, J. C. Anthony, K. A. Welsh-Bohmer, D. C. Steffens, M. C. Norton, J. T. Tschanz, . . . A. Khachaturian, "APOE-ε4 Count Predicts Age When Prevalence of AD Increases, Then Declines: The Cache County Study," *Neurology* 53, no. 2 (1999): 321–331.

42. E. Genin, D. Hannequin, D. Wallon, K. Sleegers, M. Hiltunen, O. Combarros, . . . D. Campion, "APOE and Alzheimer Disease: A Major Gene with Semidominant Inheritance," *Molecular Psychiatry* 16, no. 9 (2011): 903–907.

43. S. T. DeKosky, and S. W. Scheff, "Synapse Loss in Frontal Cortex Biopsies in Alzheimer's Disease: Correlation with Cognitive Severity," *Annals of Neurology* 27, no. 5 (1990): 457–464.

44. G. M. Shankar, B. L. Bloodgood, M. Townsend, M. Townsend, D. M. Walsh, D. J. Selkoe, and B. L. Sabatini, "Natural Oligomers of the Alzheimer Amyloid-β Protein Induce Reversible Synapse Loss by Modulating an NMDA-Type Glutamate Receptor–Dependent Signaling Pathway," *Journal of Neuroscience* 27, no. 11 (2007): 2866–2875.

45. J. J. Palop, and L. Mucke, "Amyloid-β-Induced Neuronal Dysfunction in Alzheimer's Disease: From Synapses Toward Neural Networks," *Nature Neuroscience* 13, no. 7 (2010): 812–818.

46. R. Sherrington, S. Froelich, S. Sorbi, D. Campion, H. Chi, E. A. Rogaeva, . . . P. H. St. George-Hyslop, "Alzheimer's Disease Associated with Mutations in Presenilin 2 Is Rare and Variably Penetrant," *Human Molecular Genetics* 5, no. 7 (1996): 985–988.

47. W. T. Greenough, H-M. F. Hwang, and C. Gorman, "Evidence for Active Synapse Formation or Altered Postsynaptic Metabolism in Visual Cortex of Rats Reared in Complex Environments," *Proceedings of the National Academy of Sciences USA* 82, no. 13 (1985): 4549–4552.

48. R. D. Terry, E. Masliah, D. P. Salmon, N. Butters, R. DeTeresa, R. Hill, L. A. Hansen, and R. Katzman, "Physical Basis of Cognitive Alterations in Alzheimer's Disease: Synapse Loss Is the Major Correlate of Cognitive Impairment," *Annals of Neurology* 30, no. 4 (1991): 572–580.

49. P. Penzes, M. E. Cahill, K. A. Jones, J. E. VanLeeuwen, and K. M. Woolfrey, "Dendritic Spine Pathology in Neuropsychiatric Disorders," *Nature Neuroscience* 14, no. 3 (2011): 285–293.

50. J. Hardy, and D. J. Selkoe, "The Amyloid Hypothesis of Alzheimer's Disease: Progress and Problems on the Road to Therapeutics," *Science* 297, no. 5580 (2002): 535–356.

51. P. Penzes, M. E. Cahill, K. A. Jones, J. E. VanLeeuwen, and K. M. Woolfrey, "Dendritic Spine Pathology in Neuropsychiatric Disorders," *Nature Neuroscience* 14, no. 3 (2011): 285–293.

52. H. Braak, and E. Braak, "Neuropathological Staging of Alzheimer Related Changes," *Acta Neuropathologica* 82, no. 4 (1991): 239–259.

7. THE DISCONNECTED MIND

1. D. Robinson, *The Mind: Oxford Readers* (Oxford: Oxford University Press, 1988).

2. P. Fusar-Poli and I. P. Polit, "Paul Eugen Bleuler and the Birth of Schiozophrenia," *American Journal Psychiatry* 165 (2008): 1407.

3. T. Salthouse, *Major Issues in Cognitive Aging* (Oxford: Oxford University Press, 2010).

4. P. Rabbitt, "Does It All Go Together When It Goes?" *Quarterly Journal of Experimental Psychology A* 46, no. 3 (1993): 385–434.

5. M. S. Gazzaniga, R. B. Ivry, and G. R. Mangun, *Cognitive Neuroscience: The Biology of the Mind*, 3rd ed. (New York and London: Norton & Co., 2009).

6. P. Verhaeghen and T. A. Salthouse, "Meta-analyses of Age-Cognition Relations in Adulthood: Estimates of Linear and Non-linear Age Effects and Structural Models," *Psychological Bulletin* 122, no. 3 (1997): 231–249.

7. W. J. Ma, M. Husain, and P. M. Bays, "Changing Concepts of Working Memory," *Nature Neuroscience* 17, no. 3 (2014): 347–356.

8. E. A. Maylor, S. Allison, and A. M. Wing, "Effects of Spatial and Non-spatial Cognitive Activity on Postural Stability," *British Journal of Psychology* 92, Part 2 (2001): 319–338.

9. P. Boyle, R. S. Wilson, L. Yu, A. M. Barr, W. G. Honer, J. A. Schneider, and D. A. Bennett, "Much of Late-Life Cognitive Decline Is Not Due to Common Neurodegenerative Pathologies," *Annals of Neurology* 74, no. 3 (2013): 478–489.

10. B. J. Baars, "Metaphors of Consciousness and Attention in the Brain," *Trends in Neurosciences* 21 (1998): 58–62.
11. R. Wilbur, "The Mind," *New and Collected Poems* (New York: Harcourt Books, 1988), 240.
12. D. E. Broadbent, "The Role of Auditory Localization in Attention and Memory Span," *Journal of Experimental Psychology* 47 (1954): 191–196.
13. K. W. Schaie, "The Course of Adult Intellectual Development," *American Psychologist* 49 (1994): 304–313.
14. M. Sliwinski, and H. Buschke, "Modeling Individual Cognitive Change in Aging Adults: Results from the Einstein Aging Studies," *Aging Neuropsychology and Cognition* 11 (2003): 196–211.
15. M. J. Sliwinski, S. Hofer, H. Buschke, and R. B. Lipton, "Modeling Memory Decline in Older Adults: The Importance of Preclinical Dementia, Attrition, and Chronological Age," *Psychology and Aging* 18, no. 4 (2003): 657–671.
16. D. Finkel, C. Reynolds, J. J. McArdle, H. Hamagami, and N. L. Pedersen, "Genetic Variance in Processing Speed Drives Variation in Aging of Spatial and Memory Abilities," *Developmental Psychology* 45, no. 3 (2009): 820–834.
17. Cited by P. Bourliere, *The Assessment of Biological Age in Man* (Geneva: World Health Organization, 1970), 47–55.
18. S. Rubial-Avarez, S. de Sola, M.-C. Machado, E. Sintas, P. Bohm, G. Sanchez-Benavides, K. Langohr, R. Muniz, and J. Pena-Casanova, "The Comparison of Cognitive and Functional Performance in Children and Alzheimer's Disease Supports the Retrogenesis Model," *Journal of Alzheimer's Disease* 33 (2013): 191–203.
19. P. Baltes, and U. M. Staudinger, "Wisdom: A Metaheuristic (Pragmatic) to Orchestrate Mind and Virtue Towards Excellence," *American Psychologist* 55, no. 1 (2000): 122–136.
20. K. Ritchie, and J. Touchon, "Mild Cognitive Impairment: Conceptual Basis and Current Nosological Status," *Lancet* 355, no. 9199 (2000): 225–228.
21. B. Winblat, K. Palmer, M. Kivipelto, V. Jelic, L. Fratiglioni, L. O. Wahlund, . . . R. C. Petersen, "Mild Cognitive Impairment—Beyond Controversies, Towards a Consensus: Report of the International Working Group on Mild Cognitive Impairment," *Journal of Internal Medicine* 256, no. 3 (2004): 240–246.
22. A. J. Holland, J. Hon, F. Huppert, and F. Stevens, "Incidence and Course of Dementia in Patients with Down's Syndrome: Findings from a Population-Based Study," *Journal of Intellectual Disability Research* 44, Part 2 (2000): 138–146.
23. L. J. Whalley, "The Dementia of Downs Syndrome and Its Relevance to Etiological Studies of Alzheimer's Disease," *Annals of the New York Academy of Sciences* 396 (1982): 39–53.
24. T. E. Goldberg, D. Weinberg, K. F. Berman, N. H. Pliskin, and M. H. Podd, "Further Evidence for Dementia of the Prefrontal Type in Schizophrenia: A Controlled Study of Teaching the Wisconsin Card Searching Test," *Archives of General Psychiatry* 44, no. 11 (1987): 1008–1014.
25. S. Corkin, T. J. Rosen, E. V. Sullivan, and R. A. Clegg, "Penetrating Head Injury in Young Adulthood Exacerbates Cognitive Decline in Later Years," *Journal of Neuroscience* 9, no. 11 (1989): 3876–3883.
26. E. J. Pellman, D. C. Viano, I. R. Casson, A. M. Tucker, J. F. Waeckerle, J. W. Powell, and H. Feuer, "Concussion in Professional Football: Repeat Injuries—Part 4," *Neurosurgery* 55, no. 4 (2004): 860–873.
27. L. Moretti, I. Cristofort, S. M. Weaver, A. Chau, J. N. Portelli, and J. Grafman, "Cognitive Decline in Older Adults with a History of Traumatic Brain Injury," *Lancet Neurology* 11, no. 12 (2012): 1103–1112.

28. K. W. Schaie, S. L. Willis, and G. I. L. Caskie, "The Seattle Longitudinal Study: Relationship Between Personality and Cognition," *Aging Neuropsychology and Cognition* 11, no. 2–3 (2004): 304–324.

29. L. J. Whalley, S. Sharma, H. C. Fox, A. D. Murray, R. T. Staff, A. C. Duthie, . . . J. M. Starr, "Anticholinergic Drugs in Late Life: Adverse Effects on Cognition but Not on Progress to Dementia," *Journal of Alzheimer's Disease* 30, no. 2 (2012): 253–261.

30. L. J. Whalley, and I. J. Deary, "Longitudinal Cohort Study of Childhood IQ and Survival up to Age 76," *British Medical Journal* 322 (2001): 819–822.

8. EMOTIONAL AGING

1. L. Fratiglioni, S. Paillard-Borg, and B. Windblad, "An Active and Socially Integrated Lifestyle in Late Life Might Protect Against Dementia," *Lancet* 3, no. 6 (2004): 343–353.

2. L. L. Barnes, C. F. Mendes de Leon, R. S. Wilson, J. L. Bienias, and D. A. Evans, "Social Resources and Cognitive Decline in a Population of Older African Americans and Whites," *Neurology* 63, no. 12 (2004): 2322–2326.

3. S. Bassuk, T. Glass, and L. Berkman, "Social Disengagement and Incident Cognitive Decline in Community Dwelling Elderly Persons," *Annals of Internal Medicine* 131, no. 3 (1999): 165–173.

4. M. Zunzunegui, B. Alvarado, T. Del Ser, and A. Otero, "Social Networks, Social Integration, and Social Engagement Determine Cognitive Decline in Community-Dwelling Spanish Older Adults," *Journal of Gerontology* 58B (2003): S93–S100.

5. A. Karen, M. Ertel, M. Glymour, and L. F. Berkman, "Effects of Social Integration on Preserving Memory Function in a Nationally Representative US Elderly Population," *American Journal of Public Health* 98, no. 7 (2008): 1215–1220.

6. M. Mather, and L. Carstensen, "Aging and Motivated Cognition: The Positivity Effect in Attention and Memory," *TRENDS in Cognitive Sciences* 9, no. 10 (2005): 496–502.

7. E. B. Palmore, *The Honorable Elders: A Cross-Cultural Analysis of Aging in Japan* (Durham, NC: Duke University Press, 1975), ch. 5.

8. D. Maeda, "Aging in Eastern Society," in *The Social Challenge of Aging*, ed. D. Hobman (London: Croon Helm, 1978).

9. L. L. Barnes, C. F. Mendes de Leon, R. S. Wilson, J. L. Bienias, and D. A. Evans, "Social Resources and Cognitive Decline in a Population of Older African Americans and Whites," *Neurology* 63, no. 12 (2004): 2322–2326.

10. K. L. Phan, T. Wager, S. F. Taylor, and I. Liberzon, "Functional Neuroanatomy of Emotion: A Meta-analysis of Emotion Activation Studies in PET and fMRI," *NeuroImage* 16, no. 2 (2002): 331–348.

11. N. Raz, and F. N. Rodrigue, "Differential Aging of the Brain: Patterns, Cognitive Correlates and Modifiers," *Neuroscience and Biobehavioral Reviews* 30, no. 6 (2006): 730–748.

12. T. A. Salthouse, "Neuroanatomical Substrates of Age-Related Cognitive Decline," *Psychological Bulletin* 137, no. 5 (2012): 753–784.

13. S. T. Charles, and L. L. Carstensen, "Social and Emotional Aging," *Annual Reviews in Psychology* 61 (2010): 383–409.

14. R. M. Sapolsky, "Depression, Antidepressants and the Shrinking Hippocampus," *Proceedings of the National Academy of Sciences*, 98 (2001): 12320–12322.

15. D. Goleman, *Emotional Intelligence: Why It Can Matter More Than IQ* (London: Random House, Bantam Books, 1996).

16. J. T. Erber, I. G. Prager, and X. Guo, "Age and Forgetfulness: Can Stereotype Be Modified?" *Educational Gerontology* 25, no. 3 (1999): 457–466.

17. D. P. McAdams, *The Stories We Live By: Personal Myths and the Making of Self* (New York: William Morrow, 1993).

18. R. R. McCrae, and P. T. Costa, "Personality Trait Structure as a Human Universal," *American Psychologist* 52, no. 5 (1997): 509–516.

19. B. W. Roberts, K. E. Walton, and W. Viechtbauer, "Patterns of Mean Level Change in Personality Traits Across the Life Course: A Meta-analysis of Longitudinal Studies," *Psychological Bulletin* 132, no. 1 (2006): 1–25.

20. E. H. Erikson, *Identity, Youth and Crisis* (New York: Norton, 1968).

21. R. Adolphs, "Cognitive Neuroscience of Human Social Behavior," *Nature Reviews Neuroscience* 4, no. 3 (2003): 165–178.

22. J. Kemp, O. Després, F. Sellal, and A. Dalfour, "Theory of Mind in Normal Ageing and Neuro-degenerative Pathologies," *Ageing Research Reviews* 11 (2012): 199–219.

23. S. Baron-Cohen, H. Ring, J. Moriarty, H. Ring, J. Moriarty, B. Schmitz, . . . P. Ell, "Recognition of Mental State Terms: Clinical Findings in Children with Autism and a Functional Neuroimaging Study of Normal Adults," *British Journal of Psychiatry* 165, no. 5 (1994): 640–649.

9. DEMENTIA SYNDROMES

1. G. E. Berrios, "Alzheimer's Disease—A Conceptual History," *International Journal of Geriatric Psychiatry* 5, no. 6 (1990): 355–365.

2. M. M. Esiri, F. Matthews, C. Brayne, P. G. Ince, F. E. Matthews, J. H. Xuereb, . . . J. H. Morris, Neuropathology Group UK Medical Research Council, "Pathological Correlates of Late-Onset Dementia in a Multicentre, Community-Based Population in England and Wales," *Lancet* 357, no. 9251 (2001): 169–175.

3. J. A. Schneider, Z. Arvanitakis, S. E. Leurgans, and D. A. Bennett, "The Neuropathology of Probable Alzheimer Disease and Mild Cognitive Impairment," *Annals of Neurology* 66, no. 2 (2009): 200–208.

4. R. C. Petersen, B. Caracciolo, C. Brayne, S. Gauthier, V. Jelic, and L. Fratiglioni, "Mild Cognitive Impairment: A Concept in Evolution," *Journal of Internal Medicine* 275 (2014): 214–228.

5. H. Chertkow, H. H. Feldman, C. Jacova, and F. Massoud, "Definitions of Dementia and Predementia States in Alzheimer's Disease and Vascular Cognitive Impairment: Consensus from the Canadian Conference on Diagnosis of Dementia," *Alzheimer's Research & Therapy* 5, Suppl. 1 (2013): S2.

6. B. Croisile, S. Auriacombe, F. Etcharry-Bouyx, M. Vercelletto; National Institute on Aging (U.S.); Alzheimer Association, "The New 2011 Recommendations of the National Institute on Aging and the Alzheimer's Association on Diagnostic Guidelines for Alzheimer's Disease: Preclinal Stages, Mild Cognitive Impairment, and Dementia," *Revue Neurologie* 168, no. 6–7 (2012): 471–82.

7. S. Ray, M. Britschgi, C. Herbert, Y. Takeda-Uchimura, A. Boxer, K. Blennow, D. R. Galasko, "Classification and Prediction of Clinical Alzheimer's Diagnosis Based on Plasma Signaling Proteins," *Nature Medicine* 13, no. 11 (2007): 1359–1362.

8. C. Davatzikos, Y. Fan, X. Wu, D. Shen, S. M. Resnick, A. F. Davatzikos, . . . S. M. Resnick, "Detection of Prodromal Alzheimer's Disease via Pattern Classification of Magnetic Resonance Imaging," *Neurobiology of Aging* 29, no. 4 (2008): 514–523.

9. H. Shi, O. Belbin, C. Medway, K. Brown, N. Kalsheker, M. Carrasquillo, . . . Genetic and Environmental Risk for Alzheimer's Disease Consortium; K. Morgan; Alzheimer's Research UK Consortium, "Genetic Variants Influencing Human Aging from Late-Onset Alzheimer's Disease (LOAD) Genome-Wide Association Studies (GWAS)," *Neurobiology of Aging* 33, no. 8 (2012): 1849.e5–18.

10. E. M. Reiman, J. B. Langbaum, A. S. Fleisher, R. J. Caselli, K. Chen, N. Ayutyanont, . . . P. N. Tariot, "Alzheimer's Prevention Initiative: A Plan to Accelerate the Evaluation of Presymptomatic Treatments," *Journal of Alzheimers Disease* 26, Suppl. 3 (2011): 321–329.

11. R. G. Will, J. W. Ironside, M. Zeidler, S. N. Cousens, K. Estibeiro, A. Alperovitch, . . . A. Hofman, "A New Variant of Creutzfeldt-Jakob Disease in the UK," *Lancet* 347 (1996): 921–925.

12. M. Bruce, R. G. Will, J. W. Ironside, I. McConnel, D. Drummond, A. Suttie,. . . C. J. Bostock, "Transmissions to Mice Indicate That 'New Variant' CJD Is Caused by the BSE Agent," *Nature* 389 (1997): 498–501.

13. L. A. Farrer, L. A. Cupples, J. L. Haines, B. Hyman, W. A. Kukull, R. Mayeux, . . . C. M. van Duijn, "Effects of Age, Sex, and Ethnicity on the Association Between Apolipoprotein E Genotype and Alzheimer Disease. A Meta-analysis. APOE and Alzheimer Disease Meta Analysis Consortium," *Journal of the American Medical Association* 278, no. 16 (1997): 1349–1356.

14. R. Andel, M. Crowe, N. L. Pedersen, L. Fratiglioni, B. Johansson, and M. Gatz, *Journals of Gerontology: Series A, Biological Sciences and Medicine* 63, no. 1 (2008): 62–66.

15. M. Gatz, C. A. Reynolds, L. Fratiglioni, B. Johansson, J. A. Mortimer, S. Berg, . . . N. L. Pedersen, "Role of Genes and Environments for Explaining Alzheimer Disease," *Archives of General Psychiatry* 63, no. 2 (2006): 168–174.

16. R. S. Wilson, E. Segawa, P. A. Boyle, S. E. Anagnos, L. P. Hizel, D. A. Bennett, "The Natural History of Cognitive Decline in Alzheimer's Disease," *Psychology and Aging* 27, no. 4 (2012): 1008–1017.

17. A. Burns, R. Jacoby, P. Luthert, and R. Levy, "Cause of Death in Alzheimer's Disease," *Age and Ageing* 19, no. 5 (1990): 341–344.

18. J. LeJeune, R. Turpin, and M. Gautier, "Mongolism: A Chromosomal Disease (Trisomy)," *Bulletin de l'Académie nationale de médecine* 143, no. 11–12 (1959): 256–265.

19. T. Arendt, B. Mosch, and M. Morawski, "Neuronal Aneuploidy in Health and Disease: A Cytomic Approach to the Molecular Individuality of Neurons," *International Journal of Molecular Sciences* 10 (2009): 1609–1627.

20. I. Y. Iourov, S. G. Vorsavanova, Y. Iourov, and B. Yuri, "Genomic Landscape of Alzheimer's Disease Brain: Chromosome Instability–Aneuploidy but not Tetraploidy–Mediates Neurodegeneration," *Neurodegenerative Disease* 8 (2011): 35–37, discussion 38–40.

21. C. S. Ivan, S. Seshadri, A. Beiser, R. Au, C. S. Kase, M. Kelly-Hayes, and P. A. Wolf, "Dementia After Stroke: The Framingham Study," *Stroke* 35, no. 6 (2004): 1264–1268.

22. R. T. Staff, A. D. Murray, T. Ahearn, S. Salarirad, D. Mowat, J. M. Starr . . . and L. J. Whalley, "Brain Volume and Survival from Age 78 to 85: The Contribution of Alzheimer-Type Magnetic Resonance Imaging Findings," *Journal of the American Geriatric Society* 58, no. 4 (2010): 688–695.

23. I. J. Deary, L. J. Whalley, G. D. Batty, and J. M. Starr, "Physical Fitness and Lifetime Cognitive Change," *Neurology* 67, no. 7 (2006): 1195–1200.

24. A. Hodges, *Alan Turing: The Enigma*, vintage ed. (London: Vantage Books, 1992), 250–252.

25. G. M. Savva, S. B. Wharton, P. G. Ince, G. Forster, F. E. Matthews, C. Brayne; Medical Research Council Cognitive Function and Ageing Study, "Age, Neuropathology, and Dementia," *New England Journal of Medicine* 360, no. 22 (2009): 2302–2309.

26. F. E. Matthews, C. Brayne, J. Lowe, I. McKeith, S. B. Wharton, and P. Ince, "Epidemiological Pathology of Dementia: Attributable-Risks at Death in the Medical Research Council Cognitive Function and Ageing Study," *PLoS Med* 6, no. 11 (2009): e1000180.

27. J. E. Selfridge, E. Lezi, J. Lu, and H. S. Russell, "Role of Mitochondrial Homeostasis and Dynamics in Alzheimer's Disease," *Neurobiology of Disease* 51 (2013): 3–12.

28. V. Villemagne, S. Burnham, P. Bourgeat, B. Brown, K. A. Ellis, O. Sakvado, . . . C. C. Rowe, C. Masters, "Amyloid β Deposition, Neurodegeneration, and Cognitive Decline in Sporadic Alzheimer's Disease: A Prospective Cohort Study," *Lancet Neurology* 12, no. 4 (2013): 357–367.

29. S. J. B. Vos, C. Xiong, P. J. Visser, M. S. Jasielec, J. Hassenstab, E. A. Grant, . . . A. Fagan, "Preclinical Alzheimer's Disease and Its Outcome: A Longitudinal Cohort Study," *Lancet Neurology* 12, no. 10 (2013): 957–965.

30. J. D. Warren, J. D. Rohrer, and M. Rossor, "Frontotemporal Dementia," *British Medical Journal* 347 (2013): f4827.

31. P. Svenningsson, E. Westamn, C. Ballard, and D. Aarsland, "Cognitive Impairment in Patients with Parkinson's Disease: Diagnosis, Biomarkers, and Treatment," *Lancet Neurology* 11, no. 8 (2012): 697–707.

32. I. Litvan, J. G. Goldman, A. I. Troster, B. A. Schmand, D. Weintraub, R. C. Petersen, . . . M. Emre, "Diagnostic Criteria for Mild Cognitive Impairment in Parkinson's Disease: Movement Disorder Society Task Force Guidelines," *Movement Disorders* 27, no. 3 (2012): 349–356.

33. N. Set-Salvia, J. Clarimon, J. Pagonabarraga, B. Pascual-Sedano, A. Campolongo, O. Combarros, . . . J. Kulisevsky, "Dementia Risk in Parkinson Disease: Disentangling the Role of MAPT Haplotypes," *Archives of Neurology* 68, no. 3 (2011): 359–364.

34. F. R. Guerini, E. Beghi, G. Riboldazzi, R. Zangaglia, C. Pianezzola, G. Bono, . . . E. Martignoni, "*BDNF* Val66Met Polymorphism Is Associated with Cognitive Impairment in Italian Patients with Parkinson's Disease," *European Journal of Neurology* 16 (2009): 1240–1245.

35. K. A. Fujita, M. Ostaszewski, Y. Matsuoka, S. Ghosh, E. Glaab, C. Trefois, . . . R. Balling, "Integrating Pathways of Parkinson's Disease in a Molecular Interaction Map," *Molecular Neurobiology* 49, no. 1 (2014): 88–102.

10. DEMENTIA RISK REDUCTION, 1: CONCEPTS, RESERVE, AND EARLY LIFE OPPORTUNITIES

1. Z. S. Khachaturian, D. Barnes, R. Einstein, S. Johnson, V. Lee, A. Roses, M. A. Sager, . . . L. J. Bain, "Developing a National Strategy to Prevent Dementia: Leon Thal Symposium 2009," *Alzheimer's & Dementia* 6, no. 1 (2010): 89–97.

2. B. C. Stephan, and C. Brayne, "Vascular Factors and Prevention of Dementia," *International Review of Psychiatry* 20, no. 4 (2008): 344–356.

3. M. Murphy, "The 'golden generation' in historical context," *British Actuarial Journal* 15, Suppl. (2009): 151–184.

4. C. Jagger, F. Gillies, E. Mascone, E. Cambois, H. Van Oyen, W. Nusselder, J. M. Robine; EHLEIS Team, "Inequalities in Healthy Life Years in the 25 Countries of the European Union in 2005: A Cross-National Meta-Regression Analysis," *Lancet* 372, no. 9656 (2008): 2124–2131.

5. N. Purandare, C. Ballard, and A. Burns, "Preventing Dementia," *Advances in Psychiatric Treatment* 11 (2005): 176–183.

6. K. Yaffe, M. Tocco, R. C. Petersen, C. Sigler, L. C. Burns, C. Cornelius, . . . M. C. T. Carrillo, "The Epidemiology of Alzheimer's Disease: Laying the Foundation for Drug Design, Conduct, and Analysis of Clinical Trials," *Alzheimers & Dementia* 8 (2012): 237–242.

7. Y. Stern, "Cognitive Reserve and Alzheimer's Disease," *Alzheimer Disease and Associated Disorders* 20, Suppl. 2 (2006): S69–S74.

8. R. T. Staff, A. D. Murray, I. J. Deary, and L. J. Whalley, "What Provides Cerebral Reserve?" *Brain* 127, no. 5 (2004): 1191–1199.

9. J. W. Lynch, G. W. Kaplan, and S. J. Shema, "Cumulative Impact of Sustained Economic Hardship on Physical, Cognitive, Psychological and Social Functioning," *New England Journal of Medicine* 337, no. 26 (1997): 1889–1895.

10. L. Melton, "Heat, Light and a Case of Vintage Reserve," Higher Education Supplement, *London Times*, June 17, 2005.

11. M. Rutter, and N. Madge, *Cycles of Disadvantage* (London: Heinemann, 1976).

12. A. Singh-Manoux, and M. Marmot, "The Role of Socialisation in Explaining Social Inequalities in Health," *Social Science & Medicine* 60, no. 9 (2005), 2129–2133.

13. S. Barral, S. Cosentine, R. Costa, A. Matteini, K. Christensen, S. L. Andersen, . . . R. Mayeux, "Cognitive Function in Families with Exceptional Survival," *Neurobiology of Aging* 33 (2012): 619e1–619e7.

14. B. Modin, I. Koupil, and D. Vagero, "The Impact of Early Twentieth Century Illegitimacy Across Three Generations: Longevity and Intergenerational Health Correlates," *Social Science & Medicine* 68, no. 9 (2009): 1633–1640.

15. I. J. Deary, M. D. Taylor, C. L. Hart, V. Wilson, G. Davey Smith, and D. Blane, "Intergenerational Social Mobility and Mid-life Status Attainment: Influences of Childhood Intelligence, Childhood Social Factors, and Education," *Intelligence* 33 (2005): 455–472.

16. G. Persson, and I. Skoog, "A Prospective Population Study of Psychosocial Risk Factors for Late Onset Dementia," *International Journal of Geriatric Psychiatry* 11 (1996): 15–22.

17. M. C. Norton, K. R. Smith, T. Østbye, J. T. Tschanz, S. Schwartz, C. Corcoran, . . . K. A. Welsh-Bohmer; Cache County Investigators, "Early Parental Death and Remarriage of Widowed Parents as Risk Factors for Alzheimer Disease: The Cache County Study," *American Journal of Geriatric Psychiatry* 9, no. 19 (2011): 814–824.

18. L. J. Whalley, R. T. Staff, A. D. Murray, I. J. Deary, and J. M. Starr, "Genetic and Environmental Factors in Late Onset Dementia: Possible Role for Early Parental Death," *International Journal Geriatric Psychiatry* 28, no. 1 (2013): 75–81.

19. R. Ravona-Springer, M. S. Beeri, and U. Goldbourt, "Younger Age at Crisis Following Parental Death in Male Children and Adolescents Is Associated with Higher Risk for Dementia at Old Age," *Alzheimers Disease & Associated Disorders* 26 (2012): 68–73.

20. D. A. Evans, L. E. Hebert, L. A. Beckett, P. A. Scherr, M. S. Albert, M. J. Chown, . . . J. O. Taylor, "Education and Other Measures of Socioeconomic Status and Risk of Incident Alzheimer Disease in a Defined Population of Older Persons," *Archives of Neurology* 54 (1997): 1399–1405.

21. R. S. Wilson, P. A. Scherr, G. Hoganson, J. L. Bienias, D. A. Evans, and D. A. Bennett, "Early Life Socioeconomic Status and Late Life Risk of Alzheimer's Disease," *Neuroepidemiology* 25 (2005): 8–14.

22. D. K. Lahiri, B. Maloney, and N. H. Zawia, "The LEARN Model: An Epigenetic Explanation for Idiopathic Neurobiological Diseases," *Molecular Psychiatry* 14 (2009): 992–1003.

23. A. T. Berg, H. R. Pardoe, R. K. Fulbright, S. U. Schuele, and G. D. Jackson, "Hippocampal Size Anomalies in a Community-Based Cohort with Childhood-Onset Epilepsy," *Neurology* 76 (2011): 1415–1421.

24. L. Melton, "Heat, Light and a Case of Vintage Reserve," Higher Education Supplement, *London Times*, June 17, 2005.

25. A. R. Borenstein, C. I. Copenhaver, and J. A. Mortimer, "Early-life Risk Factors for Alzheimer Disease," *Alzheimer Disease & Associated Disorders* 20 (2006): 63–72.

26. M. Dik, D. J. Deeg, M. Visser, and C. Jonker, "Early Life Physical Activity and Cognition at Old Age," *Journal of Clinical and Experimental Neuropsychology* 25 (2003): 643–653.

27. Y. Stern, "What Is Cognitive Reserve? Theory and Research Application of the Reserve Concept," *Journal of the International Neuropsychological Society* 8 (2002): 448–460.

28. C. E. Coffey, J. A. Saxton, G. Ratcliff, R. N. Bryan, and J. F. Lucke, "Relation of Education to Brain Size in Normal Aging: Implications for the Reserve Hypothesis," *Neurology* 53 (1999): 189–196.

29. E. Mori, N. Hirono, H. Yamashita, T. Imamura, Y. Ikejiri, M. Ikeda, . . . Y. Yoneda, "Premorbid Brain Size as a Determinant of Reserve Capacity Against Intellectual Decline in Alzheimer's Disease," *American Journal of Psychiatry* 154 (1997): 18–24.

30. C. Qiu, E. von Strauss, L. Backman, B. Winblad, and L. Fratiglioni, "Twenty-Year Changes in Dementia Occurrence Suggest Decreasing Incidence in Central Stockholm, Sweden," *Neurology* 80 (2013): 1888–1894.

31. L. J. Whalley, J. M. Starr, R. Athawes, D. Hunter, A. Pattie, and I. J. Deary, "Childhood Mental Ability and Dementia," *Neurology* 55 (2000): 1455–1459.

32. G. G. Potter, M. J. Helms, and B. L. Plassman, "Associations of Job Demands and Intelligence with Cognitive Performance Among Men in Late Life," *Neurology* 70 (2008): 1803–1808.

33. R. N. Kalaria, R. Akinyemi, and M. Ihara, "Does Vascular Pathology Contribute to Alzheimer Changes?" *Journal Neurological Sciences* 322, no. 1–2 (2012): 141–147.

34. E. M. Reiman, J. B. Langbaum, A. S. Fleisher, R. J. Caselli, K. Chen, N. Ayutyanont, . . . P. N. Tariot, "Alzheimer's Prevention Initiative: A Plan to Accelerate the Evaluation of Presymptomatic Treatments," *Journal of Alzheimers Disease* 26, Suppl. 3 (2011): 321–329.

35. E. M. Reiman, Y. T. Quiroz, A. S. Fleisher, K. Chen, C. Velez-Pardo, M. Jimenez-Del-Rio, . . . F. Lopera, "Brain Imaging and Fluid Biomarker Analysis in Young Adults at Genetic Risk for Autosomal Dominant Alzheimer's Disease in the Presenilin 1 E280A Kindred: A Case-Control Study," *Lancet Neurology* 11 (2012): 1048–1056.

36. J. B. Langbaum, A. S. Fleisher, K. Chen, N. Ayutyanont, F. Lopera, Y. T. Quiroz, . . . E. M. Reiman, "Ushering in the Study and Treatment of Preclinical Alzheimer Disease," *Nature Reviews Neurology* 9, no. 7 (2013): 371–381.

37. G. Kaplan, "What's Wrong with Social Epidemiology and How We Can We Make It Better?" *Epidemiological Reviews* 26 (2004): 124–135.

38. M. Rutter, and N. Madge, *Cycles of Disadvantage* (London: Heinemann, 1976).

39. S. Barral, S. Cosentine, R. Costa, A. Matteini, K. Christensen, S. L. Andersen, . . . R. Mayeux, "Cognitive Function in Families with Exceptional Survival," *Neurobiology of Aging* 33 (2012): 619e1–619e7.

40. A. Singh-Manoux, and M. Marmot, "The Role of Socialisation in Explaining Social Inequalities in Health," *Social Science & Medicine* 60 (2005): 2129–2133.

41. B. Modin, I. Koupil, and D. Vagero, "The Impact of Early Twentieth Century Illegitimacy Across Three Generations: Longevity and Intergenerational Health Correlates," *Social Science & Medicine* 68 (2009): 1633–1640.

42. I. J. Deary, M. D. Taylor, C. L. Hart, V. Wilson, G. Davey Smith, and D. Blane, "Intergenerational Social Mobility and Mid-Life Status Attainment: Influences of Childhood Intelligence, Childhood Social Factors, and Education," *Intelligence* 33 (2005): 455–472.

43. G. Persson, and I. Skoog, "A Prospective Population Study of Psychosocial Risk Factors for Late Onset Dementia," *International Journal of Geriatric Psychiatry* 11 (1996): 15–22.

44. M. C. Norton, K. R. Smith, T. V. Østbye, J. T. Tschanz, S. Schwartz, C. Corcoran, . . . K. A. Welsh-Bohmer; Cache County Investigators, "Early Parental Death and Remarriage of Widowed Parents as Risk Factors for Alzheimer Disease: The Cache County Study," *American Journal of Geriatric Psychiatry* 19 (2011): 814–824.

45. L. J. Whalley, R. T. Staff, A. D. Murray, I. J. Deary, and J. M. Starr, "Genetic and Environmental Factors in Late Onset Dementia: Possible Role for Early Parental Death," *International Journal of Geriatric Psychiatry* 28, no. 1 (2013): 75–81.

46. R. Ravona-Springer, M. S. Beeri, and U. Goldbourt, "Younger Age at Crisis Following Parental Death in Male Children and Adolescents Is Associated with Higher Risk for Dementia at Old Age," *Alzheimers Disease & Associated Disorders* 26 (2012): 68–73.

47. D. A. Evans, L. E. Hebert, L. A. Beckett, P. A. Scherr, M. S. Albert, M. J. Chown, . . . J. O. Taylor, "Education and Other Measures of Socioeconomic Status and Risk of Incident Alzheimer Disease in a Defined Population of Older Persons," *Archives of Neurology* 54 (1997): 1399–1405.

48. R. S. Wilson, P. A. Scherr, G. Hoganson, J. L. Bienias, D. A. Evans, and D. A. Bennett, "Early Life Socioeconomic Status and Late Life Risk of Alzheimer's Disease," *Neuroepidemiology* 25 (2005): 8–14.

49. D. K. Lahiri, B. Maloney, and N. H. Zawia, "The LEARN Model: An Epigenetic Explanation for Idiopathic Neurobiological Diseases," *Molecular Psychiatry* 14 (2009): 992–1003.

50. A. T. Berg, H. R. Pardoe, R. K. Fulbright, S. U. Schuele, and G. D. Jackson, "Hippocampal Size Anomalies in a Community-Based Cohort with Childhood-Onset Epilepsy," *Neurology* 76 (2011): 1415–1421.

51. R. A. Cohen, S. Grieve, K. F. Hoth, R. H. Paul, L. Sweet, D. Tate, J. Gunstad, . . . L. M. Williams, "Early Life Stress and Morphometry of the Adult Anterior Cingulate Cortex and Caudate Nuclei," *Biological Psychiatry* 59 (2006): 975–982.

52. J. D. Bremner, "Long-Term Effects of Childhood Abuse on Brain and Neurobiology," *Child and Adolescent Psychiatry Clinics of North America* 12 (2003): 271–292.

53. D. W. Hedges, and F. L. Woon, "Early-Life Stress and Cognitive Outcome," *Psychopharmacology* 214, no. 1 (2011): 121–130.

54. A. van Harmelen, M. van Tol, N. J. A. van der Wee, D. J. Veltman, A. Aleman, P. Spinhoven, . . . B. M. Elzinga, "Reduced Medial Prefrontal Cortex Volume in Adults Reporting Childhood Emotional Maltreatment," *Biolological Psychiatry* 68 (2010): 832–838.

55. P. Tomalski, and M. H. Johnson, "The Effects of Early Adversity on the Adult and Developing Brain," *Current Opinion in Psychiatry* 23 (2010): 233–238.

56. R. Katzman, "Education and the Prevalence of Dementia and Alzheimer's Disease," *Neurology* 43 (1993): 13–20.

57. D. A. Snowdon, S. J. Kemper, J. A. Mortimer, L. H. Greiner, D. R. Wekstein, and W. R. Markesbery, "Ability in Early Life and Cognitive Function and AD in Late Life: Findings from the Nun Study," *Journal of the American Medical Association* 275 (1996): 528–531.

58. L. J. Whalley, J. M. Starr, R. Athawes, D. Hunter, A. Pattie, and I. J. Deary, "Childhood Mental Ability and Dementia," *Neurology* 55 (2000): 1455–1459.

59. J. D. Bremner, "Long-Term Effects of Childhood Abuse on Brain and Neurobiology," *Child & Adolescent Psychiatry Clinics of North America* 12 (2003): 271–292.

60. L. J. Whalley, and I. J. Deary, "Longitudinal Cohort Study of Childhood IQ and Survival Up to Age 76," *British Medical Journal* 322 (2001): 1–5.

61. J. M. Starr, I. J. Deary, H. C. Fox, and L. J. Whalley, "Smoking and Cognitive Change from Age 11 to 66 Years: A Confirmatory Investigation," *Addictive Behavior* 32, no. 1 (2007): 63–68.

62. L. J. Whalley, H. C. Fox, H. A. Lemmon, S. J. Duthie, A. R. Collins, H. Peace, . . . I. J. Deary, "Dietary Supplement Use in Old Age: Associations with Childhood IQ, Current Cognition and Health," *International Journal of Geriatric Psychiatry* 18 (2003): 768–776.

63. J. M. Starr, M. D. Taylor, C. L. Hart, G. D. Smith, L. J. Whalley, D. J. Hole, . . . I. J. Deary, "Childhood Mental Ability and Blood Pressure at Midlife: Linking the Scottish Mental Survey 1932 and the Midspan Studies," *Journal of Hypertension* 22 (2004): 893–897.

64. G. D. Batty, I. J. Deary, I. Schoon, and C. R. Gale, "Mental Ability Across Childhood in Relation to Risk Factors for Premature Mortality in Adult Life: The 1970 British Cohort Study," *Journal of Epidemiology & Community Health* 61 (2007): 997–1003.

65. C. L. Hart, M. D. Taylor, G. D. Smith, L. J. Whalley, J. M. Starr, D. J. Hole, . . . I. J. Deary, "Childhood IQ and Cardiovascular Disease in Adulthood: Prospective Observational Study Linking the Scottish Mental Survey 1932 and the Midspan Studies," *Social Science & Medicine* 59 (2004): 2131–2138.

66. T. W. Teasdale, T. I. Sorensen, and A. J. Stunkard, "Intelligence and Educational Level in Relation to Body Mass Index of Adult Males," *Human Biology* 64 (1992): 99–106.

67. K. Yaffe, M. Han, T. Blackwell, E. Cherkasova, R. A. Whitmer, and N. West, "Metabolic Syndrome and Cognitive Decline in Elderly Latinos: Findings from the Sacramento Area Latino Study of Aging Study," *Journal of the American Geriatrics Society* 55 (2007): 758–762.

68. M. Richards, S. Black, and G. Mishra, "IQ in Childhood and the Metabolic Syndrome in Middle Age: Extended Follow-Up of the 1946 British Birth Cohort Study," *Intelligence* 37 (2009): 567–572.

69. J. R. Flynn, *Are We Getting Smarter? Rising IQ in the Twenty-First Century* (Cambridge: Cambridge University Press, 2012), 190–236, app. 1.

70. C. Qiu, E. von Strauss, L. Backman, B. Winblad, and L. Fratiglioni, "Twenty-Year Changes in Dementia Occurrence Suggest Decreasing Incidence in Central Stockholm, Sweden," *Neurology* 80 (2013): 1888–1894.

71. F. E. Matthews, A. Arthur, L. E. Barnes, J. Bond, C. Jagger, L. Robinson, . . . C. Brayne, "A Two-Decade Comparison of Prevalence of Dementia in Individuals Aged 65 Years and Older from Three Geographical Areas of England: Results of the Cognitive Function and Ageing Study I and II," *Lancet* 382, no. 9902 (2013): 1405–1412.

72. L. J. Whalley, and K. A. Smyth, "Crisis or Crossroads: Human Culture and the Dementia Epidemic," *Neurology* 80 (2013): 1824–1825.

73. V. Skirbekk, M. Stinawski, and E. Bonsang, "The Flynn Effect and Population Aging," *Intelligence* 41 (2013): 169–177.

74. F. E. Matthews, A. Arthur, L. E. Barnes, J. Bond, C. Jagger, L. Robinson, . . . C. Brayne, "A Two-Decade Comparison of Prevalence of Dementia in Individuals Aged 65 Years and Older from Three Geographical Areas of England: Results of the Cognitive Function and Ageing Study I and II," *Lancet* 382, no. 9902 (2013): 1405–1412.

75. W. J. Bowers, X. O. Breakefield, and M. Sena-Esteves, "Genetic Therapy for the Nervous System," *Human Molecular Genetics* 20 (2011): R21–R48.

76. D. J. Selkoe, "Preventing Alzheimer's Disease," *Science* 337, no. 6101 (2012): 1488–1491.

77. L. Jarvik, A. Larue, D. Blacker, M. Gatz, C. Kawas, J. J. McArdle, . . . A. B. Zonderman, "Children of Persons with AD: What Does the Future Hold?" *Alzheimer's Disease & Associated Disorders* 22, no. 1 (2008): 6–20.

78. N. Fillipini, B. J. Mackintosh, M. Hough, G. M. Goodwin, G. B. Frisoni, S. M. Smith, . . . C. E. Mackay, "Distinct Patterns of Brain Activity in Young Carriers of the APOE-ε4 Allele," *Proceedings of the National Academy of Sciences* 106, no. 17 (2009): 7209–7214.

79. W. J. Bowers, X. O. Breakefield, and M. Sena-Esteves, "Genetic Therapy for the Nervous System," *Human Molecular Genetics* 20 (2011): R21–R48.

80. J. Jiang, Y. Jing, G. J. Cost, J. C. Chiang, H. J. Kolpa, A. M. Cotton, . . . J. B. Lawrence, "Translating Dosage Compensation to Trisomy 21," *Nature* 500, no. 7462 (2013): 296–300.

81. E. S. Sharp, and M. Gatz, "Relationship Between Education and Dementia: An Updated Systematic Review," *Alzheimer's Disease & Associated Disorders* 25, no. 4 (2011): 289–304.

82. G. A. Kaplan, E. R. Pamuk, J. W. Lynch, R. D. Cohen, and J. L. Balfour, "Inequality in Income and Mortality in the United States: Analysis of Mortality and Potential Pathways," *British Medical Journal* 312, no. 7041 (1996): 999–1003.

83. M. Huisman, A. E. Kumst, M. Bopp, J. K. Borgan, C. Borrell, G. Costa, . . . J. P. Mackenbach, "Educational Inequalities in Cause-Specific Mortality in Middle-Aged and Older Men and Women in Eight Western European Populations," *Lancet* 365, no. 9458 (2006): 493–500.

84. L. S. Gottfedson, "Intelligence: Is It the Epidemiologist's Elusive 'Fundamental Cause' of Social Class Inequalities in Health?" *Journal of Personality & Social Psychology* 86 (2004): 174–199.

85. L. J. Whalley, and K. A. Smyth, "Human Culture and the Future Dementia Epidemic: Crisis or Crossroads?" *Neurology* 80, no. 20 (2013): 1824–1825.

86. P. Hirst, and W. Carr, "Philosophy and Education—A Symposium," *Journal of Philosophy & Education* 39 (2005): 4.

87. I. McDowell, G. Guoliang Xi, J. Lindsay, . . . M. Tierney, "Mapping the Connections Between Education and Dementia," *Journal of Clinical & Experimental Neuropsychology* 29, no. 2 (2007): 127–141.

88. J. R. Flynn, *Are We Getting Smarter? Rising IQ in the Twenty-First Century* (Cambridge: Cambridge University Press, 2012), 190–236, app. 1.

89. V. Skirbekk., M. Stinawski, and E. Bonsang, "The Flynn Effect and Population Aging," *Intelligence* 41 (2013): 169–177.

90. I. McDowell, G. Guoliang Xi, J. Lindsay, . . . M. Tierney, "Mapping the Connections Between Education and Dementia," *Journal of Clinical & Experimental Neuropsychology* 29, no. 2 (2007): 127–141.

91. A. Mechelli, J. T. Crinion, U. Noppeney, J. O'Doherty, J. Ashburner, R. S. Frackoviak, and C. J. Price, "Neurolinguistics: Structural Plasticity in the Bilingual Brain," *Nature* 431, no. 7010 (2004): 757.

92. T. A. Schweizer, J. Ware, C. E. Fischer, F. I. Craik, and E. Bialystok, "Bilingualism as a Contributor to Cognitive Reserve: Evidence from Brain Atrophy in Alzheimer's Disease," *Cortex* 48, no. 8 (2012): 991–996.

93. H. Chertkow, V. Whitehead, N. Phillips, J. Atherton, and H. Bergman, "Multilingualism (but not Always Bilingualism) Delays the Onset of Alzheimer Disease: Evidence from a Bilingual Community," *Alzheimer's Disease & Associated Disorders* 24: 118–125.

94. L. Ossher, E. Bialystok, F. I. Craik, K. J. Murphy, and A. K. Troyer, "The Effect of Bilingualism on Amnestic Mild Cognitive Impairment," *Journals of Gerontology: Series B, Psychological Science & Social Sciences* 68, no. 1 (2013): 8–12.

95. A. E. Sanders, C. B. Hall, M. J. Katz, and R. B. Lipton, "Non-Native Language Use and Risk of Incident Dementia in the Elderly," *Journal of Alzheimers Disease* 29, no. 1 (2012): 99–108.

96. E. Bialystok, F. I. Craik, and G. Luk, "Bilingualism: Consequences for Mind and Brain," *Trends in Cognitive Sciences* 16, no. 4 (2012): 240–250.

97. F. E. Matthews, A. Arthur, L. E. Barnes, J. Bond, C. Jagger, L. Robinson, and C. Brayne; Medical Research Council Cognitive Function and Ageing Collaboration, "A Two-Decade Comparison of Prevalence of Dementia in Individuals Aged 65 Years and Older from Three Geographical Areas of England: Results of the Cognitive Function and Ageing Study I and II," *Lancet* 382, no. 9902 (2013): 1405–1412.

98. L. J. Whalley, "Spatial Distribution and Secular Trends in the Epidemiology of Alzheimer's Disease," *Neuroimaging Clinics of North America* 22, no. 1 (2012): 1–10.

99. G. K. Hulse, N. T. Lautenschlager, R. J. Tait, and O. P. Almeida, "Dementia Associated with Alcohol and Other Drug Use," *International Psychogeriatrics* 17 (2005): S109–S127.

100. M. H. Meier, A. Caspi, A. Ambler, H. Harrington, R. Houts, R. S. Keefe, . . . T. E. Moffitt, "Persistent Cannabis Users Show Neuropsychological Decline from Childhood to Midlife," *Proceedings of the National Academy of Sciences of the United States* 109, no. 40 (2012): E2657–E2664.

101. T. Schilt, M. W. J. Koeter, J. P. Smal et al., "Long-Term Neuropsychological Effects of Ecstasy in Middle-Aged Ecstasy/Polydrug Users," *Psychopharmacology* 207 (2010): 583–591.

102. P. M. Thompson, K. M. Hayashi, S. L. et al., "Structural Abnormalities in the Brains of Human Subjects Who Use Methamphetamine," *Journal of Neuroscience* 24 (2004): 6028–6036

103. J. W. Langston, L. S. Forno, J. Tetrud, A. G. Reeves, J. A. Kaplan, and D. Karluk, "Evidence of Active Nerve Cell Degeneration in the Substantial Nigra of Humans Years After 1-Methyl-4-phenyl-1,2,3,6-tetrahydropyridine Exposure," *Annals of Neurology* 46, no. 4 (1999): 598–605.

104. C. M. Beynon, B. Roe, P. Duffy, et al., "Self Reported Health Status, and Health Service Contact, of Illicit Drug Users Aged 50 and Over: A Qualitative Interview Study in Merseyside, United Kingdom," *BMC Geriatrics* 9 (2009): 1–9.

105. A. S. Reece, "Evidence of Accelerated Ageing in Clinical Drug Addiction from Immune, Hepatic and Metabolic Biomarkers," *Immunity & Ageing* 4 (2007): article number 6.

106. C. Qiu, E. von Strauss, L. Backman, B. Winblat, L. Fratiglioni, "Twenty-Year Changes in Dementia Occurrence Suggest Decreasing Incidence in Central Stockholm, Sweden," *Neurology* 80, no. 20 (2013): 1888–1894. doi:10.1212/WNL.0b013e318292a2f9

107. F. E. Matthews, A. Arthur, L. E. Barnes, J. Bond, C. Jagger, L. Robinson, . . . C. Brayne, "A Two-Decade Comparison of Prevalence of Dementia in Individuals Aged 65 Years and Older from Three Geographical Areas of England: Results of the Cognitive Function and Ageing Study I and II," *Lancet* 382, no. 9902 (2013): 1405–1412.

108. L. J. Whalley, and K. A. Smyth, "Crisis or Crossroads: Human Culture and the Dementia Epidemic," *Neurology* 80 (2013): 1824–1825.

109. J. R. Flynn, *Are We Getting Smarter? Rising IQ in the Twenty-First Century* (Cambridge: Cambridge University Press, 2012), 190–236, app. 1.

110. M. D. Hilchey, and R. M. Klein, "Are There Bilingual Advantages on Nonlinguistic Interference Tasks? Implications for the Plasticity of Executive Control Processes," *Psychonomics Bulletin & Review* 18 (2011): 625–658.

111. T. Bak, J. J. Nissan, M. M. Allerhand, and I. J. Deary, "Does Bilingualism Influence Cognitive Aging?" *Annals of Neurology* 75 (2014): 955–963.

112. L. B. Zahodne, P.W. Schofield, M. T. Farrell, Y. Stern, J. J. Manly, "Bilingualism Does Not Alter Cognitive Decline or Dementia Risk Among Spanish-Speaking Immigrants," *Neuropsychology* 28 (2014): 238–246.

113. I. B. Meier, J. J. Manly, F. A. Provenzano, K. S. Louie, B. T. Wasserman, E. Y. Griffith, J. T. Hector, E. Allocco, A. M. Brickman, "White Matter Predictors of Cognitive Functioning in Older Adults," *Journal of the International Neuropsychological Society*, 18 (2012): 414–427.

114. A. M. Brickman, N. Schupf, J. J. Manly, J. A. Luchsinger, H. Andrews, M. X. Tang . . . T. R. Brown, "Brain Morphology in Older African Americans, Caribbean Hispanics, and Whites from Northern Manhattan," *Archives of Neurology*, 65 (2008): 1053–1061.

115. P. S. Sachdev, D. M. Lipnicki, N. A. Kochan, J. D. Crawford, K. Rockwood, S. Xiao, . . . J. Santabárbara, "COSMIC (Cohort Studies of Memory in an International Consortium): An International Consortium to Identify Risk and Protective Factors and Biomarkers of Cognitive Ageing and Dementia in Diverse Ethnic and Sociocultural Groups," *BMC Neurology*, 13 (2013): 165.

116. I. J. Deary, "Teaching Intelligence," *Intelligence* 42 (2014): 142–147.

11. DEMENTIA RISK REDUCTION, 2: MIDLIFE OPPORTUNITIES TO DELAY DEMENTIA ONSET

1. C. Qiu, E. von Strauss, L. Backman, B. Winblad, and L. Fratiglioni, "Twenty-Year Changes in Dementia Occurrence Suggest Decreasing Incidence in Central Stockholm, Sweden," *Neurology* 80 (2013): 1888–1894.

2. F. E. Matthews, A. Arthur, L. E. Barnes, J. Bond, C. Jagger, L. Robinson, . . . C. Brayne, "A Two-Decade Comparison of Prevalence of Dementia in Individuals Aged 65 Years and Older from Three Geographical Areas of England: Results of the Cognitive Function and Ageing Study I and II," *Lancet* 382, no. 9902 (2013): 1405–1412.

3. W. A. Rocca, R. C. Petersen, D. S. Knopman, L. E. Hebert, D. A. Evans, K. S. Hall, . . . L. White, "Trends in the Incidence and Prevalence of Alzheimer's Disease, Dementia, and Cognitive Impairment in the United States," *Alzheimers & Dementia* 7 (2011): 80–93.

4. K. M. Langa, E. B. Larson, J. H. Karlawish, D. M. Cutler, M. U. Kabeto, S. Y. Kim, and A. B. Rosen, "Trends in the Prevalence and Mortality of Cognitive Impairment in the United States: Is There Evidence of a Compression of Cognitive Morbidity?" *Alzheimers & Dementia* 4 (2008): 134–144.

5. E. M. Schrijvers, B. F. Verhaaren, P. J. Koudstaal, A. Hofman, M. A. Ikram, and M. M. Breteler, "Is Dementia Incidence Declining? Trends in Dementia Incidence Since 1990 in the Rotterdam Study," *Neurology* 78 (2012): 1456–1463.

6. K. Christensen, M. Thinggaard, A. Oksuzyan, T. Steenstrup, K. Andersen-Ranberg, B. Jeune, . . . J. W. Vaupel, "Physical and Cognitive Functioning of People Older Than 90 Years: A Comparison of Two Danish Cohorts Born 10 Years Apart," *Lancet* 382, no. 9903 (2013): 1507–1513.

7. A. Lobo, P. Saz, G. Marcos, J. L. Dia, C. De-la-Camara, T. Ventura, . . . S. Aznar; ZARADEMP Workgroup. *Acta Psychiatrica Scandinavica* 116: 299–307.

8. A. Sekita, T. Ninomiya, Y. Tanizaki, Y. Doi, J. Hata, K. Yonemoto, . . . Y. Kiyohara, "Trends in Prevalence of Alzheimer's Disease and Vascular Dementia in a Japanese Community: The Hisayama Study," *Acta Psychiatrica Scandinavica* 122 (2010): 319–325.

9. K. Y. Chan, W. Wang, J. J. Wu, L. Liu, E. Theodoratou, J. Car, I. Rudan; Global Health Epidemiology Reference Group (GHERG), "Epidemiology of Alzheimer's Disease and Other Forms of Dementia in China, 1990–2010: A Systematic Review and Analysis," *Lancet* 381 (2013): 2016–2023.

10. J. R. Flynn, "Searching for Justice: The Discovery of IQ Gains over Time," *American Psychologist* 54 (1999): 5–16.

11. J. R. Flynn, *Are We Getting Smarter? Rising IQ in the Twenty-First Century* (Cambridge: Cambridge University Press, 2012), 190–236, app. 1.

12. J. Birns, and L. Kalra, "Cognitive Function and Hypertension," *Journal of Hypertension* 23 (2009): 86–96.

13. C. Qi Qiu, E. von Strauss, L. Backman, B. Winblad, and L. Fratiglioni, "Twenty-Year Changes in Dementia Occurrence Suggest Decreasing Incidence in Central Stockholm, Sweden," *Neurology* 80 (2013): 1888–1894.

14. M. L. Daviglus, C. C. Bell, W. Berrettini, P. E. Bowen, E. S. Connolly, N. J. Cox, . . . M. Trevisan, "National Institutes of Health State-of-the-Science Conference Statement: Preventing Alzheimer Disease and Cognitive Decline," *Annals of Internal Medicine* 153 (2010): 176–181.

15. C. Qiu, and B. Winblad, "The Age Dependent Relation of Blood Pressure to Cognitive Function and Dementia," *Lancet Neurology* 4 (2005): 487–499.

16. C. Qiu, M. Kivipleto, and L. Fratiglioni, "Preventing Alzheimer Disease and Cognitive Decline," *Annals of Internal Medicine* 154 (2011): 211–212.

17. P. Soros, S. Whitehead, J. D. Spence, and V. Hachinski, "Opinion: Antihypertensive Treatment Can Prevent Stroke and Cognitive Decline," *Nature Reviews Neurology* (2012): 174–178.

18. K. Shah, Q. Salah, M. Johnson, N. Parikh, P. Schulz, and M. Kunik, "Does Use of Antihypertensive Drugs Affect the Incidence or Progression of Dementia? A Systematic Review," *American Journal of Geriatric Pharmacotherapy* 7 (2009): 250–261.

19. B. McGuinness, S. Todd, P. Passmore, and R. Bullock, "Blood Pressure Lowering in Patients Without Prior Cerebrovascular Disease for Prevention of Cognitive Impairment and Dementia," *Cochrane Database Systematic Review* 4 (2009): CD004034.

20. B. L. Plassman, J. W. Williams, J. R. Burke, T. Holsinger, S. Benjamin, "Systematic Review: Factors Associated with Risk for and Possible Prevention of Cognitive Decline in Later Life," *Annals of Internal Medicine* 153 (2010): 182–193.

21. P. B. Gorelick, D. Nyenhuis; American Society of Hypertension Writing Group; B. J. Materson, D. A. Calhoun, W. J. Elliott, . . . R. R. Townsend, "Blood Pressure and Treatment of Persons with Hypertension as It Relates to Cognitive Outcomes Including Executive Function," *Journal of the American Society of Hypertension* 6 (2012): 309–315.

22. P. B. Gorelick, A. Scuteri, S. E. Black, C. Decarli, S. M. Greenberg, C. Iadecola, and S. Seshadri, "Vascular Contributions to Cognitive Impairment and Dementia: A Statement for Healthcare Professionals from the American Heart Association/American Stroke Association," *Stroke* 42 (2011): 2672–2713.

23. M. Robertson, A. Seaton, and L. J. Whalley, "Can We Reduce the Risk of Dementia?" *Quarterly Journal of Medicine* 108 (2015): 93–97.

24. C. W. Chen, C. C. Lin, K.-B. Chen, C. B. Chen, Y. C. Kuo, Y. C. Li, and C. J. Chug, "Increased Risk of Dementia in People with Previous Exposure to General Anesthesia: A National Population-Based Case-Control Study," *Alzheimer's & Dementia* 10, no. 2 (2014): 196–204.

25. J. Witlox, L. S. M. Eurelings, J. F. M. de Jonge, K. J. Kalisvaart, P. Eikelenboom, and W. A. van Gool, "Delirium in Elderly Patients and the Risk of Postdischarge Mortality, Institutionalization, and Dementia," *Journal of the American Medical Association* 304 (2010): 443–451.

26. D. H. J. Davis, G. Muniz Terrara, H. Keage, T. Rahkonen, M. Oinas, F. E. Matthews, . . . C. Brayne, "Delirium Is a Strong Risk Factor for Dementia in the Oldest Old: A Population-Based Study," *Brain* 135 (2012): 2809–2816.

27. J. A. Luchsinger, C. Reitz, L. S. Honig, M. X. Tang, S. Shea, and R. Mayeux, "Aggregation of Vascular Risk Factors and Risk of Incident Alzheimer Disease," *Neurology* 65, no. 4 (2005): 545–551.

28. S. Pendlebury, and P. M. Rothwell, "Prevalence, Incidence, and Factors Associated with Pre-stroke and Post-stroke Dementia: A Systematic Review and Meta-analysis," *Lancet Neurology* 8 (2009): 1006–1018.

29. C. Qiu, E. von Strauss, L. Backman, B. Winblad, and L. Fratiglioni, "Twenty-Year Changes in Dementia Occurrence Suggest Decreasing Incidence in Central Stockholm, Sweden," *Neurology* 80 (2013): 1888–1894.

30. K. V. Allen, B. M. Frier, and M. W. J. Strachan, "The Relationship Between Type 2 Diabetes and Cognitive Dysfunction: Longitudinal Studies and Their Methodological Limitations," *European Journal of Pharmacology* 490 (2004): 169–175.

31. G. J. Biessels, S. Staekenborg, E. Brunner, C. Brayne, and P. Scheltens, "Risk of Dementia in Diabetes Mellitus: A Systematic Review," *Lancet* 5 (2006): 64–74.

32. J. A. Luchsinger, C. Reitz, L. S. Honig, M. X. Tang, S. Shea, and R. Mayeux, "Aggregation of Vascular Risk Factors and Risk of Incident Alzheimer Disease," *Neurology* 65, no. 4 (2005): 545–551.

33. L. J. Whalley, F. Dick, and G. McNeill, "A Life-Course Approach to the Aetiology of Late Onset Dementia," *Lancet Neurology* 5 (2006): 87–96.

34. A. J. Sommerfield, I. J. Deary, and B. M. Frier, "Acute Hyperglycemia Alters Mood State and Impairs Cognitive Performance in People with Type 2 Diabetes," *Diabetes Care* 27, no. 10 (2004): 2335–2340.

35. T. F. Hughes, R. Andel, B. J. Small, A. R. Borenstein, J. A. Mortimer, A. Wolk, and M. Gatz, "Midlife Fruit and Vegetable Consumption and Risk of Dementia in Later Life in Swedish Twins," *American Journal of Geriatric Psychiatry* 18, no. 5 (2010): 413–420.

36. M. Richards, R. Hardy, D. Kuh, M. E. Wadsworth, "Birth Weight and Cognitive Function in the British 1946 Birth Cohort: Longitudinal Population-Based Study," *British Medical Journal* 322 (2001): 199–203.

37. C. J. Wilson, C. E. Finch, and H. J. Cohen, "Cytokines and Cognition—The Case for a Head-to-Toe Inflammatory Paradigm," *Journal of the American Geriatrics Society* 50, no. 12 (2002): 2041–2056.

38. P. L. McGeer, and E. G. McGeer, "Inflammation and the Degenerative Diseases of Aging," *Annals of the New York Academy of Sciences* 1035 (2006): 104–116.

39. W. Xu, C. Qiu, M. Gatz, N. L. Pedersen, B. Johansson, and L. Fratiglioni, "Mid- and Late-Life Diabetes in Relation to the Risk of Dementia a Population-Based Twin Study," *Diabetes* 58, no. 1 (2009): 71–77.

40. K. Yaffe, K. Lindquist, B. W. Peninx, E. M. Simonsick, M. Pahor, S. Kritchevsky, . . . T. Harris, "Inflammatory Markers and Cognition in Well-Functioning African-American and White Elders," *Neurology* 61 (2003): 76–80.

41. K. Ya Yaffe, K. Lindquist, B. W. Peninx, E. M. Simonsick, M. Pahor, S. Kritchevsky, . . . T. Harris, "Inflammatory Markers and Cognition in Well-Functioning African-American and White Elders," *Neurology* 61 (2003): 76–80.

42. S. P. Mooijaart, N. Sattar, S. Trompet, J. Lucke, D. J. Stott, I. Ford, . . . A. J. de Craen; PROSPER Study Group, "C-Reactive Protein and Genetic Variants and Cognitive Decline in Old Age: The PROSPER Study," *PLoS*, 274, no. 1 (2013): 77–85.

43. K. Ya Yaffe, K. Lindquist, B. W. Peninx, E. M. Simonsick, M. Pahor, S. Kritchevsky, . . . T. Harris, "Inflammatory Markers and Cognition in Well-Functioning African-American and White Elders," *Neurology* 61 (2003): 76–80.

44. T. Jonsson, H. Stefansson, S. Steinberg, I. Jonsdottir, P. V. Jonsson, J. Snaedal, . . . K. Stefansson, "Variant of TREM2 Associated with the Risk of Alzheimer's Disease," *New England Journal of Medicine* 368 (2013): 107–116.

45. J. D. Gonzalez Murcia, C. Schmutz, C. Munger, A. Perkes, A. Gustin, and M. Peterson, "Assessment of TREM2 rs75932628 Association with Alzheimer's Disease in a Population-Based Sample: The Cache County Study," *Neurobiology of Aging* 34, no. 12 (2013): 2889.

46. M. P. Mattson, and T. Magnus, "Ageing and Neuronal Vulnerability," *Nature Reviews Neuroscience* 7 (2006): 278–294.

47. L. J. Whalley, I. J. Deary, K. M. Starr, K. W. Wahle, K. A. Rance, V. J. Bourne, and H. C. Fox, "n-3 Fatty Acid Erythrocyte Membrane Content, APOEε4 and Cognitive Variation: An Observational Follow-Up Study in Late Adulthood," *American Journal of Clinical Nutrition* 87, no. 2 (2008): 449–454.

48. L. J. Whalley, H. C. Fox, K. W. Wahle, J. M. Starr, and I. J. Deary, "Cognitive Aging, Childhood Intelligence, and the Use of Food Supplements: Possible Involvement of n-3 Fatty Acids," *American Journal of Clinical Nutrition* 80, no. 6 (2004): 1650–1657.

49. E. M. Reiman, K. Chen, J. B. S. Langbaum, W. Lee, C. Reschke, D. Brandy, . . . R. J. Caselli, "Higher Serum Total Cholesterol Levels in Late Middle Age Are Associated with Glucose Hypometabolism in Brain Regions Affected by Alzheimer's Disease and Normal Aging," *NeuroImage* 49, no. 1 (2010): 169–176.

50. E. M. Reiman, K. Chen, J. B. S. Langbaum, W. Lee, C. Reschke, D. Brandy, . . . R. J. Caselli, "Higher Serum Total Cholesterol Levels in Late Middle Age Are Associated with Glucose Hypometabolism in Brain Regions Affected by Alzheimer's Disease and Normal Aging," *NeuroImage* 49, no. 1 (2010): 169–176.

51. L. Jones, P. A. Holmans, M. L. Hamshere, D. Harold, V. Moskvina, D. Ivanov, . . . J. Williams, "Genetic Evidence Implicates the Immune System and Cholesterol Metabolism in the Aetiology of Alzheimer's Disease," *PLoS One* 15, no. 5(11) (2011): e13950.

52. P. B. Gorelick, "Role of Inflammation in Cognitive Impairment: Results of Observational Epidemiological Studies and Clinical Trials," *Annals of the New York Academy of Sciences* 1207 (2010): 155–162.

53. R. L. Buckner, "Memory and Executive Function in Aging and AD: Multiple Factors That Cause Decline and Reserve Factors That Compensate," *Neuron* 44 (2004): 195–208.

54. B. Bulic, M. Pickhardy, and E. Mandelkow, "Progress and Development in Tau Aggregation Inhibitors in Alzheimer's Disease," *Journal of Medicinal Chemistry* 56 (2013): 4135–4155.

55. G. Bu, "Apolipoprotein E and Its Receptors in Alzheimer's Disease: Pathways, Pathogenesis and Therapy," *Nature Reviews Neuroscience* 10 (2009): 335–344.

56. R. S. Doody, S. I. Gavrilova, M. Sano, and R. G. Thomas, "Effect of Dimebon on Cognition, Activities of Daily Living, Behaviour, and Global Function in Patients with Mild-to-Moderate Alzheimer's Disease: A Randomised, Double-Blind, Placebo-Controlled Study," *Lancet* 372 (2008): 207–215.

57. V. Krishnan, and E. J. Nestler, "The Molecular Neurobiology of Depression," *Nature* 455 (2008): 894–902.

58. R. M. Sapolsky, L. C. Krey, and B. S. McEwen, "The Neuroendocrinology of Stress and Aging: The Glucocorticoid Cascade Hypothesis," *Endocrine Reviews* 7 (1986): 284–301.

59. A. L. Byers, and K. Yaffe, "Depression and Risk of Developing Dementia," *Nature Reviews Neurology* 7 (2011): 323–331.

60. V. W. Henderson, "Alzheimer's Disease: Review of Hormone Therapy Trials and Implications for Treatment and Prevention After Menopause," *Journal of Steroid Biochemistry and Molecular Biology* 142C (2013): 99–106.

61. J. Holland, S. Bandelow, and E. Hogervorst, "Testosterone Levels and Cognition in Elderly Men," *Maturitas* 69 (2010): 322–327.

62. E. Rönnemaa, B. Zethelius, J. Sundelöf, J. Sundström, M. J. Degerman-Gunnarsson, L. Lannfelt, . . . L. Kilander, "Glucose Metabolism and the Risk of Alzheimer's Disease and Dementia: A Population-Based 12 Year Follow-Up Study in 71-Year-Old Men," *Diabetologia* 52 (2009): 1504–1510.

63. B. Cholerton, L. D. Baker, E. H. Trittschuh, P. K. Crane, E. B. Larson, M. Arbuckle, and S. Craft, "Insulin and Sex Interact in Older Adults with Mild Cognitive Impairment," *Journal of Alzheimer's Disease* 31 (2012): 401–410.

64. Y-H. Rhee, M. Choi, H.-S. Lee, C. H. Park, S. M. Kim, . . . S. H. Lee, "Insulin Concentration Is Critical in Culturing Human Neural Stem Cells and Neurons," *Cell Death and Disease* 4 (2013): e766.

65. C. McMillen, and J. S. Robinson, "Developmental Origins of the Metabolic Syndrome: Prediction, Plasticity, and Programming," *Physiological Reviews* (2004): 571–633.

66. D. Sitzer, E. Twamley, and D. Jeste, "Cognitive Training in Alzheimer's Disease: A Meta-Analysis of the Literature," *Acta Psychiatrica Scandinavica* 114 (2006): 75–90.

67. P. Heyn, B. Abreu, and K. Ottenbacher, "The Effects of Exercise Training on Elderly Persons with Cognitive Impairment and Dementia: A Meta-Analysis," *Archives of Physical Medicine and Rehabilitation* 85 (2004): 1694–1704.

68. D. E. Barnes, W. Santos-Modesitt, G. Poelke, A. F. Kramer, C. Castro, L. E. Middleton, and K. Yaffe, "The Mental Activity and eXercise (MAX) Trial: A Randomized Controlled Trial to Enhance Cognitive Function in Older Adults," *Journal of the American Medical Association Internal Medicine* 173 (2013): 797–804.

69. R. Brookmeyer, S. Gray, and C. Kawas, "Projections of Alzheimer's Disease in the United States and the Public Health Impact of Delaying Disease Onset," *American Journal of Public Health* 88 (1998): 1337–1342.

70. C. Qiu, E. von Strauss, L. Backman, B. Winblad, and L. Fratiglioni, "Twenty-Year Changes in Dementia Occurrence Suggest Decreasing Incidence in Central Stockholm, Sweden," *Neurology* 80 (2013): 1888–1894.

71. F. E. Matthews, A. Arthur, L. E. Barnes, J. Bond, C. Jagger, L. Robinson, . . . C. Brayne, "A Two-Decade Comparison of Prevalence of Dementia in Individuals Aged 65 Years and Older from Three Geographical Areas of England: Results of the Cognitive Function and Ageing Study I and II," *Lancet* 382, no. 9902 (2013): 1405–1412.

72. K. Ritchie, I. Carrie're, C. W. Ritchie, C. Berr, S. Artero, and M. Ancelin, "Can We Design Prevention Programs to Reduce Dementia Incidence? A Prospective Study of Modifiable Risk Factors," *British Medical Journal* 5, no. 341 (2010): c3885.

73. D. Wilson, R. Peters, K. Ritchie, and C. W. Ritchie, "Latest Advances on Interventions That May Prevent, Delay or Ameliorate Dementia," *Therapeutic Advances in Chronic Disease* 2 (2011): 161–173.

74. L. J. Whalley, and K. A. Smyth, "Human Culture and the Future Dementia Epidemic: Crisis or Crossroads?" *Neurology* 80 (2013): 1824–1825.

75. J. W. Lynch, G. A. Kaplan, R. D. Cohen, J. Tuomilehto, and J. T. Salonen, "Do Cardiovascular Risk Factors Explain the Relation Between Socioeconomic Status, Risk of All Cause Mortality, Cardiovascular Mortality and Acute Myocardial Infarction?" *American Journal of Epidemiology* 144, no. 10 (1996): 934–942.

76. R. T. Staff, A. D. Murray, T. S. Ahearn, N. Mustafa, H. C. Fox, and L. J. Whalley, "Childhood Socioeconomic Status and Adult Brain Size," *Annals of Neurology* 71 (2012): 653–660.

77. J. W. Williams, B. L. Plassman, J. Burke, T. Holsinger, and S. Benjamin, "Preventing Alzheimer's Disease and Cognitive Decline." Evidence Report/Technology Assessment No. 193. Prepared by the Duke Evidence-based Practice Center under Contract No. HHSA 290-2007-10066-I. AHRQ Publication No. 10-E005 (Rockville, MD: Agency for Healthcare Research and Quality, 2010).

78. T. A. Salthouse, *Major Issues in Cognitive Aging* (Oxford: Oxford University Press), 127–167.

12. DEMENTIA RISK REDUCTION, 3: MULTIDOMAIN APPROACHES

1. Z. S. Khachaturian, R. C. Petersen, S. Gauthier, N. Buckholtz, J. P. Corey-Bloom, B. Evans, H. Fillit, . . . J. Touchon, "A Roadmap for the Prevention of Dementia II: The Inaugural Leon Thal Symposium," *Alzheimer's & Dementia* 4, no. 3 (2008): 156–163.

2. F. Mangialasche, A. Solomon, B. Winblad, P. Mecocci, M. Kivipelto, "Alzheimer's Disease: Clinical Trials and Drug Development," *Lancet Neurology* 9, no. 3 (2010): 702–716.

3. R. A. Sperling, P. S. Aisen, L. A. Beckett, D. A. Bennett, S. Craft, A. M. Fagan, . . . C. H. Phelps, "Toward Defining the Preclinical Stages of Alzheimer's Disease: Recommendations from the National Institute on Aging–Alzheimer's Association Workgroups on Diagnostic Guidelines for Alzheimer's Disease," *Alzheimer's & Dementia* 7, no. 3 (2011): 280–292.

4. J. W. Williams, B. L. Plassman, J. Burke, T. Holsinger, and S. Benjamin, "Preventing Alzheimer's Disease and Cognitive Decline." Evidence Report/Technology Assessment No. 193. Prepared by the Duke Evidence-based Practice Center under Contract No. HHSA 290-2007-10066-I. AHRQ Publication No. 10-E005 (Rockville, MD: Agency for Healthcare Research and Quality, 2010).

5. F. Forett, M. L. Seux, J. A. Staessen, L. Thijs, W. H. Birkenhager, M. R. Babarskiene, "Prevention of Dementia in Randomised Double-Blind Placebo Controlled Systolic Hypertension in Europe (Syst-Eur) Trial," *Lancet* 352, no. 9137 (1998): 1347–1351.

6. National Research Council, Committee on Future Directions for Cognitive Research on Aging, Commission on Behavioral and Social Sciences and Education, *The Aging Mind: Opportunities in Cognitive Research*, ed. Paul C. Stern and Laura L. Carstensen (Washington, DC: National Academies Press, 2000).

7. A. D. Dangour, P. J. Whitehouse, K. Rafferty, S. A. Mitchell, L. Smith, S. Hawkesworth, and B. Vellas, "B-Vitamins and Fatty Acids in the Prevention and Treatment of Alzheimer's Disease and Dementia: A Systematic Review," *Journal of Alzheimer's Disease* 22, no. 1 (2010): 205–224.

8. E. Reynolds, "Vitamin B12, Folic Acid, and the Nervous System," *Lancet Neurology* 5, no. 11 (2006): 949–960.

9. S. J. Duthie, L. J. Whalley, A. R. Collins, S. Leaper, K. Berger, and I. J. Deary, "Homocysteine, B Vitamin Status, and Cognitive Function in the Elderly," *American Journal of Clinical Nutrition* 75, no. 5 (2002): 908–913.

10. J. A. Luchsinger, M.-X. Tang, S. Shea, J. Miller, J. Green, and R. Mayeux, "Plasma Homocysteine Levels and Risk of Alzheimer Disease," *Neurology* 62, no. 11 (2004): 1972–1976.

11. L. J. Whalley, S. J. Duthie, A. R. Collins, J. M. Starr, I. J. Deary, H. Lemmon, . . . R. T. Staff, "Homocysteine, Antioxidant Micronutrients and Late Onset Dementia," *European Journal of Nutrition* 53, no. 1 (2013): 277–285.

12. J. Durga, M. P. J. van Boxtel, E. G. Schouten, F. J. Kok, J. Jolles, M. B. Katan, and P. Verhoef, "Effect of 3-Year Folic Acid Supplementation on Cognitive Function in Older Adults in the FACIT Trial: A Randomised, Double Blind, Controlled Trial," *Lancet* 369, no. 9557 (2007): 208–216.

13. C. A. de Jager, A. Oulhaj, R. Jacoby, H. Refsum, and A. D. Smith, "Cognitive and Clinical Outcomes of Homocysteine-Lowering B-Vitamin Treatment in Mild Cognitive Impairment: A Randomized Controlled Trial," *International Journal of Geriatric Psychiatry* 27 (2012): 592–600.

14. I. Denis, B. Potier, S. Vancassel, C. Heberden, and M. Lavialle, "Omega-3 Fatty Acids and Brain Resistance to Ageing and Stress: Body of Evidence and Possible Mechanisms," *Ageing Research Reviews* 12 (2013): 579–594.

15. L. J. Whalley, H. C. Fox, K. W. Wahle, J. M. Starr, I. J. Deary, "Cognitive Aging, Childhood Intelligence, and the Use of Food Supplements: Possible Involvement of n-3 Fatty Acids," *American Journal of Clinical Nutrition* 80, no. 6 (2004): 1650–1657.

16. L. J. Whalley, I. J. Deary, J. M. Starr, K. W. Wahle, K. A. Rance, V. J. Bourne, and H. C. Fox, "n-3 Fatty Acid Erythrocyte Membrane Content, APOE Varepsilon4, and Cognitive Variation: An Observational Follow-Up Study in Late Adulthood," *American Journal of Clinical Nutrition* 87, no. 2 (2008): 449–454.

17. D. Harman, "Aging: A Theory Based on Free Radical and Radiation Chemistry," *Journal of Gerontology* 11 (1956): 298–300.

18. S. Melov, J. Ravenscroft, S. Malik, M. S. Gill, D. W. Walker, P. E. Clayton, . . . G. J. Lithgow, "Extension of Life-Span with Superoxide Dismutase/Catalase Mimetics," *Science* 289, no. 5484 (2000): 1567–1569.

19. A. Smith, R. Clark, and D. Nutt, "Anti-oxidant Vitamins and Mental Performance of the Elderly," *Human Psychopharmacology Clinical and Experimental* 14 (1999): 459–471.

20. J. H. Kang, N. Cook, and J. Manson, "A Randomized Trial of Vitamin E Supplementation and Cognitive Function in Women," *Archives of Internal Medicine* 166, no. 22 (2006): 2462–2468.

21. G. A. McNeill, A. Avenell, M. K. Campbell, J. A. Cook, P. C. Hannaford, . . . L. D. Vale, "Effect of Multivitamin and Multimineral Supplement on Cognitive Function in Men and Women Aged 65 Years and Over: A Randomised Controlled Trial," *Nutrition Journal* 6 (2007): 10.

22. J. H. Kang, N. R. Cook, J. E. Manson, J. E. Buring, J. E. Albert, and F. Grodstein, "Vitamin E, Vitamin C, Beta Carotene, and Cognitive Function Among Women with or at Risk of Cardiovascular Disease: The Women's Antioxidant and Cardiovascular Study," *Circulation* 119, no. 21 (2009): 2772–2780.

23. I. Denis, B. Potier, S. Vancassel, C. Heberden, and M. Lavialle, "Omega-3 Fatty Acids and Brain Resistance to Ageing and Stress: Body of Evidence and Possible Mechanisms," *Ageing Research Reviews* 12, no. 2 (2013): 579–594.

24. L. J. Whalley, I. J. Deary, J. M. Starr, K. W. Wahle, K. A. Rance, V. J. Bourne, and H. C. Fox, "n-3 Fatty Acid Erythrocyte Membrane Content, APOE Varepsilon4, and Cognitive Variation: An Observational Follow-Up Study in Late Adulthood," *American Journal of Clinical Nutrition* 87, no. 2 (2008): 449–454.

25. L. J. Whalley, H. C. Fox, H. A. Lemmon, S. J. Duthie, A. R. Collins, H. Peace, . . . I. J. Deary, "Dietary Supplement Use in Old Age: Associations with Childhood IQ, Current Cognition and Health," *International Journal Geriatric Psychiatry* 18, no. 9: 769–776.

26. F. Chollet, J. Tardy, J.-F. Albucher, C. Thalamas, E. Berard, C. Lamy, . . . I. Loubinoux, "Fluoxetine for Motor Recovery After Acute Ischaemic Stroke (FLAME): A Randomised Placebo-Controlled Trial," *Lancet Neurology* 10, no. 2 (2011): 123–130.

27. M. Aboukhatwa, L. Dosanjh, and Y. Luo, "Antidepressants Are a Rational Complementary Therapy for the Treatment of Alzheimer's Disease," *Molecular Neurodegeneration* 5 (2010): 10.

28. W. Chadwick, N. Mitchell, J. Caroll, Y. Zhou, S. S. Park, L. Wang,. . . S. Maudsley, "Amitriptyline-Mediated Cognitive Enhancement in Aged 3xtg Alzheimer's Disease Mice Is Associated with Neurogenesis and Neurotrophic Activity," *PLoS ONE* 6 (2011): e21660.

29. X. Han, J. Tong, J. Zhang, A. Faravah, E. Wang, J. Yang, . . . J. J. Huang, "Imipramine Treatment Improves Cognitive Outcome Associated with Enhanced Hippocampal Neurogenesis After Traumatic Brain Injury in Mice," *Journal of Neurotrauma* 28, no. 6 (2011): 995–1007.

30. I. Lourida, M. Soni, J. Thompson-Coon, N. Purandare, I. A. Lang, O. C. Ukoumunne, and D. J. Llewellyn, "Mediterranean Diet, Cognitive Function, and Dementia: A Systematic Review," *Epidemiology* 24 (2013): 479–489.

31. N. Scarmeas, J. A. Luchsinger, N. Schupf, A. M. Brickman, S. Cosentino, M. X. Yang, and Y. Stern, "Physical Activity, Diet, and Risk of Alzheimer Disease," *Journal of the American Medical Association* 302 (2009): 627–637.

32. N. Scarmeas, Y. Stern, R. Mayeux, J. J. Manly, N. Schupf, and J. A. Luchsinger, "Mediterranean Diet and Mild Cognitive Impairment," *Archives of Neurology* 66 (2009): 216–225.

33. C. Feart, C. Samieri, V. Rondeau, H. Amieva, F. Portet, J. F. Dartigues, . . . P. Barberger-Gateau, "Adherence to a Mediterranean Diet, Cognitive Decline, and Risk of Dementia," *Journal of the American Medical Association* 302 (2009): 638–648.

34. T. F. Hughes, R. Andel, B. J. Small, A. R. Borenstein, J. A. Mortimer, A. Wolk, . . . M. Gatz, "Midlife Fruit and Vegetable Consumption and Risk of Dementia in Later Life in Swedish Twins," *American Journal of Geriatric Psychiatry* 18, no. 5 (2010): 413–420.

35. M. Kivipelto, A. Solomon, S. Ahtiluoto, S. Ahtiluoto, T. Ngandu, J. Lehtisalo, . . . H. Soinenen, "The Finnish Geriatric Intervention Study to Prevent Cognitive Impairment and Disability (FINGER): Study Design and Progress," *Alzheimer's & Dementia* 9, no. 6 (2013): 657–665.

INDEX

Aβ42, 142

Aberdeen 1921 and 1936 Birth Cohorts Study, 187–188, 339, 340, 347

acetylcholine, 136–137

acetylcholinesterase inhibitors, 277

actors, older adults as, 219–222

AD. *See* Alzheimer's disease

adaptation to changing environments, 85–86

adolescence, illicit drug use in, 300

adrenal cortex, secretion of cortisol by, 131

advanced glycation end products (AGEs), 133

adversity, effects of, 22–23, 32, 290

AEDs (antiepileptic drugs), 201–202

affective theory of mind, 230

AGEMAP project, 102

AGEs (advanced glycation end products), 133

age stereotypes, 218

aging: focus on biology of, 151–158; nature of, 108–112; theories, 151, 152

aging mind. *See* brain aging

Allen Human Brain Atlas Project, 100, 112

Alzheimer, Alois, 135

Alzheimer's disease (AD): abnormal synaptic remodeling, 90; acetylcholinesterase inhibitor drugs for, 137–139; amyloid in diagnosis of, 117–118; animal studies, 80–83; *APOE* as risk factor, 157–158; association with Down syndrome, 154, 249–250; β-amyloid research, 141–146; blood flow studies, 260; and brain injury, 251–252; causes of death, 248–249;

clinical genetics of, 154; course of illness, 248; descriptions of, 140; diagnosis of, 234–235, 237–238; as distinct diagnostic category, 135; early diagnosis of, 252–254, 259; evolutionary perspective, 78–80; family history of, 246–247; gene mutations linked to, 142, 143; gene therapies for, 294–295; genetic causes, 78; genetics research, 139; implantation of neural stem cells, 127; inflammatory hypothesis of AD, 315–318; as leading cause of death, 137; maternal transmission of, 155; and MCI, 247–248; metabolic hypothesis of, 318–319; molecular pathology, 11–12, 14; MRI scans, 258; multifactorial nature of, 276–279; NSAIDs and, 317; number of cases, 244–245; patterns of brain cell loss in, 64–65; protein aggregation, 104; protein synthesis, 151–153; reactivation of developmental programs in course of, 91–103; regenerative neurology, 15; selective neuronal loss in, 140–141; SP, 115, 116; and stroke, 252; subcategories of, 134; symptoms of, 252–254; synapse loss, 160; tau proteins, 147–150; vascular disease and, 150; vulnerability of evolved neural networks to, 86–90. *See also* amyloid hypothesis of Alzheimer's disease; neurofibrillary tangles; neurovascular hypothesis of Alzheimer's disease

amyloid: amyloid-binding compounds, 117–118; diagnosis of AD, 115, 117; imaging, 117; late-onset dementias, 262; plaque, 116, 141 amyloid hypothesis of Alzheimer's disease: biology of, 142–145; criticism of, 148–149; dementia prevention, 262, 319–321; discussion of, 78–79

amyloid precursor protein. *See* APP

anatomy of emotion, 215

anesthesia-related dementia, 309–310, 311–312

animals: embryology of vertebrates, 91; gene therapy studies, 293–294; lateralization of brain functions, 87–88; studies on AD and brain aging, 80–83

antagonistic pleiotropy, 42, 102

antiamyloid therapy, 320

anticholinesterase drugs, 137–139

antidementia drugs, 248, 276

antidementia interventions, challenges in, 2

antidepressant drugs, 348

antiepileptic drugs (AEDs), 201–202

antioxidants, 127–128, 342–344

antitau aggregation drugs, 321

APOE (apolipoprotein E), 142; *APOEε4*, 158, 245, 246; cholesterol metabolism, 318–319; development of pharmacological approaches to AD based on, 321–322; discussion of, 157–158

apoptosis, 30

APP (amyloid precursor protein): dementia prevention, 319–320; discussion of, 142, 143–144; familial AD, 156; functions of, 90; genealogical studies, 139; testing known carriers of, 284

appearance of aging brain, 112–118

Arendt, Tom, 90, 250

Aricept (donepezil), 139

association studies, limitations of, 8–9

α-synuclein, 265, 267

athletes, brain injury in, 251–252, 354

attention: and balance, 174; divided, 180; driving skills, 172–173; information processing theory, 179–180; selective, 179–180; sustained, 180–181; theater metaphor, 177–179

Audrey Brown story, 169–170

Australian Imaging Biomarkers and Lifestyle Research Group, 262–263

autoregulation of cerebral blood flow (CBF), 122, 123–124

axons, 48, 49

Baer's Laws of Embryology, 98

Bak, Thomas, 302–303

balance, effects of aging on, 173–174

Baltes, Paul, 194

β-amyloid plaques: in AD, 81–82, 104; in animals, 81, 83; Arne Brun's findings, 140; biology of dementias, 141–146; criticism of amyloid hypothesis of AD, 148–149; diagnosis of dementia syndromes, 237; gene therapies, 295; historically linked to AD, 136; in normal brain, 279; possible dementia treatments, 262; production of, 319–320

Barker, David, 32

Bateman, Randall, 146

BBB (blood–brain barrier), 306–307

Beck, Harry, 11

behavior, and brain aging, 126–128

Bennett, David, 117, 175–176

Benson, Deanna, 65

Berkman, Lisa, 207

Beyreuther, Konrad, 142

Bialystok, Ellen, 299, 302

bilingualism, 299, 302–303

biological determinism, 9

biological systems, understanding of, 11–12

biology: of aging mind and brain, 13–16; of critical periods, 29–30; plausibility of risk reduction strategies, 283–287; programmed development, 31

biology of dementias: β-amyloid plaques, 141–146; focus on biology of aging, 151–158; historical trends, 134–141; neural network breakdown, 158–161; tau proteins, 147–149; vascular cognitive impairment, 149–151

biomarkers, identifying, 277

Birns, Jonathan, 307

Blane, D., 39

Blessed, Gary, 136

Bleuler, Paul Eugen, 163–164

blood–brain barrier (BBB), 306–307

blood flow studies, 260

blood pressure: and cognitive impairment, 307–308; control of hypertension, 123, 124; fetal growth and, 24

blood vessels: balance of cognitive reserve with blood–brain vessel disease, 278–279; brain aging, 121–125; dementia associated with diseases of, 270–271; diffuse white matter disease, 270–271; HS, 271; secondary dementias, 243; subcortical dementia, 270; VCI and dementia, 149–151. *See also* neurovascular hypothesis of Alzheimer's disease

body weight and reproductive status, 23

bottom-up attention, 177

Bouchard, Denis, 88

boundaries between dementia syndromes, 240–242

bovine spongioform encephalopathy (BSE), 156, 244, 354

Boyce, Thomas, 7

Braak, E., 89

Braak, H., 89

brain: biology of, 13–16; cells, 2, 119–120; connectome, 14–15; cortisol system, 34; development, 67–69; effect of childhood stress on hippocampus, 41; energy use, 122; size and complexity, 66–67, 96; weight of, 113. *See also* brain aging; brain structure; cerebral cortex

brain activity hypothesis, 325–326

brain aging, 107–108; appearance of, 112–118; biology of, 13–16; blood vessels, 121–125; brain cells, 119–120; cortisol secretion, 130–132; developmental origins of, 38–40; emotional aging and, 214–215; evolution of neurological disease, 103–105; frontal lobes, 120–121; genes, diet, and behavior, 126–128; genetic mutations linked to, 102–103; in humans and animals, 80–83; information processing model, 15–16; interaction between emotional and cognitive aging, 336; major influences on cognitive performance, 347; in MRI scans, 255–256; nature of, 108–112; overall effects, 132–133; role of cognitive reserve, 16–17; stress responses and, 128–132; structural analyses

of, 58–59; structure and function related to, 62–65; volumetric studies, 118; vulnerability to effects of, 199–205

brain banks, 259

brain-blood vessel disease, 235, 270–271, 278–279

brain imaging: in early dementia, 255–263; future of, 258–259. *See also specific imaging techniques*

brain injury: and AD, 251–252; in athletes, 251–252, 354; and cognitive impairment, 204–205; study of language-related neural circuits after, 89

brain maps, 72–73

brain structure: brain development and complexity, 65–70; brain maps, 72–73; cognitive functions of cortical regions, 166; DTI, 61–62; effects of aging, 132–133; functional organization, 53–60; gross neuroanatomy, 45–48; and language, 70–72; locationist view of, 45; microscopic neuroanatomy, 48–51; neural plasticity and critical periods, 52–53; neurons and glial cells, 51; relevant to aging, 62–65

brain training, 348–349

Brayne, Carol, 260

breast milk, 77–78

British National Survey of Health and Development, 3

Broadbent, David, 179

Brody, Harold, 119

Brookmeyer, Ron, 327

Brun, Arne, 140

BSE (bovine spongioform encephalopathy), 156, 244, 354

Bulic, B., 321

Buschke, Herman, 186

Byers, Amy, 322

Cache County study, 289–290

CADASIL (cerebral autosomal dominant arteriopathy with subcortical infarcts and leukoencephalopathy), 150

Caenorhabditis elegans (roundworm), 57, 59, 104, 108

Canadian Study of Health and Aging, 298

cannabis, 300

caregivers, counseling for, 278

Carroll, Sean, 93–94, 101

Carstensen, Laura, 209, 210, 214–215

cataracts, 172

Cattell, Raymond, 181–182, 186

CBF (cerebral blood flow), autoregulation of, 122, 123–124

cells: apoptosis, 30; differentiation, 97, 99; division, regulation of, 29–30; glial, 51; implantation of neural stem, 127; pluripotent stem, 67, 69; programmed death, 30

centenarians, dementia in, 257–258

cerebral autosomal dominant arteriopathy with subcortical infarcts and leukoencephalopathy (CADASIL), 150

cerebral blood flow (CBF), autoregulation of, 122, 123–124

cerebral cortex: anatomy of, 47–50; cellular structure of, 15; diseases of, 15; effect of AD on, 258; functional organization in, 54–56; higher cognitive functions, 56–60; insula, 215; layers in, 50; quantifying complexity of cortical layers, 58–59; thickness of, 59

cerebral hemispheres, 46

cerebral infarcts, 256

Charles, Susan, 214–215

chickens, imprinting in, 27–28

childhood: brain development during, 67; and cognitive function, 41–42; early education and dementia prevention, 296–299, 301; early parental death, 289–290; education and cognitive reserve, 280; effect on adult brain structure, 59; emotional intelligence, 216–217; familial clustering of dementia risk, 287–289; fetal growth and adult blood pressure, 24; gene therapies for late-onset dementia, 293–295; illicit drug use in adolescence, 300; infections, effect on health in later life, 33–34; intelligence, 290–292; neural plasticity in, 52; origins of common adult diseases in, 7; Piaget's stages of cognitive development, 189–192; social epidemiology, 286–287; socioeconomic status, 19, 20, 39, 40, 290; stages of development, 26; stress during, 19, 20, 41

chimpanzees: historic relationship between humans and, 84–85; studies on AD and brain aging, 81–82, 83

Chinese famine, 39

cholesterol metabolism, 318–319, 321

cholinergic drugs, 137

cholinergic synapse, 138

cholinesterase inhibitors, 137–139, 267

chromosomal instability, 249–250

chromosome 21, 250

CJD (Creutzfeld–Jakob disease), 156

classification of dementia syndromes: late-onset dementias, 239–242; limitations of, 233–238

clinical trial design for dementia prevention, 331–332

cocktail party effect, 179

coding genes, 76

cognitive aging: attention, 176–181; influences on rates of, 175–176; intelligence, 181–188; interaction between emotional aging and, 336; language processing, 181; memory, 195–199; occupational complexity, 188–189; personality and, 205; Piaget's stages of cognitive development, 189–192; processing speed, 186; research on, 164–171; role of health status in, 192–193; SLS, 182–188; vulnerability to effects of, 199–205; wisdom, 193–194

cognitive development stages (Piaget), 189–192

cognitive functions: blood pressure and, 307–308; of cortical regions, 166; effect of early life experiences on, 41–42; impairments in Parkinson's disease, 268–269; organization of, 166–168

cognitive reserve, 16–17; balance with blood-brain vessel disease, 278–279; brain activity hypothesis, 326; components of, 282; discussion of, 279–280; education and, 280; reductionism, 283; social influences, 280–282; usefulness of concept, 282–283

cognitive theory of mind, 230

cognitive training, 326, 348–349

Colman, David, 65

connectome, 14–15, 60

contact sports, risk of brain injury from, 251–252

control engineering, programming in, 31

Corkin, Susan, 204

Corsellis, Nick, 237

cortex. *See* cerebral cortex

cortical microinfarcts, 125

cortisol system, 34, 35, 36, 130–132

Costa, Paul, 221–222

Craft, Suzanne, 324

Creutzfeld– Jakob disease (CJD), 156

Crimmins, Eileen, 33–34

critical periods, 27–31, 53, 337

crystallized (Gc) intelligence, 181–182, 185–187

cultural transmission of acquired adaptive behaviors, 76

cultures, variations in dementia incidence between, 299

cytokines, 317

data maps, 12

death: brain pathology of people dying without dementia, 260–261; early parental death, 289–290; of patients with AD, causes of, 248–249

deep white matter lacunes, 125

defense mechanisms, 223–224

delays to onset, 351

delirium, 308–312

dementia pugilistica, 251

dementias: anesthesia and risk of, 311–312; brain pathology of people dying without, 260–261; delays to onset, 351; dietary changes and increased risk of, 77–78; empathy disrupted by, 56–57; epidemic of, 274–276; future concerns, 352–356; hereditary or familial, 155–157; neural networks related to, 58; patterns of brain cell loss in subtypes of, 64–65; postmortem studies, 260; protein synthesis, 151–153; reports of declining prevalence, 304–306; role of cognitive reserve, 16–17; role of epigenetics, 3; spatial patterning of, 160–161. *See also* biology of dementias; risk of dementia, reducing

dementia syndromes: associated with brain blood vessel disease, 270–271; brain imaging in early dementia, 255–263; causes of, 242–244; classification of, 233–238; classification of late-onset dementias, 239–242; frontotemporal, 263–264; major subdivisions, 243; overlap of, 240–242; Parkinson's disease with dementia, 265–269. *See also* Alzheimer's disease

dementia with Lewy bodies (DLB), 239; discussion of, 266–267; drugs for, 267; early diagnosis of, 259; lewy bodies, 267; synucleins, 265

dendrites, 48, 49, 121

dendritic spines, cognition linked to, 159–160

depression, risk of dementia related to, 321–323

determinism, biological, 9

developmental origins of aging and adult disease, 7, 38–40

developmental stages, 24–26

DHA (docosahexaenoic acid), 341

diabetes, 314–315

diagnosis: of AD, 117–118, 237–238; of dementia, 234; early diagnosis of AD, 252–254, 259

Diagnostic and Statistical Manual (DSM), 234

Dickinson, Alan, 244

dietary habits: association studies, 8; and brain aging, 126–128; changes during Industrial Revolution, 85–86; increased risk of dementia, 77–78; Mediterranean diet, 349–351; role on brain development and aging, 17. *See also* nutrition

dietary supplements, 337–338, 345

diffuse gliosis, 125

diffuse white matter demyelinations, 125

diffuse white matter disease, 270–271

diffusion tensor imaging (DTI), 61–62

disconnected mind: attention, 176–181; cognitive aging, 164–171; effect of aging on mental life, 163–164; influences on rates of cognitive aging, 175–176; information processing, 181; memory, 195–199; personality and cognitive aging, 205; sensory systems, 171–175; vulnerability to effects of brain aging, 199–205. *See also* general mental ability

disease, developmental origins of, 38–40

displaying emotions, 212

divided attention, 180

DLB. *See* dementia with Lewy bodies

DNA: critical periods, 30; human genome project, 10; noncoding, 10

Dobzhansky, Theodosius, 75

docosahexaenoic acid (DHA), 341

Domazet-Los, Tomislav, 103

dominant hemispheres, 63

donepezil (Aricept), 139

Down syndrome, 154, 201; association with AD, 249–250; gene therapies for AD, 295

driving, age-related visual limitations, 172–173

drug industry, 275

drugs: anticholinesterase, 137–139; antidementia, 248, 276; antidepressants, 348; antiepileptic, 201–202; antitau aggregation, 321; in development for treatment and prevention, 320; illicit use of, 300, 354; L-Dopa, 136–137, 265; neuroleptics, 267; NSAIDs, 317

DSM (Diagnostic and Statistical Manual), 234

DTI (diffusion tensor imaging), 61–62

Durga, Jane, 340

Dutch famine birth cohort study, 39

E2 (estrogen) deficiency and dementia risk, 322

early life dementia prevention opportunities: childhood intelligence, 290–292; early education, 296–299, 301; early parental death, 289–290; familial clustering of dementia risk, 287–289; gene therapies for late-onset dementia, 293–295; illicit drug use in adolescence, 300

early life experiences. *See* childhood

early onset Alzheimer's disease (EOAD): in families, 238; gene mutations, 135; genetic transmission, 154

ecstasy, 300

education, 4, 59; early education and dementia prevention, 296–299, 301; effect on cognitive reserve, 280; familial clustering of dementia risk, 288; post–WWII view of, 304–305

embryology, 91, 97, 98

emotional aging: aging brain and, 214–215; anatomy of emotion, 215; cognitive reserve, 280–282; emotional experts, 211–212; emotional intelligence, 216–218; emotional life, 208–211; emotional states, 213; interaction between cognitive aging and, 336; life narratives and self-concept, 222–226; personality traits, 219–222; social integration, 207; social support, cohesion and pain, 226–228; time perspective, 228–231

emotional intelligence, 216–218

Emotional Intelligence (Goleman), 216

emotions, 209–211; control of, 211–212; displaying, 212; physical components of, 213; recognizing in others, 212

empathy, 56–57

Emx2, 68

Engel de Abreu, Pascale, 302

EOAD. *See* early onset Alzheimer's disease

epidemic of dementia, 274–276

epigenetic modifications of noncoding genes, 76

epigenetics, 3, 97–99

epilepsy, 201–202

episodic memory, 196

Erber, Joan, 218

Erikson, Erik, 225–226

Ertel, Karen, 207

essential fatty acids, 86, 317, 318, 341–342, 344

estrogen (E2) deficiency and dementia risk, 322

evo-devo, 98

evolution, 5; AD from evolutionary perspective, 78–83; human, 83–86; of neurological disease, 103–105; principles of brain design, 95–96; processing spatial information, 10; reactivation of developmental programs in course of AD, 91–103; recent dietary changes, 77–78; transmission of information through generations, 75–80; vulnerability of evolved neural networks to AD, 86–90

evolutionary biology, 94–95

evolutionary gerontology, 92–93

Exelon (rivastigmine), 139

exercise, 128, 349–350

experiences, long-term outcomes of, 5–6

explicit memory, 196

external structure of brain, 46

extrinsic aging, 109

FAD. *See* familial Alzheimer's disease

falling, fear of, 174

familial Alzheimer's disease (FAD), 155–157, 238, 293

family: familial clustering of dementia risk, 287–289; history of AD, 246–247; history of PD, 266; reproductive success of, 76

Farrer, Lindsay, 246–247

fear of falling, 174

fetal development: and adult blood pressure, 24; brain development, 67–68; cell differentiation, 97, 99; effect of undernutrition on adult health, 35, 38–40; fetal origins of adult disease hypothesis, 38–40, 337; maternal transmission of AD, 155

fight-or-flight reaction, 128, 131

Finch, Caleb, 33–34

fish oils, 86, 318, 341–342, 344

five-factor model, 221–222

fluid (Gf) intelligence, 181–182, 185–187

Flynn, James, 305

Flynn effect, 305

fMRI (functional magnetic resonance imaging), 15, 60, 214

focal gliosis, 125

folate, 338–341

Folkman, Judah, 122

formal operations, Piaget's stage of, 190–191, 193

Fratiglioni, Laura, 207, 314

free radical theory of aging, 342–343

Freud, Sigmund, 25, 223–224

frontal lobes, 54, 120–121

frontotemporal dementia (FTD), 239, 259, 263–264

functional magnetic resonance imaging (fMRI), 15, 60, 214

galantamine (Reminyl), 139

Gatz, Margaret, 154, 296

Gc (crystallized) intelligence, 181–182, 185–187

geese, imprinting in, 29

Gems, David, 92–93

gene mutations: causing AD, 142, 143, 146; early onset AD, 135; FAD, 238; linked to aging, 102

general mental ability: categories of, 181–182; occupational complexity, 188–189; Piaget's stages of cognitive development, 189–192; role of health status in cognitive aging, 192–193; SLS, 182–188; wisdom, 193–194

genes: brain aging, 126–128; causing AD, 78; controlling head size, 86–90; epigenetics, 97; linking to brain functions, 100; MAPT, 147, 264; measuring changes in expression of, 121; molecular mechanisms of aging and development, 94; mutant, 14; PD risk, 269; PGRN, 148, 264; pleiotropy, 101–102; Sonic hedgehog, 101; transmission of information through, 75–76; tree of life, 103. *See also* gene mutations; presenilin genes

gene therapies for late-onset dementia, 293–295

genetic polymorphisms, 105–106

genetic programs, 102

genetic regulatory networks (GRNs), 111

genetic screening, 355

genetic testing for hereditary dementia, 156

genome-wide association studies (GWAS), 105–106, 154

Genomics and Personalized Medicine Act, 74

Gf (fluid) intelligence, 181–182, 185–187

Glen, Iain, 137

Glenner, George, 141

glial cells, 51

Glucocorticoid Cascade Hypothesis of Aging, 322

Glymour, Maria, 207

golden generation, 275

Goleman, Daniel, 216

Golgi, Camillo, 50

goslings, imprinting in, 29

grandmothers, effect of nutritional status on following generations, 23

Grant, William, 126–127

graph theory, 59

gray matter, 48
GRNs (genetic regulatory networks), 111
gross neuroanatomy, 45–48, 112–113
growth factors: nerve, 322; vascular, 122–123
guided participation, 281
GWAS (genome-wide association studies), 105–106, 154
gyri, 46, 112

Harman, Denham, 342–343
harmful environmental agents, exposure to, 30–31
Hart, C., 39
HD (Huntington's disease), 64–65, 153, 270, 327
head size, 86–90, 96
health: affected by childhood stress, 19, 20; of golden generation, 275; inequalities due to poverty, 18–19; IQ scores and view of, 291; role in cognitive aging, 192–193
Health and Retirement Study, 207
health food paradox, 345
Healthy Life Expectancy Index, 275
hearing loss, 173
heart disease, 33
Henderson, Victor, 322
hereditary dementia, 155–157, 238
Heston, Len, 139
higher cognitive functions, 56–60, 88, 105, 120
hippocampal sclerosis (HS), 271
hippocampus: effect of childhood stress on, 41; loss in volume with AD, 256–257
Hogervorst, Eva, 323
Hole, D., 39
Holland, effects of famine in, 38–39
home visiting for disadvantaged children, 42
homocysteine hypothesis of AD, 338–341
homocysyteine reduction, 338–341, 344
Hongerwinter, 38–39
Honolulu Heart Program, 126
hormone replacement therapy, 323
hormones, 128
Horn, Gabriel, 27–28
Horn, John, 181–182, 186
Hornykiewicz, Oleh, 136–137

HS (hippocampal sclerosis), 271
Hubel, David, 29
human breast milk, 77–78
Human Connectome Map, 60
human evolution, 83–86
human genome project, 10
human new variant CJD (nvCJD), 244
Huntington's disease (HD), 64–65, 153, 270, 327
Huntley, George, 65
hypertension, 124

illegitimacy of children, 288–289
illicit drug use, 300, 354
implantation of neural stem cells, 127
implicit memory, 196
individual differences in brain aging and dementia, 5
Industrial Revolution, dietary changes during, 85–86
infections: effect on health in later life, 33–34; secondary dementias, 244
inflammatory hypothesis of AD, 315–318
information processing model of brain aging, 15–16
information processing theory of attention, 179–180
inherited dementia, 155–157, 238
injury, brain, 204–205, 251–252
ink drop analogy, 61
innateness, concept of, 9
insula, 215
insulin–insulin growth factor signaling (ISS) pathway, 104
insulin levels and dementia risk, 324–325
intellectual disabilities, 201
intelligence: categories of, 181; childhood, 290–292; crystallized, 181–182, 185–187; emotional, 216–218; fluid, 181–182, 185–187; SLS, 182–188
intelligence quotient (IQ), 291–292, 301, 305, 335
intergenerational continuities, 288
internal structure of brain, 46
intimate relationships, 232
intrinsic aging, 109
Iourov, Ivan, 250

Iourov, Yuri, 250
IQ (intelligence quotient), 291–292, 301, 305, 335
Israeli Ischemic Heart Disease study, 289
ISS (insulin–insulin growth factor signaling) pathway, 104

Jablonka, Eva, 10, 31
Japanese-American men, Alzheimer's disease in, 126
Jarvik, Lissy, 295
Jiang, Jun, 295
job skills, 188–189
Johnson, M. H., 290
judgements, social, 217–218
Jung, Carl Gustav, 224–225

Kalaria, Raj, 122–123, 284
Kalra, Lalit, 307
Kaplan, George, 285–286
Katzman, Robert, 137, 296
Kay, David, 296
Klunk, William, 118
Kosik, Kenneth, 14

lacunar infarcts, 123, 124, 126
Lamb, Marion, 10, 31
Lambert, Jean-Claude, 154
language: bilingualism, 299, 302–303; clinical studies on, 71–72; evolutionary perspective, 88–89; processing, 181; semantic paraphasias, 72; speech processing, 70–71
late-onset Alzheimer's disease (LOAD), 135, 154, 158
late-onset dementias: childhood intelligence linked to risk for, 290–292; classification of, 239–242; early parental death linked to risk for, 289; gene therapies for, 293–295
lateralization of brain functions, 63, 87–88
L-Dopa, 136–137, 265
learning, in critical periods, 27–31
left hemisphere, 63
Lejeune, Jerome, 249
lesions, microvascular, 125, 150
Lewy, Frederick H., 267

lewy bodies, 267
life chart, 44
life course approach: Adolf Meyer, 21–24; clinical trial design for dementia prevention, 333–337; critical periods, 27–31; defined, 43; developmental origins of aging and adult disease, 38–40; developmental stages, 24–26; effect of early life experiences on cognitive function, 41–42; general discussion, 1–2; influential ideas, 18–24; observational studies, 7; poverty and mortality, 18–19; programmed development, 31–36; social class and health, 4; survivorship, 4; Terman Life-Cycle Study, 20–21; terminology, 42–44; three lives, 36–38
life cycle, 43
life expectancy, 43, 274–276, 288
life history, 43
Lifelines: Biology, Freedom, Determinism, 9
life narratives, 222–226
life span, 42
life span construct, 42
life span perspective, 42
life span psychology, 42
life story, 43
Lindsay, Joan, 298
LOAD. *See* late-onset Alzheimer's disease
locationist view of brain, 45
locus of control, 228–229
London underground analogy, 11, 345
long-term memory, 196
Lopera, Francisco, 156–157, 238, 284–285
Lorenz, Konrad, 29

macular degeneration, 172
mad cow disease, 156, 244, 354
Madge, Nicola, 287–288
magnetic resonance imaging (MRI): brain structure, 46; effects of brain aging, 113, 114; graph theory aligned with, 59–60; imaging for early dementia, 255; limitations of, 61; signs of AD, 256–257, 258; signs of normal aging, 255–256; WMHs, 124–125
mapping neuronal circuits, 57–58

MAPT (microtubule-associated protein tau) gene, 147, 264

Masters, Colin, 142

Mathis, Chester, 118

Matthews, Fiona, 260, 292

Mayeux, Dick, 288, 314, 338–339, 349

Maylor, Elizabeth, 174

McAdams, Dan, 219

McCrae, Robert, 221–222

McDowell, Ian, 298

McEwen, Bruce, 7, 322

MCI. *See* mild cognitive impairment

Medical Research Council's Neurosciences Board, 283–284

Mediterranean diet, 349–351

memantine, 138, 139

memory: changes with aging, 16; critical periods, 27–31; effect of aging on, 195–199; long-term, 196; memories, retrieval of, 325; self-rating of, 198–199; short-term, 196

men: insulin levels and dementia risk, 324; intimate relationships and health, 232; testosterone levels and dementia risk, 323; weight of brain, 113

mental ability, linked to risk for dementia, 290–292

mental exercise, benefits of, 3

mentalizing, 230

mental state attribution, 230

metabolic hypothesis of Alzheimer's disease, 318–319

metabolic reprogramming, 153

metabolic syndrome, 291, 293

methamphetamines, 300

Meyer, Adolf, 21–24

Meyer's Life Chart, 21–24

microarray technology, 121

microbleeds, 125, 256

microencephaly, 86, 87

microglia, 316, 317–318

micronutrients, 337–338

microscopic neuroanatomy, 48–51, 112–113

microtubule-associated protein tau (MAPT) gene, 147, 264

microvascular lesions, 125

middle-age adults: amyloid and related therapeutic approaches hypothesis, 319–321; brain activity hypothesis, 325–326; diabetes and dementia risk, 314–315; inflammatory hypothesis of AD, 315–318; metabolic hypothesis of AD, 318–319; midlife crisis, 25; positive aspects seen in, 25; postoperative delirium, 308–312; reports of declining dementia prevalence, 304–306; stress, depression, and role of growth factors in dementia risk, 321–325; stroke and dementia risk, 312–314

mild cognitive impairment (MCI), 200; and AD, 234, 247–248; caregivers, 278

mind reading, 230

MIRAGE (Multi-Institutional Research in Alzheimer's Genetic Epidemiology program), 246–247

mirror neurons, 88

mitochondrial dysfunction in postoperative delirium, 311–312

mitochondrial theory of aging, 343

molecular changes in aging brain, 111

molecular genetics, 15, 150

molecular mechanisms of aging and development, 92–94

molecular pathology, 11–12

Moreno, Jacob, 228

Mori, Susomi, 61

morphogenesis, 97

mosaic pleiotropy, 101

mothers. *See* women

motor cortex, 54

Mountcastle, Vernon, 69–70

MPTP (1-methyl-4-phenyl-1,2,3,6-tetrahydropyridine), 300

MRI. *See* magnetic resonance imaging

MS (multiple sclerosis), 202–203

multidomain approach to dementia prevention, 305–306; antioxidants, 342–344; background, 330–333; discussion of, 345–351; fish oils, 341–342; nutritional studies, 337–338, 344–345; randomized controlled clinical trials, 333–337; vitamin B12 and folate with homocysteine reduction, 338–341 multifactorial nature of Alzheimer's disease, 276–279

Multi-Institutional Research in Alzheimer's Genetic Epidemiology program (MIRAGE), 246–247

multiple sclerosis (MS), 202–203

mutant genes, 14

myelin, 48

myelination, 89, 90

National Institute of Health (NIH) Human Connectome Map, 60

nature of brain aging, 108–112

navigation, 9–10

Neill, David, 89, 90

neocortex, 56

nervous system: diagram of, 47; gross neuroanatomy, 45–48; microscopic neuroanatomy, 48–51; neural plasticity and critical periods, 52–53; neurons and glial cells, 51

networks, social, 226–227

neural networks: breakdowns and dementias, 158–161; disruptions in pathways, 182, 183; evolution and vulnerability to AD, 86–90; functional organization, 58, 59–60; maintenance of, 127

neural plasticity, 29, 52–53

neural stem cells, implantation of, 127

neuroanatomy: of aging brain, 112–118; of emotion, 214; gross, 45–48; microscopic, 48–51; neural plasticity and critical periods, 52–53; neurons and glial cells, 51

neuroangiogenesis, 122

neurofibrillary tangles (NFTs), 83, 89; diagnosis of AD, 115; diagnosis of dementia syndromes, 237; as distinguishing pathological feature of AD, 146; formation of, 115–116, 148–149; tau proteins, 147

neurogenesis, 51

neuroleptics, 267

neurological disease, evolution of, 103–105

neuronal networks, 51

neurons: columns of, 69–70; components of, 49; in cortex layers, 50; counting of, 119; general discussion, 51; loss of, 119–120; mirror, 88; neurogenesis, 51; selective neuronal loss in AD, 140–141; types of, 48; understanding connections between, 57–58

neuroregeneration, 348

neurotrophins, 64

neurovascular hypothesis of Alzheimer's disease, 123; and delirium, 308–312; and diabetes, 314–315; discussion of, 306–308; plausibility, 284; and stroke, 312–314

NFTs. See neurofibrillary tangles

NIH (National Institute of Health) Human Connectome Map, 60

noncoding DNA, 10

noncommunicable diseases, increase in, 76

nonpenetrating head injury, 251

nonsteroidal anti-inflammatory drugs (NSAIDs), 317

Norton, Maria, 289

Novartis Symposium on cognitive reserve, 282

NSAIDs (nonsteroidal anti-inflammatory drugs), 317

nuclei, 47

Nun Study, 270, 291

nutrition, 17; antioxidants, 342–344; and brain aging, 126–128; changes during Industrial Revolution, 85–86, 341; dietary changes and dementia, 77–78; effect of nutritional status of grandmother on following generations, 23; effects of undernutrition on fetus, 35; fish oils, 341–342; Mediterranean diet, 349–351; nutritional studies, 337–338, 344–345; omega-3 and omega-6 essential fatty acids, 317, 318, 341–342, 344; role on brain development and aging, 17, 38–39; vitamin B12 and folate with homocysteine reduction, 338–341

nutritional studies, 337–338, 344–345

nvCJD (human new variant CJD), 244

obesity, 337, 355

observational studies, 332, 334

occipital lobes, 54

occupational complexity, 188–189

ocular dominance, 53

oldest-old adults, dementia in, 257–258

omega-3 and omega-6 essential fatty acids, 317, 318, 341–342, 344

1-methyl-4-phenyl-1,2,3,6-tetrahydropyridine (MPTP), 300

openness to experience, 205, 281

OPTIMA project, 340–341

organization of cognitive functions, 166–168

pain, emotional perception of, 215

pain, social, 226

paired helical filaments (PHF-tau), 147

parents, early death of, 289–290

parietal lobes, 54

Parkinsonism, 265

Parkinson's disease: cognitive impairments in, 268–269; with dementia, 265–269, 270; DLB, 266–268; L-Dopa treatment, 136–137; patterns of brain cell loss in, 64–65; selective neuronal loss in, 140–141

Partridge, Linda, 92–93

paternal death in childhood, linked to dementia, 289

Pax6, 68

Pedersen, Nancy, 154, 186, 247

Pendlebury, Sarah, 313

penetrating head injury, 251

personality: and cognitive aging, 205; disorders, 219–221; openness to experience, 205, 281; resilience, 280–281; stability of traits, 219–222

personalized and preventive medicine, 12

PET (positron emission tomography) scans, 117–118, 214, 259

Petersen, Nancy, 326

PGRN (progranulin) gene, 148, 264

Phan, Luan, 214

PHF-tau (paired helical filaments), 147

physical activity, 128, 349–350

physostigmine, 137

Piaget, Jean, 189–192, 193

Pick's disease, 263

Pittsburgh Compound B (PiB), 117–118

plasticity, neural, 29, 52–53

pleiotropy, 101–102

pluripotent stem cells, 67, 69

population aging, 2

positron emission tomography (PET) scans, 117–118, 214, 259

possible selves, 223

postanesthetic dementia syndromes, 309–310, 311–312

postmortem studies, 113, 117, 118, 260

posttraumatic stress disorder, 131

poverty and poor health, 18–19

PoWs (prisoners of war), 203

prefrontal cortex: activation induced by emotional states, 214; functional organization, 54

pregnancy: breast milk, 77–78; effect of maternal cortisol and maternal psychosocial stress on fetus, 36; effect of stress on fetal cortisol system, 34; effect of undernutrition on fetus and adult health, 35, 38–40; factors that increase risk of heart disease and stroke, 33; harm to fetus during, 26; modern changes in reproductive cycle, 86

presenilin genes: in developing brain, 90; familial AD, 156; linked to AD, 142, 143, 145, 146; prevention testing, 284–285

prevention of dementia. See risk of dementia, reducing

prevention testing, 331–337

primary dementias, 243

primary prevention, 277

primates: historic relationship between humans and, 84–85; studies on AD and brain aging, 81–82, 83

principles of brain design, 95–96

prion diseases, 156

prisoners of war (PoWs), 203

processing speed, 186

programmed cell death, 30

programmed development, 31–36

programming, 102

progranulin (PGRN) gene, 148, 264

prose recall tests, 169–170

prostaglandins, 317, 318

protein aggregation, 103–104

protein synthesis, 151, 152

psychodynamic theory, 223–224

psychological tests, 164
psychological traits, effect on health, 19, 20
psychological trauma, severe, 203
punch drunk syndrome, 251
Putnam, Robert, 6

Rakic, Pasko, 66
Ramón y Cajal, Santiago, 50, 52
randomized controlled clinical trials, 333–337
Rapoport, Stephen, 89
reactive oxygen species (ROS), 312, 316
recurrent stroke, 313–314
reductionism: cognitive reserve, 283; dementia
    prevention, 277; human genome project, 10;
    problems with, 8–9
regenerative neurology, 15, 29
Reiman, Eric, 284–285
relationships, social, 227–228, 229, 232
reliance on others, 281–282
reminiscence therapy, 326
Reminyl (galantamine), 139
resilience, 5–7, 280–281
respiratory disorders, death caused by, 249
restorative neurology, 52, 127
retina, effects of age on, 172
retirement, 188–189
reverse causality, 8
Reynolds, Chandra, 154
Richards, Marcus, 291
right hemisphere, 63
risk of dementia, reducing: amyloid and related
    therapeutic approaches hypothesis, 319–321;
    biological plausibility of strategies, 283–287;
    brain activity hypothesis, 325–326; childhood
    intelligence, 290–292; cognitive reserve,
    279–283; diabetes, 314–315; early education,
    296–299, 301; early parental death, 289–290;
    epidemic of dementia, 274–276; familial clus-
    tering of dementia risk, 287–289; gene thera-
    pies for late-onset dementia, 293–295; illicit
    drug use in adolescence, 300; inflammatory
    hypothesis of AD, 315–318; metabolic hypoth-
    esis of AD, 318–319; multifactorial nature
    of AD, 276–279; postoperative delirium,
308–312; reports of declining dementia prevalence, 304–306; social epidemiology, 285–
287; stress, depression, and role of growth
factors, 321–325; stroke, 312–314. *See also* multidomain approach to dementia prevention;
neurovascular hypothesis of Alzheimer's
disease
rivastigmine (Exelon), 139
Ronnemaa, Elina, 324
ROS (reactive oxygen species), 312, 316
Rose, Steven, 9
Roses, Allen, 157–158
Roth, Martin, 136
Rothwell, Peter, 313
roundworm (*Caenorhabditis elegans*), 57, 59,
    104, 108
Royal Academy of Engineering, 13
Rutter, Michael, 287–288

Sagan, C., 56–60
Salthouse, Timothy, 4, 111, 164–165, 170
Sapolsky, Robert, 322
schizophrenia, 163–164, 202
Seattle Longitudinal Study (SLS), 182–188
secondary dementias, 243–244
secondary prevention, 277
second languages, 299, 302–303
selective attention, 179–180
selective neuronal loss, 140
self-concept, 222–226
self-organizing principle, 96, 110
semantic dementia, 71, 72, 263
semantic memory, 196
semantic paraphasias, 72
senile dementia, 134–135
senile plaques (SP), 115–116, 141–142
sensory systems, effects of aging on, 171–175
severe psychological trauma, 203
sheep, brain disease of, 244
Sherwood, Chet, 81–82, 83
Shonkoff, Jack, 7
short-term memory, 196
Skirbek, Vergard, 292
Skoog, Ingmar, 289

Sliwinski, Martin, 186
SLS (Seattle Longitudinal Study), 182–188
smell, effects of aging on sense of, 174–175
Smith, David, 340–341
Smith, G. D., 39
Snowdon, David, 270, 291
social capital, 5–7, 296
social cognition, 16–17, 229–230
social cognitive neuroscience, 56–57
social epidemiology, 285–287
social integration, 207, 211, 226–228, 229
social judgements, 217–218
socioeconomic status: and dietary habits, 339; early parental death, 289–290; health food paradox, 345; health linked to childhood stress from, 19, 20, 39, 40; and resilience, 281
sociograms, 227–228
sociomotivational theory of emotional aging, 210
Sonic hedgehog gene, 101
SP (senile plaques), 115–116, 141–142
spatial information, processing, 10
spatial patterning of dementias, 160–161
spectrotemporal analysis, 71
spinocerebellar degeneration, 270
Sporns, Olaf, 60
sports, risk of brain injury from, 251–252, 354
stereotypes, 217–218
stress: early parental death, 289–290; effect on hippocampus, 41; long-term health outcomes, 19, 20, 32, 39–40; risk of dementia related to, 321–325
stress responses, 108, 128–132
stroke: dementia following, 252; diffuse white matter disease, 270–271; fetal development and risk for, 33; implantation of neural stem cells, 127; and neurovascular hypothesis of AD, 312–314; study of language-related neural circuits after, 89
subcortical dementia, 270
subcortical gray matter, 125
subcortical nuclei, 49, 50
substantia nigra, 265, 266
sulci, 112

survivorship, 4
sustained attention, 180–181
synapses, 49, 51, 81
synaptic plasticity, stimulating, 348
synaptic remodeling, 89, 90
synthetic heroin, 300
synucleinopathies, 239
synucleins, 265, 267
systems biology, 13, 22, 130

tacrine (tetrahydroaminoacridine), 138
TAF (tumor angiogenesis factor), 122
taste, effects of aging on sense of, 174–175
tauopathies, 239
tau proteins, 147–149, 262, 320, 321
Tautz, Diethard, 103
TBI (traumatic brain injury), 204–205
temporal lobes, 54
Terman, Lewis M., 20–21
Terman Life-Cycle Study, 20–21
terminology, 42–44
tertiary prevention, 277
testamentary capacity, 230–231
testosterone levels and dementia risk, 323
tests for cognitive aging, 165
tetrahydroaminoacridine (tacrine or THA), 138
theater metaphor, 177–179
theory of mind (TOM), 230–231
Theory of Psychosocial Development (Erikson), 225–226
thickness of cortex, 59
Thompson, Godfrey, 21
thrifty phenotype hypothesis, 35
time perspective, 228–231
TOM (theory of mind), 230–231
Tomalski, P., 290
Tomlinson, Bernard, 136
Tommasi, Luca, 10
top-down attention, 177
touch, effects of aging on sense of, 173–174
tractography, 62
transmission of information through generations, 75–76
traumatic brain injury (TBI), 204–205

tree of life, 82, 103

trisomy 21. *See* Down syndrome

tumor angiogenesis factor (TAF), 122

twin studies: of AD, 247; brain activity hypothesis, 326; diabetes linked to dementia, 314

U.K. MRC Cognitive Aging Follow-up Studies, 292

U.S. Adverse Child Experience Study, 40

U.S. Department of Health and Human Services, 328

U.S. Dictionary of Occupational Titles, 189

useful field of view, limitations in, 172

U.S. Nurses' Health Study cohort, 40

Van Essen, David, 60

van Gool, Willem, 310

Vantaa 85+ study, 310

van Veluw, Susanne, 118

vascular cognitive impairment (VCI), 121; and dementia, 149–151; depressive symptoms, 322; hypertension and dementia, 307, 308

vascular dementia, 121

vascular endothelial growth factor (VEGF), 122

verbal recall testing, 168–170

veterans of war, 203, 204

veterinary neuropathology, 80–83

vigilance, 180–181

vision changes with age, 172

visual cortex, 29

visual hallucinations, 267–268

visualization, value of, 10

vitamin B12, 338–341

volumetric studies on brain aging, 118

von Baer, Karl Ernst, 98

Waddington, Conrad, 97

walking, effects of aging on, 173–174

Warner Schaie, K., 182, 184–185

war veterans, 203, 204

Watson, Lyall, 57

weight of brain, 113

Weisel, Torsten, 29

White, Joseph, 329

white matter: diffuse white matter disease, 270–271; discussion of, 48–49; lost in aging, 80

white matter hyperintensities (WMH), 114, 124–125, 150, 256

Willis, Sherrey, 182, 184–185

Winblad, Bengt, 207, 304

wisdom, 193–194

WM (working memory), 170–171

WMH (white matter hyperintensities), 114, 124–125, 150, 256

Wolf, Phil, 252

women: body weight and reproductive status, 23; breastfeeding, 77–78; effect of nutritional status of grandmother on following generations, 23; estrogen deficiency and dementia risk, 322; intimate relationships and health, 232; maternal transmission of AD, 155; modern changes in reproductive cycle, 86; weight of brain, 113. *See also* pregnancy

Wong, Caine, 141

working memory (WM), 170–171

work skills, 188–189

World War II, view of education after, 304–305

Yaffe, Kristine, 322

Zhang, Jiangyang, 61